ADVANCES IN CLINICAL CHEMISTRY

VOLUME 39

Advances in CLINICAL CHEMISTRY

Edited by

GREGORY S. MAKOWSKI

University of Connecticut Health Center
Farmington, Connecticut

VOLUME 39

ELSEVIER
ACADEMIC
PRESS

AMSTERDAM • BOSTON • HEIDELBERG • LONDON
NEW YORK • OXFORD • PARIS • SAN DIEGO
SAN FRANCISCO • SINGAPORE • SYDNEY • TOKYO

Elsevier Academic Press
525 B Street, Suite 1900, San Diego, California 92101-4495, USA
84 Theobald's Road, London WC1X 8RR, UK

This book is printed on acid-free paper. ∞

For all information on all Elsevier Academic Press publications visit our Web site at www.books.elsevier.com

ISBN: 0-12-010339-7

PRINTED IN THE UNITED STATES OF AMERICA
05 06 07 08 9 8 7 6 5 4 3 2 1

CONTENTS

Clinical and Analytical Review of Ischemia-Modified Albumin Measured by the Albumin Cobalt Binding Test

Fred S. Apple

Human Tissue Kallikreins: From Gene Structure to Function and Clinical Applications

George M. Yousef, Christina V. Obiezu, Liu-Ying Luo, Angeliki Magklara, Carla A. Borgoño, Tadaaki Kishi, Nader Memari, Iacovos P. Michael, Michael Sidiropoulos, Lisa Kurlender, Katerina Economopolou, Carl Kapadia, Nahoko Komatsu, Constantina Petraki, Marc Elliott, Andreas Scorilas, Dionyssios Katsaros, Michael A. Levesque, and Eleftherios P. Diamandis

Cholesterol and Lipids in Depression: Stress, Hypothalamo-Pituitary-Adrenocortical Axis, and Inflammation/Immunity

TIAO-LAI HUANG AND JUNG-FU CHEN

Advances on Biological Markers in Early Diagnosis of Alzheimer Disease

ALESSANDRO PADOVANI, BARBARA BORRONI, AND MONICA DI LUCA

Endothelial Microparticles (EMP) as Vascular Disease Markers

JOAQUIN J. JIMENEZ, WENCHE JY, LUCIA M. MAURO, LAURENCE L. HORSTMAN, CARLOS J. BIDOT, AND YEON S. AHN

Proteomics in Clinical Laboratory Diagnosis

STACY H. SHOSHAN AND ARIE ADMON

Molecular Determinants of Human Longevity

FRANCESCO PANZA, ALESSIA D'INTRONO, ANNA M. COLACICCO, CRISTIANO CAPURSO, ROSA PALASCIANO, SABRINA CAPURSO, ANNAMARIA GADALETA, ANTONIO CAPURSO, PATRICK G. KEHOE, AND VINCENZO SOLFRIZZI

Laboratory Findings of Caloric Restriction in Rodents and Primates
YOSHIKAZU HIGAMI, HARUYOSHI YAMAZA, AND ISAO SHIMOKAWA

Immuno-PCR as a Clinical Laboratory Tool
MICHAEL ADLER

CONTRIBUTORS

Numbers in parentheses indicate the pages on which the authors' contributions begin.

MICHAEL ADLER (239), *Chimera Biotec GmbH, Biomedicinecenter Dortmund, 44227 Dortmund, Germany*

ARIE ADMON (159), *Department of Biology, Technion-Israel Institute of Technology, Haifa 32000, Israel*

YEON S. AHN (131), *Department of Medicine, Division of Hematology Oncology, Wallace H. Coulter Platelet Laboratory, University of Miami School of Medicine, Miami, Florida 33136*

FRED S. APPLE (1), *Hennepin County Medical Center; Department of Laboratory Medicine and Pathology, University of Minnesota School of Medicine, Minneapolis, Minnesota 55415*

CARLOS J. BIDOT (131), *Department of Medicine, Division of Hematology Oncology, Wallace H. Coulter Platelet Laboratory, University of Miami School of Medicine, Miami, Florida 33136*

CARLA A. BORGOÑO (11), *Department of Pathology and Laboratory Medicine, Mount Sinai Hospital, Toronto, Ontario M5G 1X5, Canada; Department of Laboratory Medicine and Pathobiology, University of Toronto, Toronto, Ontario M5G 1L5, Canada*

BARBARA BORRONI (107), *Department of Neurological Sciences, University of Brescia, 25100 Brescia, Italy*

ANTONIO CAPURSO (185), *Department of Geriatrics, Center for the Aging Brain, Memory Unit, University of Bari, 11-70124 Bari, Italy*

CRISTIANO CAPURSO (185), *Department of Geriatrics, Center for the Aging Brain, Memory Unit, University of Bari, 11-70124 Bari, Italy; Department of Geriatrics, University of Foggia, 71100 Foggia, Italy*

SABRINA CAPURSO (185), *Department of Geriatrics, Center for the Aging Brain, Memory Unit, University of Bari, 11-70124 Bari, Italy*

JUNG-FU CHEN (81), *Department of Internal Medicine, Chang Gung Memorial Hospital, Kaohsiung 833, Taiwan, Republic of China*

ANNA M. COLACICCO (185), *Department of Geriatrics, Center for the Aging Brain, Memory Unit, University of Bari, 11-70124 Bari, Italy*

ALESSIA D'INTRONO (185), *Department of Geriatrics, Center for the Aging Brain, Memory Unit, University of Bari, 11-70124 Bari, Italy*

ELEFTHERIOS P. DIAMANDIS (11), *Department of Pathology and Laboratory Medicine, Mount Sinai Hospital, Toronto, Ontario M5G 1X5, Canada; Department of Laboratory Medicine and Pathobiology, University of Toronto, Toronto, Ontario M5G 1L5, Canada*

KATERINA ECONOMOPOLOU (11), *Department of Pathology and Laboratory Medicine, Mount Sinai Hospital, Toronto, Ontario M5G 1X5, Canada; Department of Laboratory Medicine and Pathobiology, University of Toronto, Toronto, Ontario M5G 1L5, Canada*

MARC ELLIOTT (11), *Department of Pathology and Laboratory Medicine, Mount Sinai Hospital, Toronto, Ontario M5G 1X5, Canada; Department of Laboratory Medicine and Pathobiology, University of Toronto, Toronto, Ontario M5G 1L5, Canada*

ANNAMARIA GADALETA (185), *Department of Geriatrics, Center for the Aging Brain, Memory Unit, University of Bari, 11-70124 Bari, Italy*

YOSHIKAZU HIGAMI (211), *Department of Pathology and Gerontology, Nagasaki University Graduate School of Biomedical Science, Nagasaki 852-8523, Japan*

LAURENCE L. HORSTMAN (131), *Department of Medicine, Division of Hematology Oncology, Wallace H. Coulter Platelet Laboratory, University of Miami School of Medicine, Miami, Florida 33136*

TIAO-LAI HUANG (81), *Department of Psychiatry, Chang Gung Memorial Hospital, Kaohsiung 833, Taiwan, Republic of China*

JOAQUIN J. JIMENEZ (131), *Department of Medicine, Division of Hematology Oncology, Wallace H. Coulter Platelet Laboratory, University of Miami School of Medicine, Miami, Florida 33136*

WENCHE JY (131), *Department of Medicine, Division of Hematology Oncology, Wallace H. Coulter Platelet Laboratory, University of Miami School of Medicine, Miami, Florida 33136*

CARL KAPADIA (11), *Department of Pathology and Laboratory Medicine, Mount Sinai Hospital, Toronto, Ontario M5G 1X5, Canada; Department of Laboratory Medicine and Pathobiology, University of Toronto, Toronto, Ontario M5G 1L5, Canada*

DIONYSSIOS KATSAROS (11), *Department of Gynecology, Gynecologic Oncology Unit, University of Turin, 10060 Turin, Italy*

PATRICK G. KEHOE (185), *Department of Care of the Elderly, University of Bristol, Bristol BS16 1LE, England*

TADAAKI KISHI (11), *Department of Pathology and Laboratory Medicine, Mount Sinai Hospital, Toronto, Ontario M5G 1X5, Canada*

NAHOKO KOMATSU (11), *Department of Pathology and Laboratory Medicine, Mount Sinai Hospital, Toronto, Ontario M5G 1X5, Canada*

LISA KURLENDER (11), *Department of Pathology and Laboratory Medicine, Mount Sinai Hospital, Toronto, Ontario M5G 1X5, Canada; Department of Laboratory Medicine and Pathobiology, University of Toronto, Toronto, Ontario M5G 1L5, Canada*

MICHAEL A. LEVESQUE (11), *Department of Pathology and Laboratory Medicine, Mount Sinai Hospital, Toronto, Ontario M5G 1X5, Canada*

MONICA DI LUCA (107), *Institute of Pharmacological Sciences, University of Milan, 20133 Milan, Italy*

LIU-YING LUO (11), *Department of Pathology and Laboratory Medicine, Mount Sinai Hospital, Toronto, Ontario M5G 1X5, Canada; Department of Laboratory Medicine and Pathobiology, University of Toronto, Toronto, Ontario M5G 1L5, Canada*

Angeliki Magklara (11), *Department of Pathology and Laboratory Medicine, Mount Sinai Hospital, Toronto, Ontario M5G 1X5, Canada*

Lucia M. Mauro (131), *Department of Medicine, Division of Hematology Oncology, Wallace H. Coulter Platelet Laboratory, University of Miami School of Medicine, Miami, Florida 33136*

Nader Memari (11), *Department of Pathology and Laboratory Medicine, Mount Sinai Hospital, Toronto, Ontario M5G 1X5, Canada; Department of Laboratory Medicine and Pathobiology, University of Toronto, Toronto, Ontario M5G 1L5, Canada*

Iacovos P. Michael (11), *Department of Pathology and Laboratory Medicine, Mount Sinai Hospital, Toronto, Ontario M5G 1X5, Canada; Department of Laboratory Medicine and Pathobiology, University of Toronto, Toronto, Ontario M5G 1L5, Canada*

Christina V. Obiezu (11), *Department of Pathology and Laboratory Medicine, Mount Sinai Hospital, Toronto, Ontario M5G 1X5, Canada; Department of Laboratory Medicine and Pathobiology, University of Toronto, Toronto, Ontario M5G 1L5, Canada*

Alessandro Padovani (107), *Department of Neurological Sciences, University of Brescia, 25100 Brescia, Italy*

Rosa Palasciano (185), *Department of Geriatrics, Center for the Aging Brain, Memory Unit, University of Bari, 11-70124 Bari, Italy*

Francesco Panza (185), *Department of Geriatrics, Center for the Aging Brain, Memory Unit, University of Bari, 11-70124 Bari, Italy*

Constantina Petraki (11), *Department of Pathology, Evangelismos Hospital, 11528 Athens, Greece*

Andreas Scorilas (11), *Department of Biochemistry and Molecular Biology, University of Athens, 15701 Athens, Greece*

Isao Shimokawa (211), *Department of Pathology and Gerontology, Nagasaki University Graduate School of Biomedical Science, Nagasaki 852-8523, Japan*

Stacy H. Shoshan (159), *Department of Biology, Technion-Israel Institute of Technology, Haifa 32000, Israel*

MICHAEL SIDIROPOULOS (11), *Department of Pathology and Laboratory Medicine, Mount Sinai Hospital, Toronto, Ontario M5G 1X5, Canada; Department of Laboratory Medicine and Pathobiology, University of Toronto, Toronto, Ontario M5G 1L5, Canada*

VINCENZO SOLFRIZZI (185), *Department of Geriatrics, Center for the Aging Brain, Memory Unit, University of Bari, 11-70124 Bari, Italy*

HARUYOSHI YAMAZA (211), *Department of Pathology and Gerontology, Nagasaki University Graduate School of Biomedical Science, Nagasaki 852-8523, Japan*

GEORGE M. YOUSEF (11), *Department of Pathology and Laboratory Medicine, Mount Sinai Hospital, Toronto, Ontario M5G 1X5, Canada; Department of Laboratory Medicine and Pathobiology, University of Toronto, Toronto, Ontario M5G 1L5, Canada*

PREFACE

This volume marks the first issue of a biannual publication for the Advances in Clinical Chemistry series. Starting this year, the series will be published twice a year to keep the readership abreast of the rapid changes and latest developments in clinical chemistry, clinical laboratory science, and laboratory diagnostics.

This volume, number 39 in the series, contains chapters on topics of wide interest, including advances in assay methods such as immuno–polymerase chain reaction technology and proteomic assessment. Other chapters focus on the development and potential applications of novel biomarkers of chronic conditions such as Alzheimer's disease, cancer, cardiovascular disease, and depression. In addition to these long-term debilitating diseases, several chapters specifically address molecular and biochemical findings in the aging process. I hope these reviews will provide insight into the evolving role of the clinical laboratory in meeting the healthcare challenges of 21st-century diagnostics.

I personally thank each of the authors in making this volume a continuing resource for clinical laboratory scientists and practitioners. I also extend my deepest appreciation to colleagues who participated in the peer review process. I would like to acknowledge Ms. Pat Gonzalez of the Elsevier staff for her effort and fine attention to detail throughout the publication of this volume.

I hope the readership will enjoy the first volume of 2005 and actively use it. As always, I warmly welcome comments and solicit participation in keeping subsequent volumes of the Advances in Clinical Chemistry series in the forefront of clinical laboratory sciences.

As customary and in keeping with the tradition of the series, I would like to dedicate this volume to my wife Melinda for her support and scientific interest in making this series a continuing reality.

Gregory S. Makowski

CLINICAL AND ANALYTICAL REVIEW OF ISCHEMIA-MODIFIED ALBUMIN MEASURED BY THE ALBUMIN COBALT BINDING TEST

Fred S. Apple

Hennepin County Medical Center; Department of Laboratory Medicine and Pathology, University of Minnesota School of Medicine, Minneapolis, Minnesota 55415

1. Introduction

Seven to eight million patients with chest pain present annually to urgent care centers and emergency departments (EDs) [1]. Five million members of this group are judged to have suspected acute coronary syndromes (ACSs) or unstable ischemic heart disease and are admitted to the hospital to rule out acute myocardial infarction (MI). Less than half of those admitted ultimately are found to have a cardiac diagnosis, resulting in costly unnecessary admissions. Two to three million members of this group are discharged from the ED each year, and approximately 50,000 of these patients are inadvertently discharged without the diagnosis of MI. Mortality rates of missed diagnoses in patients sent home are twofold greater than those patients admitted. Therefore, there is a considerable clinical interest and clinical research effort underway to identify biomarkers of myocardial ischemia that could be monitored during the early, reversible stage of ACSs, to assist in the appropriate triage of patients presenting with symptoms indicative of ACS.

0065-2423/05 $35.00
DOI: 10.1016/S0065-2423(04)39001-3

Given the understanding that release of cardiac troponin into the circulation reflects myocardial cell death, it is understandable why, clinically, therapies are oriented toward inhibiting the pathophysiologic processes of thrombosis, fibrinolysis, platelet aggregation, and inflammation leading to ischemia and, ultimately, myocardial cell death. Several markers of ischemia have been proposed [2], and research and development studies are underway to adequately validate their clinical usefulness or lack of evidence. These include the topic of this chapter (ischemia-modified albumin [IMA]), as well as choline, unbound free fatty acids, and nourin.

At present, cardiac troponin, a marker of myocardial cell necrosis, is the biomarker defined by the European Society of Cardiology/American College of Cardiology/American Heart Association as the "standard" for detection of myocardial injury [3, 4]. Further, in the clinical setting of ischemia, an increased cardiac troponin is the basis of the diagnosis of MI. However, as it may take 2–6 hours before cardiac troponin increases above a predetermined reference limit in the circulation, the quest to identify an early and, ideally, cardiac tissue–specific marker of myocardial ischemia continues. The purpose of this chapter is to present a general overview of the clinical and analytical issues pertaining to the albumin–cobalt binding (ACB) test, which measures IMA—the first Food and Drug Administration (FDA)–cleared assay to rule out myocardial ischemia [5].

2. IMA and the ACB Test

The observation that serum albumin in patients with myocardial ischemia produced a lower metal binding capacity for cobalt than did serum albumin in nonischemic normal controls lead to the development of the recently FDA-cleared ACB test [5–7]. The precise mechanisms for production of IMA during coronary ischemia are not known. Modifications have been localized to the NH_2-Asp-Ala-His-Lys sequence of human albumin and are postulated to be caused by the production of free radicals during ischemia or reperfusion, acidosis, reduced oxygen, and cellular electrolyte pump disorders [8, 9]. Bhagavan and coworkers have also shown that a deletion defect of the NH_2 terminus was responsible for reduced cobalt binding, leading to a false-positive test for ischemia [10]. The ACB test is a quantitative assay that measures IMA in human serum. In principle, in the serum of patients with ischemia, cobalt added to serum does not bind to the N-terminus of IMA, leaving more free cobalt to react with dithiothreitol and form a darker color. At present, the assay is configured (and FDA cleared) to be measured on the COBAS Mira Plus instrument with an absorbance read at 500 nm [5]. Protocols are being developed by the manufacturer (Ischemia Technologies,

Denver, CO) on additional instruments, using the colorimetric assay as well as investigations into the use of an immunoassay technology (personal communication with manufacturer, Donna Edmonds).

According to the manufacturer's package insert [5], specific preanalytical requirements need to be followed, including avoiding the use of collection tubes with chelators, performing assay analysis within 2.5 hours or freezing at $\leq 20\ ^{\circ}C$, and avoiding sample dilutions. In addition, ACB test results should be interpreted with caution when serum albumin concentrations are <2.0 g/dl or >5.5 g/dl. Because of specificity issues, increased IMA values may be found in patients with cancer, infections, end-stage renal disease, liver disease, and brain ischemia. The analytical characteristics of the assay appear to be good, with no known drug interferences (including several commonly used medications: acetaminophen, aspirin, heparin, ibuprofen, and fluoxetine), acceptable assay total imprecision (<9% coefficient of variation [CV]) at the medical decision cutoff, a lower limit of detection at 14 U/ml, and linearity to 200 U/ml. Expected (normal) values, determined from a population of 283 healthy subjects, ranging in age from 35 to 98 years and equally distributed between males and females, ranged from 52 to 116 U/ml, with a 95th percentile at 85 U/ml. The manufacturer recommends that each laboratory establish its own reference values and clinical cutoff concentrations, which may vary depending on geographic, dietary, and environmental factors.

3. Clinical Studies Review

This review includes several clinical studies that have evaluated the performance of the ACB test for monitoring IMA in various ischemic populations. Studies were typically performed in small numbers of patients with various study designs. As a consequence, published data are preliminary and still need to be confirmed in larger studies and trials. The goals of these studies have been to demonstrate that the ACB test is a sensitive, early marker of reversible and irreversible cardiac ischemia and to demonstrate that IMA increases before markers of myocardial necrosis (cardiac troponin). However, many questions remain unanswered. The ACB test needs to be evaluated by incorporating it into decision-making algorithms under ED conditions. The highest expected benefit of the test would be to rule out ACS in low to moderate pretest probability conditions with negative necrosis markers and a negative ECG (electrocardiogram). This was the language for which the ACB test was cleared by the FDA for clinical use.

Table 1 summarizes the clinical studies reviewed. In the first descriptive clinical study published, IMA was measured in 99 patients with cardiac chest pain and 40 patients with noncardiac chest pain [7]. Samples were collected

TABLE 1

SUMMARY OF CLINICAL STUDIES MEASURING ISCHEMIA-MODIFIED ALBUMIN BY THE ALBUMIN–COBALT BINDING TEST

Reference	Setting	Number of patients	NPV for ischemia	PPV for ischemia	Outcome measure	Type of study
Bar-Or, 2000 [7]	ACS	139	NA	NA	Final diagnosis of ischemia	Prospective enrollment
Christenson, 2001 [9]	ACS	224	96%	33%	cTnI at 6–24 hours	Prospective enrollment
Bar-Or, 2001 [8]	PTCA	54	NA	NA	Change in albumin–cobalt binding assay value after procedure	Prospective enrollment
Bhagavan, 2003 [10]	ACS	167	91%	92%	Final diagnosis of ischemia	Retrospective enrollment
Sinha, 2003 [11]	PTCA	30	NA	NA	Change in albumin–cobalt binding assay value after procedure	Prospective enrollment
Quiles, 2003 [12]	PCI	34	NA	NA	Ischemia-modified albumin changes related to number and duration of inflation during PCI	Prospective enrollment
Garrido, 2004 [13]	PCI	90	NA	NA	Ischemia-modified albumin increases related to PCI and collaterals present	Prospective enrollment
Roy, 2004 [15]	DCCV	24	NA	NA	DCCV for A fib caused increases in ischemia-modified albumin	Prospective enrollment
Sinha, 2004 [16]	ACS	245	72%	59%	Compare PPV and NPV of ischemia-modified albumin vs. other biomarkers and electrocardiogram	Prospective enrollment

NPV = negative predictive value; PPV = positive predictive value; ACS = acute coronary syndrome; PTCA = percutaneous transluminal coronary angioplasty; PCI = percutaneous coronary intervention; DCCV = direct-current cardioversion; A fib = atrial fibrillation; NA = not applicable.

within 4 hours of presentation to the ED in patients with the primary complaint of chest pain. Ninety-five of 99 patients with cardiac chest pain had increased IMA levels, and 37 of 40 patients with noncardiac chest pain had normal values. However, no other biochemical markers or ECG status were reported at the time of presentation for comparison.

A separate multicenter study involving 224 ED patients with signs and symptoms indicative of ACS [9] examined the ability of the ACB test to predict a cardiac troponin (cTn) positive or negative result within 6–24 hours after presentation. All patients had a negative cTnI result at presentation. Patients were considered troponin positive if one or more cTnI values were above the upper limit of normal within 6–24 hours. At the optimum cut-off for the ACB test, sensitivity and specificity were 70% and 80%, with a negative predictive value of 96%. There were 6 false negatives and 131 true negatives. cTnI alone was used as the outcome measure, and ECG status at presentation was not considered in the design of the study. A control group of 109 healthy adults (ages 20–85 years) was also tested to determine reference limits. The cut-off value of 75 U/ml for the study group, derived by the receiver–operator characteristic curve, was lower than the 80.2 U/ml value that represented the 95th percentile of the control population, demonstrating overlap between normal and increased values. This resulted in a positive predictive value of 33%.

In another study, by Bhagavan and coworkers [10], ACB assay results were correlated with final discharge diagnoses in 75 ED patients with myocardial ischemia and 92 patients with general good health. A clinical diagnosis of myocardial ischemia was based on clinical signs and symptoms, imaging, ECG, and other biochemical markers. A subgroup had MI, the diagnosis that was based on the new European Society of Cardiology/American College of Cardiology criteria [3, 4]. The sensitivity and specificity for myocardial ischemia were 88% and 94%, and the positive and negative predictive values were 92% and 91%, respectively. The ACB test, however, was a poor discriminator between ischemic patients with and without MI.

The inflation of a balloon during angioplasty causes transient myocardial ischemia in humans. On the basis of previous reports demonstrating that percutaneous transluminal coronary angioplasty (PTCA) can be used as an *in vivo* model of mild transient myocardial ischemia in humans, a fourth study examined 41 patients undergoing elective PTCA [8]. The ACB test, CK-MB, myoglobin, and cTnI were monitored before, immediately after, and 6 and 24 hours after PTCA. During the procedure, 37 of 41 patients had stents placed, and 33 had signs, symptoms, or ECG signs of ischemia. Of 41 patients, 34 had increased IMA values immediately after the procedure, representing a 10.1% mean difference for IMA concentrations between those with and without a change. There was no significant mean percentage change of IMA from baseline at either 6 or 24 hours, however. There also was no significant difference between the patients who did and who did not have signs and symptoms of ischemia during the procedure.

CK-MB, myoglobin, and cTnI showed no changes immediately after the procedure, but at 6 and 24 hours there were significant IMA increases. However, no patients showed an increase above their respective upper reference limits. Further, there was no correlation between ACB test results and the other biomarkers. In addition, a control group of 13 patients was also tested; they had coronary angiography without angioplasty and stenting. None had symptoms and signs of ischemia during the procedure and no significant mean percentage changes from baseline were found for any biochemical markers. This study demonstrated that changes in IMA occurred minutes after the onset of transient ischemia and returned to baseline at 6 hours in the PTCA setting. No samples were taken, however, between the procedure and 6 hours postprocedure. Thus, kinetic data, which were not demonstrated in this study, are needed for a better understanding of the mechanisms of IMA formation, release, and clearance.

In a recent study by Sinha and coworkers [11], IMA and cTnT were compared before, immediately after, 30 minutes after, and 12 hours after elective PTCA. The study group consisted of 19 patients who had >70% single-vessel disease and who had chest pain or ischemic ECG changes during the procedure. Stents were deployed as required. IMA levels were elevated from baseline (72 U/ml) in 18 of 19 patients both immediately after (101 U/ml) and 30 minutes after (87 U/ml) the procedure and returned to below baseline at 12 hours. None of the patients had cTnT levels above the upper limit of normal. A control group of 11 patients undergoing diagnostic angiography were also included who did not have significant changes in IMA levels.

Quiles and coworkers also confirmed and expanded previous reports that IMA was a marker of ischemia in the setting of percutaneous coronary intervention (PCI) [12]. In 34 patients who underwent elective single-vessel PCI for the management of stable angina, IMA was monitored both 10 minutes before and within 5 minutes after the last balloon inflation. All serum specimens collected were frozen within 2 hours and were batch analyzed. Before PCI, IMA levels were 59.9 U/ml (SD19) and increased to 80.9 U/ml (SD22) after angioplasty (both values are within the normal reference interval of the assay). IMA levels were higher in patients with more balloon inflations ($\geq$5; 88.8 vs. 70.9 U/ml), higher-pressure inflations, and longer inflation duration (duration of inflations 151.6 ± 1.01 seconds). Overall, these data support IMA as a marker of the occurrence of balloon-induced myocardial ischemia. Similar observations were also made by Garrido and coworkers, who demonstrated that IMA levels increased 10 minutes after

PCI, with larger increases occurring in patients without collateral vessels (4.8% with vs. 14% without) [13]. Although these studies indicate that IMA is an early marker of transient ischemia in the PTCA setting, larger trials are necessary to validate this hypothesis.

In an abstract presented at the 2003 meeting of the Society of Academic Emergency Medicine, the utility of IMA in making risk stratification decisions at ED presentation was projected by reviewing 251 records in a prospective design [14]. A risk level was assigned to each patient based on the reviewer's clinical practice, using demographics, cardiac risk profile, signs and symptoms, ECG, and biochemical marker status. After 2 weeks, risk levels were reassigned, adding the IMA concentration measured at presentation into the assessment. The study population was at low risk, with a 10% frequency of ACS. Without knowledge of the IMA result, 66 "very low risk" risk assessments (consistent with expeditious discharge home) were made. With the knowledge of IMA, 236 cases were identified as very low risk. No patients with negative IMA results were found to have ACS, demonstrating a negative predictive value (NPV) of 100%.

Roy and coworkers studied 24 patients who underwent elective direct current cardioversion for atrial fibrillation [15]. Serum collected before and at both 1 and 6 hours after direct current cardioversion was frozen at 70 °C within 2 hours and batch analyzed after thawing. At baseline, IMA was 62 U/ml and increased to 80 U/ml 1 hour after direct current cardioversion in patients with ECG changes. However, in patients with ECG changes, no changes were observed (pre: 61 U/ml; post 63 U/ml). Overall, the authors postulate that IMA release was indicative of cardiac ischemia relating to ECG changes. As noted earlier, the manufacturer's package insert of the ACB assay states that a combination of the ACB test with a negative cardiac troponin, a nondiagnostic ECG, and appropriate risk stratification may bring about a near 100% negative predictive value [5]. These observations were recently reported by Sinha and coworkers [16] in their study of 208 patients suggestive of ACS presenting to the ED within 3 hours of acute chest pain. In this study, IMA, in conjunction with ECG and cTnT, was monitored. Results, alone and in combination, were correlated with final diagnoses (nonischemic chest pain, unstable angina, MI). Sensitivities and specificities at presentation were as follows: IMA alone 82%, 46%; ECG alone 45%, 91%; cTnT alone at >0.05 μg/l cutoff 20%, 99%; ECG and cTnT 53%, 90%; IMA and cTnT 90%, 44%; IMA and ECG 92%, 43%; IMA, ECG and cTnT 95%, 42%. These data showed that the sensitivity of IMA alone was no different than IMA and ECG combined and that IMA was highly sensitive for the diagnosis of myocardial ischemia.

The role of IMA, even with the low specificity of 46%, was thought to be useful, pending larger trials, to improve diagnostic strategies for patients with acute chest pain presenting to EDs. However, this study did not monitor IMA prospectively (all samples were frozen and batch analyzed), as would be necessary to be implemented in an ED decision algorithm. To be truly effective in a rapid triage ED chest pain evaluation protocol, near-bedside (point of care) testing, with a 24 hours a day, 7 days a week turn-around time of <30 minutes would be desirable. The test had limited specificity with many false positives, with considerable overlap between normal and ischemic IMA levels. A positive ACB test result also did not appear to discriminate between unstable angina and early myocardial necrosis. However, a negative ACB test effectively differentiated these two groups from noncardiac patients with angina-like symptoms.

4. Future Needs and Unanswered Questions

IMA, which appears to be an indicator of oxidative stress, may not be specific for cardiac ischemia. There is limited data about IMA levels in noncardiac ischemia. There is anecdotal evidence indicating that IMA increases in stroke, end-stage renal disease, and some neoplasms [12]. In a group of marathon runners, IMA did not increase immediately after a marathon run, indicating that skeletal muscle ischemia during exercise does not change IMA levels [17]. However, there were significant increases at 24–48 hours after the run, which were attributed to exercise-induced latent gastrointestinal ischemia. This latent increase is an issue that may potentially complicate the use of the test in clinical practice. Further, resolving the influence of fluid shifts, effects of increased lactic acid concentrations, and albumin concentration changes on the ACB test, which occur following strenuous exercise and other pathologies, need to be more fully understood.

Future studies that are needed for the evolution of this new assay include normal population distributions by gender and ethnicity; an optimum cut-off value for a high versus normal concentration in ACS patients, comparing IMA levels in common disease states both with or without accompanying cardiac disease; and added information in common diseases that coexist with cardiac ischemia, such as congestive heart failure, diabetes mellitus, chronic renal disease, hypertension (18). A better understanding of IMA kinetics over the early hours after the onset of an ACS is essential to understanding the interpretative value, if any, of serial monitoring of IMA after presentation.

Of the numerous questions that remain, the most prominent is how clinicians will interpret a positive IMA finding. The positive predictive value of the ACB test seems to be too low for use in ruling in ischemia, a use that clinicians hope the laboratory could provide. It is not known whether patients with negative ECG and necrosis markers (cardiac troponin) and a positive IMA result might benefit from early triage and intervention according to stratified pretest probabilities. Therefore, outcome and risk stratification studies monitoring IMA alone and in combination with other clinical and biomarkers findings are needed. In clinical practice, this lack of information can potentially lead to overtreatment of low-risk patients, with a positive (IMA) result. Whether positive (IMA) results in noncardiac patients may be associated with significant clinical conditions that justify admission for a more detailed examination needs to be explored.

5. Summary

IMA measured by the ACB test is proposed as a novel marker that appears sensitive to cardiac ischemia. It has the potential to become a triage tool in suspected ACS patients, especially to rule out ACS, and might also find utility in stroke, stress testing, nuclear imaging, and states of noncardiac ischemia and oxidative stress. Rapid-testing platforms will be necessary to achieve the optimal goal of assisting in the triage of chest (ACS) patients that present to EDs. However, it does not appear to be highly tissue or clinically specific.

There is a continued need to improve the clinical and analytical evidence base of IMA to substantiate its clinical use in diagnostics and outcomes assessment. The search for a marker, whether IMA or another marker that establishes an evidence base, that would effectively rule in as well as rule out early cardiac ischemia continues. The ACB test has barely begun the exploration of the exciting challenges and discoveries that lie ahead in assisting clinicians in the early detection of myocardial ischemia to assist and improve patient triage, therapy, and management.

Acknowledgment

Dr. Apple has received research funding and honorarium and expenses for consulting from Ischemia Technologies.

References

[1] Storrow AB, Gibler WB. Chest pain centers: Diagnosis of acute coronary syndromes. Ann Emerg Med 2000; 35:449–461.

[2] Wu AHB, editor. Clinical Markers: Pathology and Laboratory Medicine, second ed. Totowa, NJ: Humana, 2003.

[3] Alpert J, Thygesen K, Antman E, et al. Joint European Society of Cardiology/American College of Cardiology Committee. Myocardial infarction redefined—a consensus document of the Joint European Society of Cardiology/American College of Cardiology Committee for the redefinition of myocardial infarction. J Am Coll Cardiol 2000; 36:959–969.

[4] Jaffe AS, Ravkilde J, Roberts R, et al. It's time for a change to a troponin standard. Circulation 2000; 102:1216–1220.
[5] ACB® Test Reagent Pack, COBAS MIRA® Plus. Package insert. Denver, CO: Ischemia Technologies, August 2002.
[6] Bar-Or D, Curtis G, Rao N, Bampos N, Lau E. Characterization of the Co^{2+} and Ni^{2-} binding amino-acid residues of the N-terminus of human albumin. Eur J Biochem 2001; 268:42–47.
[7] Bar-Or D, Lau E, Winkler JV. Novel assay for cobalt-albumin binding and its potential as a marker for myocardial ischemia—a preliminary report. J Emerg Med 2000; 19:311–315.
[8] Bar-Or D, Winkler JV, Vanbenthuysen K, Harris L, Lau E, Hetzel FW. Reduced albumin-cobalt binding with transient myocardial ischemia after elective percutaneous transluminal coronary angioplasty: A preliminary comparison to creatine kinase-MB, myoglobin, and troponin I. Am Hear J 2001; 141:985–991.
[9] Christenson RH, Duh SH, Sanhai WR, et al. Characteristics of an albumin cobalt binding test for assessment of acute coronary syndrome patients: A multicenter study. Clin Chem 2001; 47:464–470.
[10] Bhagavan NV, Lai EM, Rios PA, et al. Evaluation of human serum albumin cobalt binding assay for the assessment of myocardial ischemia and myocardial infarction. Clin Chem 2003; 49:581–585.
[11] Sinha MK, Gaze DC, Tippins JR, Collinson PO, Kaski JC. Ischemia modified albumin is a sensitive marker of myocardial ischemia after percutaneous coronary intervention. Circulation 2003; 107:2403–2405.
[12] Quiles J, Roy D, Gaze D, et al. Relation of ischemia-modified albumin (IMA) levels following elective angioplasty for stable angina pectoris to duration of balloon-induced myocardial ischemia. Am J Cardiol 2003; 92:322–324.
[13] Garrido IP, Roy D, Calvino R, et al. Comparison of ischemia-modified albumin levels in patients undergoing percutaneous coronary intervention for unstable angina pectoris with versus without coronary collaterals. Am J Cardiol 2004; 93:88–90.
[14] Pollack CV, Peacock WF, Summers RW, et al. Ischemia-modified albumin (IMA) is useful in risk stratification of emergency department chest pain patients. Acad Emerg Med 2003; 10:555–556.
[15] Roy D, Quiles J, Sinha M, et al. Effect of direct-current cardioversion on ischemia modified albumin levels in patients with atrial fibrillation. Am J Cardiol 2004; 93:366–368.
[16] Sinha MK, Roy D, Gaze D, Collinson PO, Kaski JC. Role of ischemia modified albumin, a new biochemical marker of myocardial ischemia, in the early diagnosis of acute coronary syndromes. Emerg Med J 2004; 21:29–34.
[17] Apple FS, Quist HE, Otto AP, Mathews WE, Murakami MM. Release characteristics of cardiac biomarkers and ischemia-modified albumin as measured by the albumin cobalt-binding test after a marathon race. Clin Chem 2002; 48:1097–1100.
[18] Wu AHB. The ischemia-modified albumin biomarker for myocardial ischemia. Medical Laborator Observer June, 2003; 36–40.

HUMAN TISSUE KALLIKREINS: FROM GENE STRUCTURE TO FUNCTION AND CLINICAL APPLICATIONS

George M. Yousef,[*,†] Christina V. Obiezu,[*,†] Liu-Ying Luo,[*,†] Angeliki Magklara,[*] Carla A. Borgoño,[*,†] Tadaaki Kishi,[*] Nader Memari,[*,†] Iacovos P. Michael,[*,†] Michael Sidiropoulos,[*,†] Lisa Kurlender,[*,†] Katerina Economopolou,[*,†] Carl Kapadia,[*,†] Nahoko Komatsu,[*] Constantina Petraki,[‡] Marc Elliott,[*,†] Andreas Scorilas,[§] Dionyssios Katsaros,[||] Michael A. Levesque,[*] and Eleftherios P. Diamandis[*,†]

[*]Department of Pathology and Laboratory Medicine, Mount Sinai Hospital, Toronto, Ontario M5G 1X5, Canada
[†]Department of Laboratory Medicine and Pathobiology, University of Toronto, Toronto, Ontario M5G 1L5, Canada
[‡]Department of Pathology, Evangelismos Hospital, 11528 Athens, Greece
[§]Department of Biochemistry and Molecular Biology, University of Athens, 15701 Athens, Greece
[||]Department of Gynecology, Gynecologic Oncology Unit, University of Turin, 10060 Turin, Italy

0065-2423/05 $35.00
DOI: 10.1016/S0065-2423(04)39002-5

1. Introduction

Kallikreins are serine proteases with diverse physiological functions. Until recently, it was thought that the human kallikrein gene family included only three members, but recent studies have led to the complete characterization of the human kallikrein gene locus and identification of all 15 members of this family. Kallikreins are expressed in many organs, most prominently in endocrine-related tissues such as the prostate, breast, ovary, uterus, vagina, and testis. Many kallikreins are regulated by steroid hormones in cancer cell lines, and several lines of investigation have supported a link between kallikreins and cancer.

Prostate-specific antigen (PSA, hK3) and, more recently, human glandular kallikrein 2 (hK2) are used as tumor markers for prostate cancer. Several other kallikreins, including hK5, hK6, hK8, hK10, hK11, and hK14, are emerging as new serum biomarkers for ovarian cancer diagnosis and prognosis. Some other kallikreins are differentially expressed at both the mRNA and protein levels in various endocrine-related malignancies. The coexpression of many kallikreins in several cancer types and other information points to the possibility of their involvement in a cascade-like pathway that may be associated with cancer pathogenesis or progression. Finally, in addition to their diagnostic/prognostic utilities, kallikreins may be attractive novel therapeutic targets.

2. Discovery of the Human Tissue Kallikrein Gene Locus: A Short Historical Perspective

Until approximately 1998, it was thought that the human kallikrein gene locus included three genes: pancreatic/renal kallikrein (*KLK1*), glandular kallikrein (*KLK2*), and PSA (*KLK3*). Several early estimates of the size of the human kallikrein gene family were contradictory. Southern blot analysis indicated that the size of this family varied from just three to four genes [1] to as many as 19 genes [2]. Starting around 1996, several independent groups reported the cloning of new serine proteases that colocalized at the chromosomal region 19q13.4 and shared a high degree of homology with the known kallikreins (also called the "classical" kallikreins in this review). Extensive work from our laboratory and from other groups analyzing genomic sequences in the vicinity of chromosome 19q13.4, available through the Human Genome Project, has lead to the cloning of all 15 members of the human kallikrein gene family. Table 1 summarizes information from genomic and protein databases for all tissue kallikreins. Later, we present the detailed description of the locus and contrast these data with information derived from the independent analysis of the DNA sequence of chromosome 19 [3].

3. Kallikreins in Rodents and Other Species

Kallikreins are found in both primates and nonprimates. Kallikrein genes and proteins have been identified in six different mammalian orders: Primates, Rodentia, Carnivora, Proboscidea, Perissodactyla, and Artiodactyla [4]. The number of kallikreins varies among species, and kallikreins in rodents and other animal species have been extensively described in a number of excellent reviews [2, 5–8]. In this section, we provide only a quick overview and some recent updates about kallikrein families in different species, with special emphasis on their structural and localization relationships with the

TABLE 1

OFFICIAL AND OTHER GENE AND PROTEIN NAMES FOR MEMBERS OF THE HUMAN KALLIKREIN GENE FAMILY[a]

Official gene name	Other names/symbols	Genbank accession number	Unigene cluster	Merops ID	SwissProt ID
KLK1	Pancreatic/renal kallikrein, hPRK	M25629 M33105	Hs.123107	S01.160	Q07276
KLK2	Human glandular kallikrein 1, hGK-1	M18157	Hs.181350	S01.161	P20151
KLK3	Prostate-specific antigen, PSA, APS	X14810 M24543 M27274	Hs.171995	S01.162	P07288
KLK4	Prostase, KLK-L1, EMSP1, PRSS17, ARM1	AF113141 AF135023 AF148532	Hs.218366	S01.251	Q9Y5K2
KLK5	KLK-L2, HSCTE	AF135028 AF168768	Hs.50915	S01.017	Q9Y337
KLK6	Zyme, Protease M, Neurosin, PRSS9	AF013988 AF149289 U62801 D78203	Hs.79361	S01.236	Q92876
KLK7	HSCCE, PRSS6	L33404 AF166330	Hs.151254	S01.300	P49862
KLK8	Neuropsin; Ovasin; TADG-14, PRSS19, HNP	AB009849 AF095743 AB010780 AF055982	Hs.104570	S01.244	O60259
KLK9	KLK-L3 protein	AF135026	Hs.448942	S01.307	Q9UKQ9
KLK10	NES1, PSSSL1	AF055481 NM_002776	Hs.69423	S01.246	O43240
KLK11	TLSP/Hippostasin, PRSS20	AB012917	Hs.57771	S01.257	Q9UBX7
KLK12	KLK-L5 protein	AF135025	Hs.159679	S01.020	Q9UKR0
KLK13	KLK-L4 protein	AF135024	Hs.165296	S01.306	Q9UKR3
KLK14	KLK-L6 protein	AF161221	Hs.283925	S01.029	Q9P0G3
KLK15	prostinogen, HSRNASPH	AF303046	Hs.250770	S01.081	Q9H2R5

[a] According to the Human Gene Nomenclature Committee (http://www.gene.ucl.ac.uk/nomenclature).

human kallikreins. More detailed discussion can be found in our recent review [9].

3.1. The Mouse Kallikrein Gene Family

In the mouse, kallikreins are represented by a large multigene family, initially thought to include 25 genes. Among those, at least 14 genes were presumed to be encoding for serine proteases [5]. Recent data from the Mouse Genome Project (http://www.ncbi.nlm.nih.gov/genome/guide/mouse/) indicate the existence of 36 mouse kallikrein genes (including annotated genes). Olsson and Lundwall published the organization of the kallikrein gene family in the mouse through sequence analysis of different databases [10]. Mouse tissue kallikrein genes reside in a locus that spans approximately 310 kb at or near the Tam-1 locus on mouse chromosome 7. Data from the human/mouse homology maps (http://www.ncbi.nlm.nih.gov/projects/Homology/) show that this region is highly syntenic to the human chromosome 19q13.4, which harbors the human kallikrein genes (up to 75% sequence similarity). Sequence analysis indicated the presence of possible mouse orthologues for the human kallikreins 1 and 4 through 15, but not for the classical kallikreins *KLK2* and *KLK3*. Comparing the human and mouse kallikrein loci, however, indicated that although the distance between the human *KLK1* and *KLK15* genes is only 1.5 kb, the same area in the mouse genome is 290 kb in length and harbors the rest of the mouse kallikreins [10]. All mouse kallikrein genes are transcribed in the same direction and share a high degree of structural homology at both the mRNA and protein levels (70%–90%). They also share the same genomic organization, being formed of five coding exons and four introns, with completely conserved exon–intron splice sites. A TATA box variant "TTAAA" and consensus polyadenylation signal sequences were found in all mouse kallikreins [11–13]. All mouse kallikreins code for pre-pro-kallikreins that are 261 amino acids in length, with an 18–amino acid signal peptide followed by a profragment of six amino acids. Similar to the human family, only one mouse kallikrein (*mGK6*) has a true kininogenase activity [11]. Other mouse kallikrein proteins have different types of activities. *mGK3* and *mGK4* are nerve growth factor–binding and nerve growth factor–processing enzymes; *mGK9, mGK13*, and *mGK22* are epidermal growth factor–binding proteins; *mGK22* is a nerve growth factor–inactivating enzyme; *mGK16* is γ-renin and *mGK26* is prorenin-converting enzyme-2 [14, 15]. With the expansion of the mouse kallikrein gene family and the identification of new mouse orthologues of the human kallikreins, a revision of the mouse, human, and other species' kallikrein nomenclature, based on map sequence and evolutionary analyses, should be considered in the future.

3.2. The Rat Kallikrein Gene Family

Another large family of about 13 kallikreins was originally identified in the rat, of which at least 10 are transcriptionally active [16]. More recent data from the Rat Genome Database (http://ratmap.gen.gu.se/) show the possibility of the existence of additional kallikreins in the rat genome (see following). Rat kallikreins are clustered in the same chromosomal region and share a high degree of structural similarity. They also have the same conserved structure of five coding exons and four introns, with most of the similarity in the exonic, rather than intronic, regions [17]. Ten of the 13 rat kallikreins code for potentially active serine proteases that are 261 amino acids in length; the rest are assumed to be pseudogenes. As is the situation in the human and the mouse, only one rat kallikrein (*rKLK1*) meets the functional definition of a kallikrein [18]. *rKLK2* codes for tonin, which converts angiotensinogen to angiotensin II [19], and *rKLK10* codes for a kininogenase, which cleaves T-kininogen to release T-kinin [19]. Rat kallikrein mRNA levels were found to be responsive to hormonal manipulation, and castration of male animals resulted in a decrease of mRNA, which could be restored by testosterone [20]. Pituitary *rKLK1* has been found to be upregulated by estrogen [8]. Interestingly, rodent kallikreins are mainly expressed in the salivary gland, with very few of them having a wider tissue expression pattern [8].

Recently, Olsson *et al.* reported a more precise genomic organization of the rat kallikrein locus [21]. The rat *KLK* locus spans approximately 580 kb on chromosome 1q21, contains 22 genes and 19 pseudogenes, and is devoid of *KLK2* and *KLK3* orthologues. This locus contains nine duplications of an approximately 30-kb region harboring the *KLK1, KLK15*, and pseudogene Ψ*KLK2*, and *KLK4*, resulting in nine paralogues of each gene. However, only the *KLK1* paralogues seem to be functional. For more details, see Olsson *et al.* (2004) [21].

3.3. The Mastomys Kallikrein Gene Family

Mastomys is an African rodent that is intermediate in size and that has physical characteristics between the mouse and rat. It has been studied because of the presence of an androgen-responsive prostate in the female. Fahnestock reported the cloning of cDNAs from Mastomys. Two of these cDNAs were expressed in the kidney as well as the submandibular gland, and one is hypothesized to code for a true tissue kallikrein [22]. A third kallikrein was found only in the submandibular gland. DNA sequence analysis and hybridization studies demonstrate that Mastomys represents an interesting hybrid between mouse and rat [22].

3.4. The Monkey and Chimpanzee Kallikrein Gene Families

A cynomolgus monkey tissue kallikrein gene has been characterized from a monkey renal cDNA library and has been shown to be 90% homologous to its human counterpart at the nucleotide level [23]. The *cmKLK1* encodes a tissue kallikrein of 257 amino acids, which is 93% homologous to the human kallikrein protein. The key residues important for kininogenase activity are entirely conserved. A rhesus monkey prostatic *KLK3* cDNA encoding the simian counterpart of PSA (*KLK3*) has also been cloned [24]. It consists of 1515 nucleotides, encoding a preproenzyme of 261 amino acids, with a long 3′ untranslated region. The deduced amino acid sequence is 89% homologous to human hK3 and 71% to human hK2. Tyr^{93}, a residue important for the kininogenase activity of the human kallikrein 1, is replaced by a serine, indicating that rm*KLK3* will lack kininogenase activity, as does its human counterpart.

Recent data from the Chimpanzee Genome Project have revealed that the chimpanzee *KLK* locus is strikingly similar to the human locus, spans about 350 kb of genomic sequence on chromosome 20, and contains orthologues to all human kallikrein genes, with an overall >99% sequence similarity at the DNA and amino acid levels (our unpublished data).

3.5. The Dog Kallikrein Gene Family

Only two kallikreins have been identified in the dog: *dKLK1*, encoding a true tissue kallikrein based on functional definition, and *dKLK2*, encoding a canine arginine esterase [25]. *dKLK1* encodes a polypeptide of 261 amino acids with a typical 24-residue pre-pro-peptide, a conserved catalytic triad of serine proteases, and a tissue kallikrein substrate-binding pocket. A prostatic cDNA and the gene encoding for canine arginine esterase (*dKLK2*) have both been cloned. As for all other mammalian kallikrein genes, *dKLK2* consists of five exons and four introns, with fully conserved exon/intron boundaries, and an "AGTAAA" polyadenylation signal identical to that of human hK1. The *dKLK2* gene product shows a wide pattern of tissue expression and has less overall conservation (~50%) with kallikreins of other species.

3.6. Kallikreins in Other Species

No more than three tissue kallikreins were identified up to now in the guinea pig [26]. Two kallikreins were cloned in the horse—a renal kallikrein [27] and a horse prostate kallikrein [28], which is a homologue of human PSA. Southern blot analysis data detected *KLK2*- and *KLK3*-positive bands in several nonhuman primate species including macaque, orangutan, chimpanzee, and gorilla, but not in cows and rabbits [29]. Kallikreins were also isolated from the pancreas, colon, and submadibular gland of the cat [30]. Nonprimates do not contain any prostate-localized proteins homologous to PSA [31].

Completion of many other future genome projects will facilitate accurate comparisons of the *KLK* families in many species.

4. Characterization and Sequence Analysis of the Human Tissue Kallikrein Gene Locus

4.1. Locus Overview

The first comprehensive attempt to characterize the human kallikrein locus was reported by Riegman *et al.* [32], who proposed that the locus is formed of only three genes, *KLK1, KLK2*, and *KLK3*. These three genes were found to be clustered in a 60-kb region on chromosome 19q13.4. Their alignment in the genome is centromere-*KLK1-KLK3-KLK2*-telomere. *KLK2* and *KLK3* are transcribed in the direction from centromere to telomere. *KLK1* is transcribed in the opposite direction [32, 33].

Recently, with the discovery of all human kallikrein genes, we were able to fully characterize the human kallikrein gene locus with high accuracy and to precisely localize each of the 15 members of the human tissue kallikrein gene family, as well as to determine the distances between them and their directions of transcription [34]. The human tissue kallikrein gene locus spans a region of 261,558 bp on chromosome 19q13.4 and is formed of 15 kallikrein genes with no intervening nonkallikreins. More recently, a potential kallikrein-processed pseudogene has been cloned [34a], and the possibility exists for the presence of at least another two pseudogenes [35]. The kallikrein gene family is flanked centromerically by the testicular acid phosphatase gene [36] and telomerically by the *CAG* (cancer-associated gene) [37] and the Siglec family of genes [38]. The latest analysis of the genomic region confirmed and extended our previous findings, as shown in Fig. 1 and Table 2. Minor differences are a result of discrepancies in noncoding regions.

The kallikrein genes are clustered together, and the distances between two adjacent genes range from 1.5 (*KLK1* and *KLK15*) to 32.5 kb (*KLK4* and *KLK5*). Detailed information about the locus is presented in Fig. 1 and Table 2. The locus has been extensively analyzed for the presence of other kallikreins [34, 35, 39]. It is thus unlikely that new members will be identified inside the locus or on either end, with the possible exception of more pseudogenes.

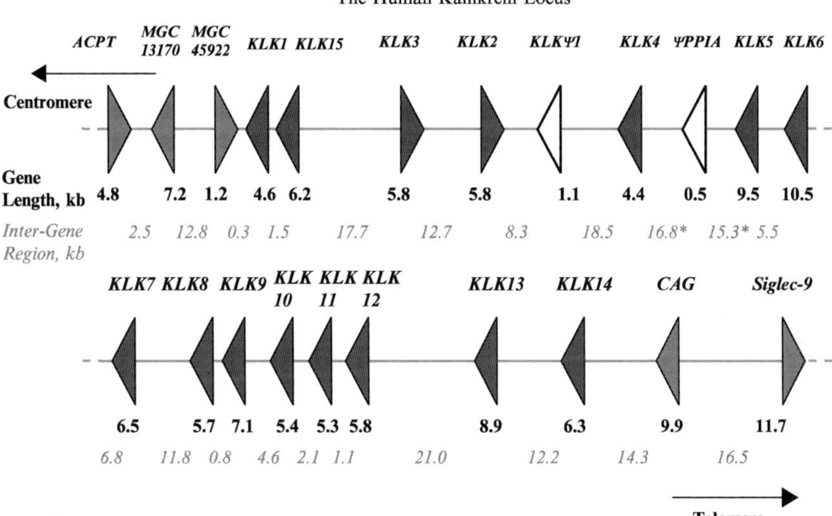

FIG. 1. Schematic representation of the human kallikrein gene locus on chromosome 19q13.4. Genes are represented by arrowheads indicating the direction of transcription. Kallikrein genes are shown in blue, and nonkallikrein genes and pseudogenes are presented by grey and white arrows, respectively. Official gene symbol is shown above each gene, and the approximate gene length is shown below each gene in kilobytes. The approximate intergenic regions are shown in red in Kb. CAG, cancer associated gene (GenBank accession AY279382); ACPT, testicular acid phosphatase (GenBank accession AF321918). The position of the PPIA pseudogene is provisional and the length of the intergenic regions (shown with asterisks) may change in the future. MGC, mammalian gene collection; these two genes have not as yet been characterized. (See Color Insert.)

4.2. Repeat Elements and Pleomorphism

The kallikrein locus was also analyzed for the presence of repeat elements [40]. The entire sequence has 49.59% GC content, which is comparable to other genomic regions. Approximately 52% of the region was found to contain various repetitive elements (on either strand). Short interspersed nuclear elements, such as ALU and MIR repeats, are the most abundant repetitive elements (22.53%), followed by the long interspersed nuclear elements, which represent 13.1% of the repetitive elements. Other repeat elements, including Tigger2, MER8, and MSR1, were also identified in *KLK4* introns [41].

The human kallikrein locus contains a unique minisatellite element that is restricted to chromosomal band 19q13, and ten clusters of this minisatellite are distributed along the kallikrein locus. These clusters are mainly located

TABLE 2

Coordinates of all Genes and Pseudogenes Contained within the Human Kallikrein Locus Numbers are given According to GenBank Contig NT_011109

Name	Strand	Start	End	Length	Number of exons	Intergene region
ACPT	+	23561862	23566671	4810	11	2479
MGC13170	–	23576164	23569151	7014	2	12962
MGC 45922	+	23589127	23590324	1198	2	269
KLK1	–	23595233	23590594	4640	5	1501
KLK15	–	23608659	23596735	11925	5	17701
KLK3	+	23626361	23632210	5850	6	12668
KLK2	+	23644879	23650051	5173	5	8299
KLKΨ1	–	23659265	23658351	915	1	18532
KLK4	–	23682184	23677798	4387	6	16794
ΨPPIA[a]	–	23699479	23698977	503	1	15269
KLK5	–	23724466	23714749	9718	6	5611
KLK6	–	23741119	23730078	11042	7	6799
KLK7	–	23755340	23747919	7422	6	12113
KLK8	–	23773148	23767454	5695	6	810
KLK9	–	23781080	23773959	7122	5	4635
KLK10	–	23791472	23785716	5757	6	2204
KLK11	–	23799480	23793677	5804	6	1057
KLK12	–	23806338	23800538	5801	6	21314
KLK13	–	23836557	23827653	8905	5	12206
KLK14	–	23855692	23848764	6929	8	14272
CAG[b]	–	23876037	23869965	6073	3	20317
SIGLEC9	+	23896355	23901757	5403	7	—

[a]Coordinates for ΨPPIA are provisional.
[b]CAG is also known as LOC90353.

in the promoters and enhancers of genes, as well as in introns and in the untranslated regions of mRNAs. Polymerase chain reaction (PCR) analysis of two clusters of these elements indicates that they are polymorphic, and thus they can be useful tools in linkage analysis and DNA fingerprinting. Interestingly, one of the clusters was found to extend from the last part of exon 3 of the *KLK14* gene. Our preliminary data show that the distribution of the different alleles of these minisatellites might be associated with malignancy [40].

5. Structural Features of the Human Tissue Kallikrein Genes and Proteins

Extensive analyses over the last few years have led to the identification of many common structural features of kallikreins. Some of these features are shared with other members of the S1 family of serine proteases (see following). Other features, however, are unique to certain kallikreins. As mentioned above, some of the kallikrein-specific features are conserved among species (e.g., all kallikreins have five coding exons, and only one member in each species has true kininogenase activity). Human tissue kallikrein gene lengths range from 4 to 10 kb, with most of the differences attributed to introns.

5.1. Common Structural Features

The common structural features of kallikreins can be summarized as follows (see also Table 3) [34, 42–44]. First, all genes possess five coding exons (except for a *KLK4* variant, which has four exons), and most of them have one or two extra 5′ untranslated exons. The first coding exon always contains a 5′ untranslated region, followed by the methionine start codon, located 37–88 bp away from the end of the exon. The stop codon is always located 150–189 bp from the beginning of the last coding exon. Second, exon sizes are very similar or identical. Third, the intron phases of the coding exons (i.e., the position where the intron starts in relation to the last codon of the previous exon) are conserved in all genes. The pattern of the intron phase is always I-II-I-0. Fourth, the positions of the residues of the catalytic triad of serine proteases are conserved, with the histidine always occurring near the end of the second coding exon, the aspartate at the middle of the third coding exon, and the serine residue at the beginning of the fifth coding exon. Fifth, all kallikrein proteins are predicted to be synthesized as pre-pro-peptides, with a signal peptide of 16–57 amino acids at the N terminus, followed by an activation peptide of about three to nine amino acids, followed by the enzymatically active (mature) protein (223–252 amino acids). Sixth, the amino acid of the substrate-binding pocket is either aspartate (11 kallikreins) or glutamate (1 kallikrein), indicating trypsin-like specificity (12 kallikreins), or another amino acid (probably conferring chymotryptic or other activity), as is the case with hK3 (serine), hK7 (asparagine), and hK9 (glycine). Seventh, in addition to the conservation of the catalytic amino acid triad, seven additional protein motifs were also found to be highly conserved in kallikreins [44]. Eighth, most, if not all, kallikrein genes are under steroid hormone regulation. Ninth, all proteins contain 10–12 cysteine residues, which will form five (in hK1, hK2, hK3, and hK13) or six (in all other kallikreins) disulphide bonds. The positions of the cysteine residues are also fully conserved. Finally, classical or variant polyadenylation signals have been found 10–20 bases away from the poly-A tail of all kallikrein mRNAs. All three classical kallikreins have the same variant polyadenylation signal AGTAAA [5, 33, 45–47]. Multiple alignments of all kallikrein proteins have been published previously [43].

TABLE 3

SUMMARY OF HUMAN TISSUE KALLIKREIN PROTEIN CHARACTERISTICS

Protein name	pI[a]	MW[b]	Full length[c]	Signal peptide[c]	Activation peptide[c]	Mature protein[c]	Cysteines[d]	Substrate specificity[e]
hK1	4.62	28.9	262	17	7	238	10	Trypsin-like
hK2	6.44	28.7	261	17	7	237	10	Trypsin-like
hK3	6.78	28.8	261	17	7	237	10	Chymotrypsin-like
hK4	4.71	27.0	254	26	4	224	12	Trypsin-like
hK5	8.64	32.0	293	57	9	227	12	Trypsin-like
hK6	7.15	26.8	244	16	5	223	12	Trypsin-like
hK7	8.82	27.5	253	22	7	224	12	Chymotrypsin-like
hK8	8.61	28.5	260	28	4	228	12[d]	Trypsin-like
hK9	7.10	27.5	250	19	3	229	12	Chymotrypsin-like
hK10	8.95	30.1	276	33	9	234	12	Trypsin-like
hK11	8.80	27.5	250	18	3	229	12[f]	Trypsin-like
hK12	7.57	26.7	248	17	4	227	12	Trypsin-like
hK13	8.78	30.6	277	20	5	252	10	Trypsin-like
hK14	9.30	27.5	251	18	6	227	12	Trypsin-like
hK15	8.27	28.1	256	16	5	235	12	Trypsin-like

[a] Isoelectric point.

[b] Molecular weight of the full-length protein in kilodaltons, excluding any post-translational modifications.

[c] Number of amino acids.

[d] The number of cysteine residues in the mature enzyme.

[e] Some substrate specificities have not been experimentally verified.

[f] Has an extra "C" residue in a nonconserved position.

5.2. Three-Dimensional Structure

The crystal structure has been revealed for some rodent kallikreins. The three-dimensional structure of a horse orthologue of the human hK3 has also been reported [28]. In contrast, hK1 and hK6 are the only human kallikreins for which crystal structures have been determined [48–50]. Most of the discussion in this section is derived from comparative model building of the human kallikrein proteins.

Kallikreins can be roughly divided into two categories, the classical kallikreins (hK1, hK2, and hK3) and the "new" kallikreins. The new kallikreins appear to be unique in their three-dimensional structure and share some features with trypsins and other features with the classical kallikreins. Comparative protein models show that the pattern of hydrophobic side-chain packing in the protein core is nearly identical in all human kallikreins, and the observed differences occur within the solvent-exposed loop segments.

An 11–amino acid residue insertion relative to the trypsin sequences in loop E (residues 91–103 in the bovine chymotrypsinogen consensus numbering), also known as "the kallikrein loop," is a unique feature of the three classical kallikreins. Loop E is located between the fifth (residues 81–90) and the sixth strand (residues 104–108), and loop G between the seventh and the eighth strand (residues 156–163). None of the new human kallikreins contains this loop in its entirety. Loop E in hK10 is longest, with an eight-residue-long insertion relative to the trypsin sequences. Loop E overhangs the substrate-binding groove on the surface of the protease molecule, and its length and sequence can directly influence substrate recognition.

The *KLK15* gene is particularly interesting, as it lies between two classical tissue kallikrein genes, *KLK1* and *KLK3* [51], yet the sequence and structure of hK15, with six disulphide bonds and no insertion in the so-called kallikrein loop (E), clearly place it among the new kallikreins. Moreover, in loop G, hK15 has an eight-residue insertion [51] that is not found in any other kallikrein. In the three-dimensional structure, the extended loop G lies on the opposite side of the active site relative to loop E, and although it is more distant to the substrate-binding groove, loop G may also participate in substrate and inhibitor recognition.

6. Sequence Variations of Human Kallikrein Genes

Sequence changes, including polymorphisms and mutations, are clinically important. In addition to medicolegal applications, they can also be indicators for susceptibility and prognosis for different malignancies [52]. *KLK3* is the most extensively studied kallikrein in this respect. Comparison of the published mRNA sequences of *KLK3* revealed infrequent and inconsistent sequence variations. Baffa *et al.* found no evidence of mutations in the *KLK3* mRNA sequence in prostate cancer compared to matched normal tissues from the same patient [53]. Similarly, no mutations were found in the coding portion of the *KLK3* gene in breast cancer tissues and cell lines, with the exception of a polymorphism in exon 2 in some breast tumors [54]. Three distinct forms of *KLK1* mRNA, differing in one or two amino acid substitutions, were identified in different tissues [1, 55, 56]. Experimental evidence, however, indicates that the protein products of these variants have no difference in their protein activity [57]. Probably the most polymorphic sequence of *KLK4* is that deposited by Hu *et al.* [41]. In addition to a large insertion in the 3′ untranslated region, there are 18 differences between their sequence and those deposited by others. These probable polymorphisms will affect the derived amino acid product [41]. We have recently identified a germline single nucleotide variation in exon 3 of the *KLK10* gene that will change the amino acid from alanine to serine. This polymorphism is less prevalent in prostate cancer patients in comparison to control subjects [58]. Also, four other polymorphisms were identified in exon 4 of the same gene.

Within a 5.8-kb promoter/enhancer region of *KLK3*, 16 different mutational hotspots (appearing more than once in nine tumors) were found in breast cancer [54]. A single nucleotide variation (G $\rightarrow$ A) was identified at position −158 within androgen response element 1 (ARE-1). Univariate Cox regression modeling showed a 28% reduction in the risk of death in patients with homozygous G genotype compared to those with homozygous A [59].

In general, the examination of sequence variation within the 300-kb kallikrein locus has not been performed in detail, and it is therefore possible that inactivating mutations within the coding region of kallikrein genes exist but await discovery. Kallikrein gene inactivation in human diseases would likely provide clues for the physiological functions of these diseases.

7. The Tissue Kallikreins in the Context of Other Serine Proteases in the Human Genome

Proteases are enzymes that cleave proteins by hydrolyzing peptide bonds. On the basis of their catalytic mechanisms, they can be classified into five main types: proteases that have an activated cysteine residue (cysteine proteases), an activated aspartate (aspartate proteases), a metal ion (metalloproteases), or an activated threonine (threonine proteases), and proteases with an active serine (serine proteases). Within each type, enzymes are separated into "clans" (also referred to as "superfamilies") based on evidence of evolutionary relationship [60, 61] from the linear order of catalytic site residues and the tertiary structure, in addition to distinctive aspects of catalytic activity such as specificity or inhibitor sensitivity. Each clan is given a two-letter identifier, of which the first letter is an abbreviation for the catalytic type. Next, proteins are classified into families (each denoted by a unique number) and subfamilies (denoted by another letter) based on sequence similarity to a chosen type example for that family.

7.1. Genomic Overview of Serine Proteases

According to our recent survey, and studies by others, there are approximately 500 confirmed, nonredundant proteases in the human genome, including nonpeptidase homologues [62–65]. This represents about 2% of all gene products in humans [60]. This number increases to about 700 when the "predicted" genes and proteins are included. Approximate figures indicate that proteases are distributed as follows: 4% are aspartate proteases, 26% are cysteine proteases, 34% are metalloproteases, 5% are threonine proteases, and 31% are serine proteases.

Serine proteases (SP) are a family of enzymes that use a uniquely activated serine residue in the substrate-binding pocket to catalytically hydrolyze peptide bonds [66]. SP carry out a diverse array of physiological functions, of which the best known are digestion, blood clotting, fibrinolysis, fertilization, and complement activation during immune responses [67]. They have also been shown to be abnormally expressed in many diseases including cancer, arthritis, and emphysema [42, 43, 67–70].

Out of the estimated 500 proteases in the human genome, 31% are predicted to be serine proteases [71]. This large family includes the digestive enzymes (e.g., trypsin, chymotrypsin), the kringle domain-containing growth factors (e.g., tissue plasminogen activator), some of the blood clotting factors, and the kallikreins. In terms of absolute numbers, a total of 150 serine proteases have been identified in the human genome and are distributed in all chromosomes except 18 and Y [62]. The density of serine proteases varies from just one to two genes in most chromosomes to up to 23 genes on chromosome 19. Most SP are localized sporadically, and relatively few clusters exist. Kallikreins represent the largest cluster of serine proteases in the human genome.

7.2. Structural and Sequence Analysis

From a structural point of view, kallikreins belong to the serine protease family of enzymes. The essential features of serine proteases that are preserved in the kallikreins can be summarized as follows: only one serine residue of the protein is catalytically active; two residues, a histidine and an aspartate, are always associated with the activated serine in the catalytic site, forming together what is known as the "catalytic triad" of serine proteases; each of the catalytic triad residues is surrounded by a highly conserved motif [the motif GD*S*GGP surrounds serine, TAA*H*C surrounds histidine, and *D*IMLL surrounds aspartate residues [60]]; the active serine is situated in an internal pocket, and the aspartate and histidine residues are closely located in the three-dimensional structure; the catalytically essential histidine and serine are almost immediately adjacent to their exon boundaries; they are initially produced in a "zymogen" form; they exhibit a high degree of sequence similarity; and they contain most of the 29 amino acids that are reported by Dayhoff to be invariable in many species [72].

Kallikreins belong to the S1 family (also known as the trypsin family), of clan SA of serine proteases. This is supported by sequence, structural, and phylogenetic analyses. We have recently performed detailed analyses for 79 protein sequences representing this family, according to the latest information from the Human Genome Project and other databases [63]. Our results show that seven residues are absolutely conserved in this family. An additional 15 showed almost complete conservation (>95%), and in total, 48 residues were found to be more than 80% conserved, and 87 residues were found to display greater than 50% conservation. Conserved residues tended to group together, likely representing certain necessary structural or functional domain elements. This conclusion is supported by the fact that in most cases of substitution, a residue of similar character (i.e., size, hydrophobicity, or polarity) was inserted. This implies that the overall character of the local region is conserved for proper function, more so than some of the individual amino acid identities. In addition to the conserved motifs around the residues of the catalytic triad, we recently identified 32 other highly conserved amino acid motifs in the S1A family of SP, including tissue kallikreins [63]. The biological significance of these motifs has yet to be determined. Multiple alignment showed the presence of a conserved domain (**R**/K)(**I**/V)(**V**/I)(**G**/N) at the N-terminal cleavage site of the zymogen (proenzyme) end of most members of this family. Most enzymes are cleaved after an Arg or Lys, indicating the need for a trypsin-like enzyme for activation. In the case of trypsin, cleavage occurs between residues Lys^{15} and Ile^{16} (chymotrypsinogen numbering). After cleavage Ile^{16} forms the new N terminus of the protein, and Asp^{194} rotates to interact with it. This rotation, and the resulting salt bridge, produces a conformational change that completes the formation of the oxyanion hole and the substrate binding pocket, both of which are necessary for proper catalytic activity. Certain sequences did not display conservation of this trypsin cleavage site, with substitutions at either the 15th or 16th positions (e.g., the granzymes). These substitutions likely result in either cleavage by a protease with different specificity or no cleavage at all. The presence of Asp at position 189 indicates that most members of the S1A family will have a trypsin-like specificity. In chymotrypsin and chymotrypsin-like proteases (e.g., hK3), there is a serine at this position.

Structurally conserved regions usually remain conserved in all members of the family and are usually composed of secondary structure elements, the immediate active site, and other essential structural framework residues of the molecule. For instance, Ser^{214} in chymotrypsin-like proteases contributes to the S1 binding pocket and appears to be a fourth member of a catalytic tetrad [73]. Between these conserved elements are highly variable stretches (also called variable regions). These are almost always loops that lie on the external surface of the protein and that contain all additions and deletions between different protein sequences. The former regions (structurally conserved regions) have been successfully used as the basis for predicting the three-dimensional structure of newly identified SP based on information on existing members [74]. The latter (variable regions) are important for studying the evolutionary history of SP.

Certain residues with a variable degree of conservation can be investigated for their usefulness as "evolutionary markers" that can provide insight into the history of each enzyme family or clan and allow comparative analysis with other families or clans. Krem and Di Cera [75] identified several such markers with proven evolutionary usefulness. In addition to the use of these markers for rooting the phylogenetic trees, attempts were made to classify serine proteases into functional groups based on these markers or their coding sequences.

Serine proteases exhibit preference for hydrolysis of peptide bonds adjacent to a particular class of amino acids. In the trypsin-like group, the protease cleaves peptide bonds following basic amino acids such as arginine or lysine, as it has an aspartate (or glutamate) in the substrate-binding pocket, which can form a strong electrostatic bond with these residues. The chymotrypsin-like proteases have a nonpolar substrate-binding pocket and thus require an aromatic or bulky nonpolar amino acid such as tryptophan, phenylalanine, tyrosine, or leucine. The elastase-like enzymes, however, have bulky amino acids (valine or threonine) in their binding pockets, thus requiring small hydrophobic residues, such as alanine [76].

Activation reactions catalyzed by serine proteases (including kallikreins) are an example of "limited proteolysis" in which the hydrolysis is limited to one or two particular peptide bonds. Hydrolysis of peptide bonds starts with the oxygen atom of the hydroxyl group of the serine residue that attacks the carbonyl carbon atom of the susceptible peptide bond. At the same time, the serine transfers a proton first to the histidine residue of the catalytic triad and then to the nitrogen atom of the susceptible peptide bond, which is then cleaved and released. The other part of the substrate is now covalently bound to the serine by an ester bond. The charge that develops at this stage is partially neutralized by the third (asparate) residue of the catalytic triad. This process is followed by "deacylation," in which the histidine draws a proton away from a water molecule and the hydroxyl ion attacks the carbonyl carbon atom of the acyl group that was attached to the serine. The histidine then donates a proton to the oxygen atom of the serine, which will then release the acid component of the substrate.

8. Tissue Expression and Cellular Localization of the Kallikrein Genes

8.1. Overview

Many kallikreins are transcribed predominantly in only a few tissues, as indicated by Northern blotting. By using the more sensitive reverse transcriptase (RT)–PCR technique, kallikreins were found to be expressed at lower amounts in several other tissues. Many kallikreins are expressed in the salivary gland—the tissue in which most of the rodent kallikreins are expressed. In addition, several kallikreins were also found in the central nervous system and endocrine-related tissues such as the prostate, breast, and testis. A more detailed review of the expression of kallikreins in different tissues by various techniques can be found elsewhere [44]. A global view of kallikrein expression in 36 different tissues, as determined by RT-PCR, is presented in Fig. 2.

8.2. Immunohistochemical Expression of Human Tissue Kallikreins

Most studies use quantitative methods such as quantitative RT-PCR to reveal the expression of human tissue kallikreins in benign and malignant tissues. The recent development by our group of monoclonal and polyclonal antibodies against many kallikrein proteins has helped in defining their distribution in tissue extracts [77–79]. Furthermore, using the streptavidin–biotin method with monoclonal and polyclonal antibodies, we have already studied the immunohistochemical expression (IE) of seven hKs (hK5, hK6, hK7, hK10, hK11, hK13, and hK14) in several normal human tissues, as well as in the corresponding malignant tissues [6, 80–85, and our unpublished data). Different antibodies used for each kallikrein (polyclonal and monoclonal) revealed a similar immunostaining pattern in many tissues. The IE was always cytoplasmic and in some tissues displayed a characteristic immunostaining pattern that was membranous, droplet-like, supranuclear, subnuclear, and luminal.

Comparison of the IE patterns of a few studied kallikreins in different tissues revealed no major differences, indicating that they may share a common mode of regulation. It is worth mentioning that our results concerning the IE of the studied hKs in different normal human tissues correspond fairly

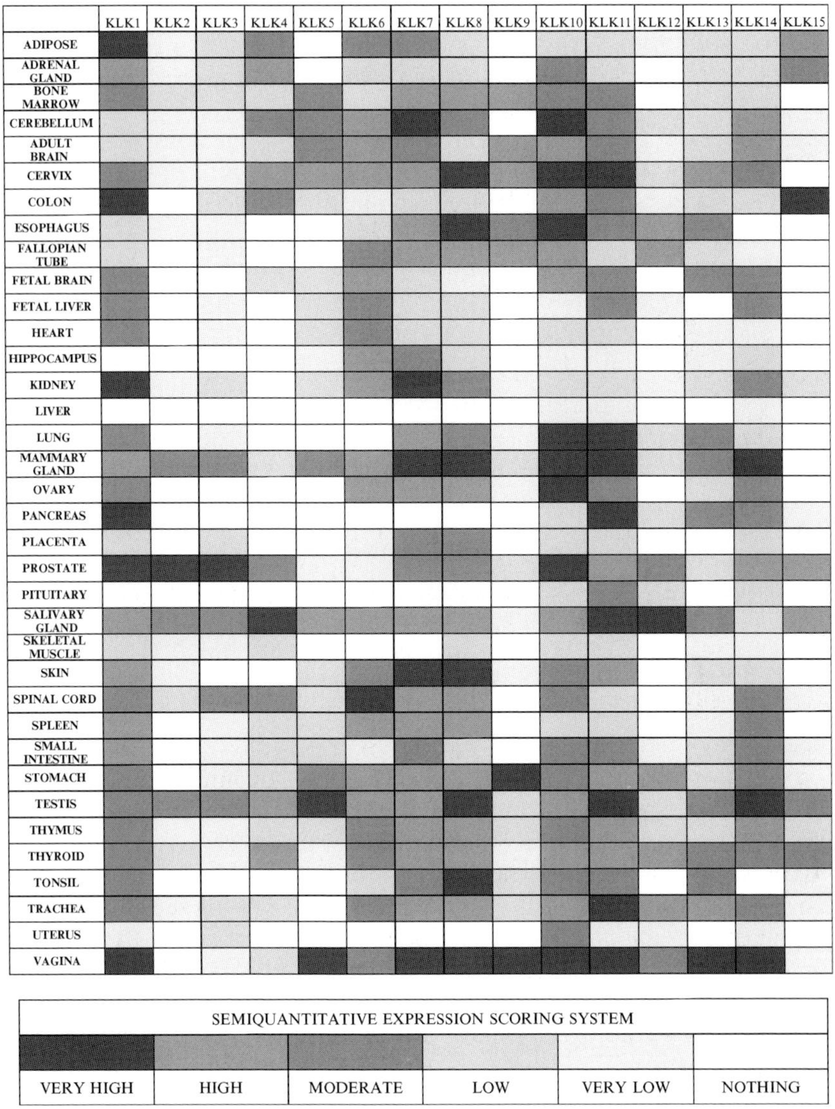

	KLK1	KLK2	KLK3	KLK4	KLK5	KLK6	KLK7	KLK8	KLK9	KLK10	KLK11	KLK12	KLK13	KLK14	KLK15
ADIPOSE															
ADRENAL GLAND															
BONE MARROW															
CEREBELLUM															
ADULT BRAIN															
CERVIX															
COLON															
ESOPHAGUS															
FALLOPIAN TUBE															
FETAL BRAIN															
FETAL LIVER															
HEART															
HIPPOCAMPUS															
KIDNEY															
LIVER															
LUNG															
MAMMARY GLAND															
OVARY															
PANCREAS															
PLACENTA															
PROSTATE															
PITUITARY															
SALIVARY GLAND															
SKELETAL MUSCLE															
SKIN															
SPINAL CORD															
SPLEEN															
SMALL INTESTINE															
STOMACH															
TESTIS															
THYMUS															
THYROID															
TONSIL															
TRACHEA															
UTERUS															
VAGINA															

SEMIQUANTITATIVE EXPRESSION SCORING SYSTEM					
VERY HIGH	HIGH	MODERATE	LOW	VERY LOW	NOTHING

FIG. 2. Expression map of human tissue kallikreins in a variety of tissues, as determined by reverse transcriptase polymerase chain reaction. The relative semiquantitative expression levels for each gene are indicated. (See Color Insert.)

well with other data based on ELISA assays and RT-PCR. According to these studies, except for *KLK2* and *KLK3*, none of the remaining KLKs is tissue-specific, although certain genes are preferentially expressed in some organs [31, 86–88]. Many hKs were immunohistochemically revealed in a variety of tissues, indicating that no protein is tissue specific. Immunohistochemistry has advantages over other methods, as it defines the protein distribution in different cell types independent from its quantity in the tissue. A tissue may therefore immunohistochemically express a kallikrein and yet yield negative results using a quantitative method. This likely explains why we did not find any immunohistochemical difference in the tissue expression among the different hKs, whereas other methods showed tissue preferences for each hK.

In recent RT-PCR studies, many kallikreins have been proposed as new biomarkers for malignancies other than prostate cancer. Breast, ovarian, and testicular cancers are the most studied. Certain kallikreins were found to be differentially expressed in various malignancies (up- or downregulated), and the increase or decrease of their expression may be associated with prognosis [43, 58, 70, 89–94]. We have immunohistochemically evaluated some kallikreins in malignant diseases, including two series of prostate and renal cell carcinoma, and have examined their prognostic values [84, 85].

In malignancy, glandular epithelia constitute the main kallikrein immunoexpression sites, and staining of their secretions indicating that these proteases are secreted. Similar to the IE pattern in normal glandular tissues, all hKs are expressed in adenocarcinomas. In the future, it is clearly worthwhile to study the relation of the IE of all hKs with prognosis of several malignancies and to correlate these results with those obtained by other methods.

In Figs. 3–5, we present immunohistochemical data for various kallikreins. For more discussion, please refer to our detailed publications [80–85].

Regarding immunohistochemical localization, hK4 appears to be a notable exception. Recently, Xi *et al.* suggested that hK4 is a predominantly nuclear protein that is overexpressed in prostate cancer [95]. It appears that the vast majority of *KLK4* mRNA lacks exon 1, which codes for the signal peptide. These preliminary data need to be reproduced.

9. Regulation of Kallikrein Activity

9.1. At the mRNA Level

Promoter analysis and hormonal stimulation experiments allowed us to obtain insights into the mechanisms that regulate expression of the human kallikrein genes. In addition to *KLK3* and *KLK2*, and more recently *KLK10*, no other kallikrein gene promoter has been functionally tested.

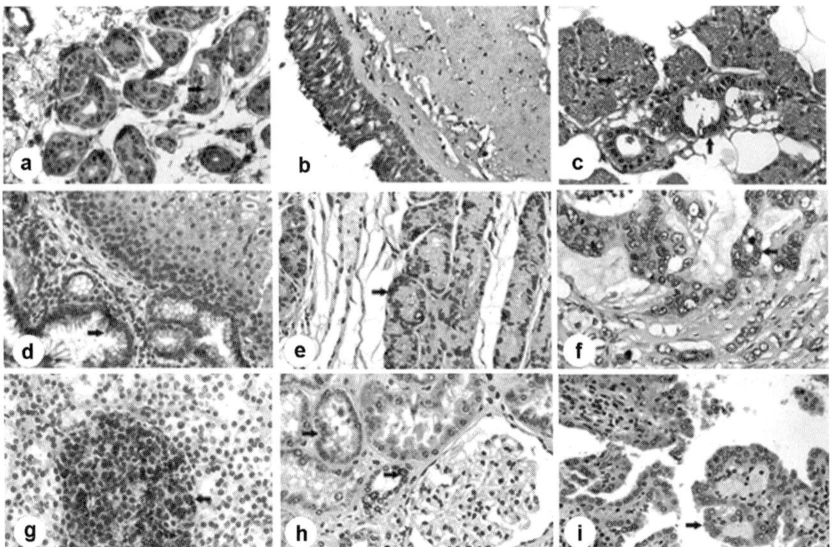

FIG. 3. Immunohistochemical expression of (a) hK7 by the epithelium of eccrine glands of the skin (monoclonal antibody, clone 73.2), (b) hK13 by the epithelium of the bronchus (monoclonal antibody, clone IIC1), (c) hK5 by the ductal epithelium of the parotid gland (polyclonal antibody), (d) hK7 by the esophageal glands (monoclonal antibody, clone 73.2), (e) hK13 by the gastric mucosa (monoclonal antibody, clone 2–17), (f) hK6 by the large intestine mucosa (polyclonal antibody), (g) hK10 in an islet of Langerhans in the pancreas (monoclonal antibody, clone 5D3), (h) hK11 by the epithelium of the urinary tubuli (monoclonal antibody), (i) hK11 by a papillary renal cell carcinoma (monoclonal antibody). (See Color Insert.)

TATA box variants are found in the three classical kallikreins. *KLK1* has the variant TTTAAA, whereas *KLK2* and *KLK3* share another variant, TTTATA. [5, 33, 45–47]. In addition, two AREs have been identified and experimentally verified [96]. Another ARE was mapped in the far upstream enhancer region of the gene and shown to be functional and tissue specific [97, 98]. More recently, five additional low-affinity AREs have been identified close to ARE-III [99], and three distinct regions surrounding ARE-III were found to bind ubiquitous and cell specific proteins. A functional ARE was also identified in the *KLK2* promoter [100]. Interestingly, a negative regulatory element was also found between −468 and −323 of *KLK2* [100], and another ARE was identified between −3819 and −3805 of the *KLK2* promoter [101].

Apart from *KLK1, KLK2*, and *KLK3*, no obvious TATA boxes were found in the promoter of other kallikreins. Two major obstacles exist in predicting the promoter response elements: the inaccurate localization of

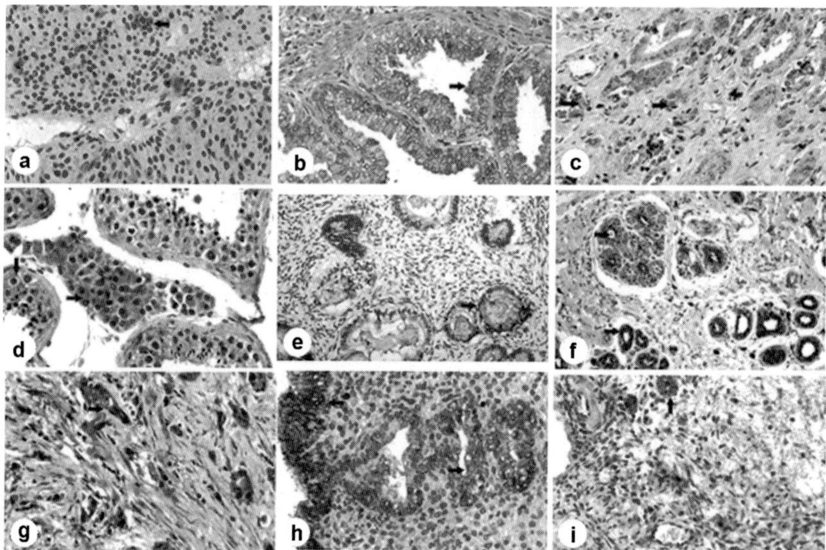

FIG. 4. Immunohistochemical expression of: (a) hK10 by a low-grade urothelial carcinoma (monoclonal antibody, clone 5D3), (b) hK11 by the secretory epithelium of the prostate gland (polyclonal antibody), (c) hK11 by a Gleason score 6 prostate carcinoma (polyclonal antibody), (d) hK14 by the spermatic epithelium and the stromal Leydig cells in the testis (polyclonal antibody), (e) hK10 by epithelial elements in a testicular immature teratoma (monoclonal antibody, clone 5D3), (f) hK14 by lobuloalveolar structures of the breast (polyclonal antibody), (g) hK14 by a ductal breast carcinoma, grade II (polyclonal antibody), (h) hK13 by the glandular epithelium of the endometrium (polyclonal antibody), (i) hK14 by luteinized stromal cells of the ovary (polyclonal antibody). (See Color Insert.)

the transcription start site and the presence of more than one splice variant with more than one transcriptional start site. Recently, by using EST analysis alone, Grimwood *et al.* extended the 5′-end of many published mRNAs from chromosome 19 by more than 50 bp [3].

9.2. At the Protein Level

There are different mechanisms for controlling serine protease activity by which unwanted activation is avoided and precise spatial and temporal regulation of the proteolytic activity is achieved. One important mechanism is by producing kallikreins in an inactive "proenzyme" (or zymogen) form, which will be activated as necessary. The N-terminal extension of the mature enzyme, or the "prosegment," sterically blocks the active site and thus prevents binding of substrates. It is also possibly implicated in folding, stability, and intracellular sorting of the zymogen. For more detailed

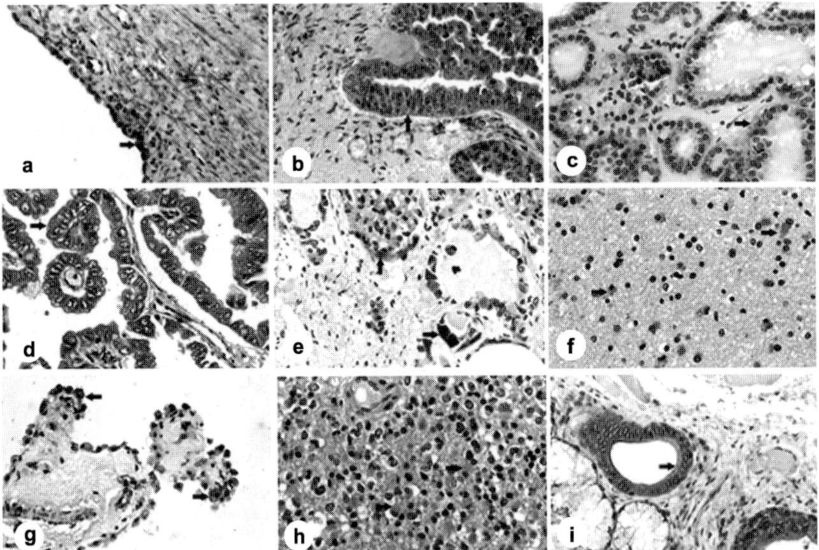

FIG. 5. Immunohistochemical expression of (a) hK14 by the ovarian surface epithelium (polyclonal antibody), (b) hK14 by a cystadenocarcinoma of the ovary (polyclonal antibody), (c) hK10 by hyperplastic follicles of the thyroid gland (monoclonal antibody, clone 5D3), (d) hK6 by a papillary thyroid carcinoma (polyclonal antibody), (e) hK13 by endocrine cells in the pituitary gland (monoclonal antibody, clone 2–17), (f) hK10 by glial cells in the brain (monoclonal antibody, clone 5D3), (g) hK13 by the choroid plexus epithelium (monoclonal antibody, clone 2–17), (h) hK5 by a glioma (monoclonal antibody, clone 6.10), (i) hK7 by the ducts of the submucosal glands of the tonsils (monoclonal antibody, clone 85.2). (See Color Insert.)

discussion, see the recent review by Khan and James [102]. The activation of the zymogen can occur intracellularly (i.e., in the trans-Golgi apparatus or in the secretory granules) or extracellularly after secretion, and it can be autolytic or dependent on the activity of another enzyme (see following). Interestingly, all of the "proforms" of the kallikrein enzymes, with the exception of hK4, are predicted to be activated by cleavage at the C-terminal end of either arginine or lysine (the preferred trypsin cleavage site), indicating that they need an enzyme with trypsin-like specificity for their activation. This observation has been experimentally demonstrated for some kallikreins. For example, hK5 and hK7 can be converted to the active enzyme by trypsin treatment [103, 104], and hK11 can be activated by enterokinase. Autoactivation is a common phenomenon among kallikreins—hK2, though not hK3, is capable of autoactivation [105]. hK3 has chymotryptic activity, but it needs a trypsin-like activating enzyme. hK4 is also autoactivated during the refolding process [106], and there is evidence that hK6 is also capable of autoactivation [107]. hK13 is also autoactivated on secretion (G. Sotiropoulou, personal communication). This autoactivation can be explained by the finding that many kallikreins have trypsin-like substrate activity and that the same type of activity is needed for their activation.

Proteolytic activation is irreversible. Hence, other means of switching off the activity of these enzymes are needed. Once activated, serine proteases are controlled by ubiquitous endogenous inhibitors. Laskowski and Qasim divide all known inhibitors into two categories: inhibitors devoid of significant class specificity, and class-specific inhibitors [108]. The former includes proteins of the α_2-macroglobulin family, which bind proteases through a molecular trap mechanism and inhibit them by steric hindrance [109]. With respect to specific serine protease inhibitors, at least 23 structurally distinct families are currently known, including the Kunitz, soybean trypsin inhibitors–Kunitz, Kazal, and hirudin families, as well as the serpins (serine proteinase inhibitors) [108]. Many of the specific inhibitors are capable of inhibiting the same serine protease, and the same inhibitor may inhibit several serine proteases [108]. Some molecular complexes of kallikreins with protease inhibitors have clinical applicability because they can improve the diagnostic sensitivity or specificity of cancer biomarkers such as PSA [110].

The majority of hK3 (PSA) in the prostate and seminal plasma is in its free form, with less than 5% complexed with protein C inhibitor [111]. In serum, the majority of hK3 binds to protease inhibitors, and only 15%–25% is in the free form. The hK3–α_1-antichymotrypsin complex constitutes 70%–85% of total serum hK3, with the hK3–α_2-macroglobulin and hK3–α_1-antitrypsin complexes representing 15% and 3%, respectively [112–114].

hK2 has been shown to complex with protease C inhibitor both in *in vitro* studies and in seminal plasma [115]. In serum, most of hK2 is in the free form, with only a small amount complexed with α_1-antichymotrypsin [116, 117]. α_2-macroglobulin [118], and antithrombin III [119] have also been shown to be able to bind hK2. hK2 was also found to be bound to α_2-antiplasmin and plasminogen activator inhibitor 1 [120]. More recently, hK5 was found to form complexes with α_1-antitrypsin [121], α_1-antichymotrypsin was identified as an inhibitor for hK6 [122], and hK13 was reported to form complexes with α_1-antichymotrypsin [123]. Another mechanism for controlling the activity of proteases is by internal cleavage and subsequent degradation. Self-digestion is reported for hK7 [104]. Around 30% of hK3 in seminal plasma is inactivated by internal cleavage between lysine 145 and lysine 146 [124], and about 25% of hK2 was also found to be internally cleaved between amino acids 145–146 (Arg-Ser) [125]. Other internal cleavage sites have also been identified for hK2 and hK3, but the enzymes responsible for such cleavages are not yet known.

9.3. Locus Control of Kallikrein Expression

The coexpression of many kallikreins in the same tissues and the parallel differential regulation of groups of kallikreins in pathological conditions (e.g., the upregulation of seven kallikreins in ovarian cancer [126]) raise the possibility of the existence of a common mechanism that controls the expression of groups of kallikrein genes in a cluster known as a locus control region (LCR). Added to this are the relatively short distances between adjacent kallikreins [which could be as short as the 1.5 kb between *KLK1* and *KLK15* [34]] and the absence of classic promoter sequences, as shown by prediction analysis, in all kallikreins except *KLK2* and *KLK33*. Clustering of coexpressed homologous genes could be explained by the evolutionary history of the genomic region. The probable mechanism in this case would include local duplication and divergence of amplified copies, resulting in an array of paralogues that may retain common regulatory elements [127].

Previous studies have shown that two sequence elements were essential for initiating DNA replication of an adjacent group of beta globin genes: the initiation region and the LCR, residing 50 kb upstream of the initiation region. The beta globin LCR is located 6–20 kb upstream of a cluster of five functional globin genes. It consists of five DNAse hypersensitive sites and numerous binding sites for transcription factors. LCRs are operationally defined by their ability to enhance the expression of linked genes to physiological levels in a tissue-specific and copy number–dependent manner [128]. Although their composition and locations relative to their cognate genes are different, LCRs have been described in a broad spectrum of mammalian gene systems, indicating that they play an important role in the control of eukaryotic gene expression. Other intergenic sequences, such as domain boundaries or barriers, and chromatin architecture might also be involved. Acquisition of knowledge about these processes is a key step toward the understanding of the role of kallikreins in normal physiology and pathobiology.

Another proposed regulatory mechanism is gene potentiation, which is the process of opening a chromatin domain that will render genes accessible to the various factors required for their expression. The formation of an open chromatin structure is central to the establishment of cell fate and tissue-specific gene expression. Many eukaryotic genes are organized into functional chromatin domains. This facilitates their coordinated regulation during development [129]. The ability of individual cells to regulate the genes contained within such chromatin domains is of paramount importance for their differentiation. Perturbations in chromatin structure can act both locally, to alter the accessibility of *trans*-acting factors to *cis*-regulatory elements, and globally, to affect the opening and closing of entire chromatin domains [130]. The potentiated state of a gene can also be influenced by alterations in the local chromatin environment. For example, many eukaryotic genes are differentially expressed by altering their methylation status. These genes are largely unmethylated in cells where they are transcribed but fully methylated in all nonexpressing cells [131]. Histone acetylation also acts on the local gene environment during the transition from the 30-nm fiber to a more open structure that can be linked to a 10-nm fiber, stabilizing the more relaxed open structure [132]. It has also been postulated that DNA methylation patterns may serve to modulate histone acetylation, thereby maintaining local chromatin states. Both DNA demethylation and histone acetylation render increased accessibility of ubiquitous and tissue-specific *trans*-acting factors to *cis*-regulatory elements, facilitating transcriptional activation [133].

9.4. Epigenetic Regulation of Kallikrein Gene Expression

In human cancers, numerous mechanisms may contribute to loss of tumor suppressor gene function, including homozygous deletions, allelic loss in combination with mutations, abnormal splicing, and CpG island methylation [134]. Methylation contributes to inactivation of numerous genes, including the cell cycle regulator *p16* [135], the growth suppressor *ER* [136], the epithelial cadherin *E-cadherin* [137], the DNA repair gene *MGMT* [138], the Ras-associated domain family 1A gene *RASSF1A* [139], the angiogenesis inhibitor *THBS1* [140], and the metastasis suppressor *TIMP3* [141]. In addition, hK10 can also be inactivated by CpG island hypermethylation in breast cancer [142] and acute lymphoblastic leukemia [143].

The physiological function of hK10 is still unclear. However, recent data indicate that *KLK10* may have a tumor suppressor function, based on its downregulation in breast and prostate cancer cell lines and the finding that overexpression of *KLK10* in nude mice can suppress tumor formation [144, 145]. The expression of the protein in certain tissues and its homology to other members of the kallikrein family should provide a starting point in the search for physiologically relevant substrates.

This putative tumor suppressor activity prompted us to speculate that this gene may be a target for either somatic mutations or hypermethylation, in analogy to other tumor suppressor genes that are inactivated by mutations or methylation and that thereby predispose to cancer development or progression. A previous study from our laboratory [58] examined in detail the polymorphic and mutational status of the *KLK10* gene, using DNA isolated from normal tissues and from cancers of the breast, ovary, prostate, and testis. Bharaj *et al.* confirmed that the *KLK10* gene does not seem to be a target for somatic mutations in either breast, ovarian, prostate, or testicular cancer. A single nucleotide variation at codon 50, however, was associated with increased prostate cancer risk. Recently, Li *et al.* demonstrated an important role for CpG island methylation in the loss of *KLK10* gene expression in breast cancer [142].

Despite the discovery of *KLK10* as a tumor suppressor through gene downregulaton, recent data indicate that one major mechanism of *KLK10* inactivation may be at the epigenetic level. The frequent loss of *KLK10* expression indicates that inactivating the function of *KLK10* may be a critical step toward carcinogenesis. By treating *KLK10* nonexpressing cells with a demethylating agent and using methylation-specific PCR and sequence analysis of sodium bisulfite-treated genomic DNA, Li *et al.* [142] demonstrated a strong correlation between *KLK10* exon 3 hypermethylation and loss of *KLK10* mRNA expression in a panel of breast cancer cell lines and primary breast tumors. Furthermore, this study supports the notion that *KLK10* expression and its methylation status can be used as a molecular marker in breast cancer. These results justify using larger follow-up studies to evaluate the methylation status of *KLK10* as a screening tool for the detection of breast cancer, as well as other malignancies such as leukemia [143].

For the remaining human kallikrein genes, in none of them has methylation been tested as a potential mechanism for inactivation in the development or progression of cancer. Accumulating data provide indirect evidence that these kallikrein genes may be involved in the development of various malignancies, as a subset of these genes are downregulated in many cancers [43]. It is possible that some of these genes are epigenetically suppressed or silenced through CpG island methylation. Clearly, there is a need to characterize the CpG islands within these genes and to better understand the mechanism and role that epigenetic regulation plays within the human kallikrein locus.

10. Hormonal Regulation of Kallikreins

Steroid hormones, acting through their receptors, play important roles in the normal development and function of many organs. In addition, they are involved in the pathogenesis of many types of cancer [146]. Several reports confirmed that many kallikreins are under steroid hormone regulation in endocrine-related tissues and cell lines [100, 147–155].

An interesting observation is the tissue-specific pattern of hormonal regulation of several of these genes in different tissues. For example, *KLK4* is upregulated by androgen in prostate and breast cancer cell lines [148] and by estrogen in endometrial cancer cell lines [147]. Also, *KLK12* was found to be upregulated by androgens and progestins in prostate cancer cell lines and by estrogens and progestins in breast cancer cell lines [156]. *KLK14* and *KLK15* are mainly regulated by androgens. Recent preliminary kinetic and blocking experiments indicate that this upregulation is mediated through the androgen receptor [155, 157].

In general, it can be concluded that most, if not all, kallikrein genes are regulated by steroid hormones, either predominantly by androgens or by estrogens/progestins/glucocorticoids. Because most of the data have been generated in cell lines, which contain variable amounts of various steroid hormone receptors, it will be very interesting to delineate the hormonal regulation of these genes *in vivo*. Manipulation of kallikrein gene expression by steroids may have therapeutic potential in some diseases such as cancer, psoriasis, or others.

Recently, Palmer *et al.*, working with human colon cancer cell lines, reported dramatic up-regulation of kallikreins 6 and 10 by 1α, 25-dihydroxyvitamin D3 [158]. This finding raises the possibility that some kallikreins could be regulated by a multitude of nuclear receptors.

11. Evolution of Kallikreins

The glandular kallikreins are simple secreted serine proteases. Some studies have investigated the phylogenetic relationship between different serine proteases, but no definitive conclusions regarding the glandular kallikreins could be drawn [62, 63, 159]. A more thorough phylogenetic analysis of the serine proteases is needed to elucidate the origin of the glandular kallikrein family.

A comparison of the genes in the glandular kallikrein family of man and mouse, combining phylogenetic analysis with a structural analysis of the loci, provided important clues about the evolution of the genes within this family [10]. The kallikreins from *KLK4* through to *KLK15* are well conserved in the two species, but the classical glandular kallikreins (*KLK1, KLK2*, and *KLK3*) differ significantly, and only *KLK1* (encoding tissue kallikrein) is present in both species. When the separation of the murine lineages from mammals of higher orders took place approximately 115 million years ago [160], 14 glandular kallikrein genes existed. These genes were *KLK4* to *KLK15* and two classical glandular kallikreins, *KLK1* and a progenitor of the human genes encoding PSA and hK2. After the division of the two species, this progenitor was silenced in the mouse lineage (this pseudogene is referred to as Ψ*mGK*) and in the lineage leading to humans it was duplicated, resulting in the genes encoding PSA and hK2. *KLK1* was kept unaltered in human, but in mouse, it was extensively duplicated, resulting in another 24 paralogues. When studying the evolution of glandular kallikreins, it is tempting to speculate that the evolutionarily older and conserved kallikreins from *KLK4* to *KLK15* may be involved in processes that are more fundamental than the younger classical glandular kallikreins. Further, the classical glandular kallikreins, which vary dramatically between species, appear to be involved in physiological processes that are more species specific.

12. Cross-Talk Between Kallikreins: A Possible Novel Enzymatic Cascade Pathway

Interactions between serine proteases are common, and substrates of serine proteases are usually other serine proteases that are activated from an inactive precursor [66]. The involvement of serine proteases in cascade pathways is well documented. One important example is the blood coagulation cascade. Blood clots are formed by a series of zymogen activations. In this enzymatic cascade, the activated form of one factor catalyzes the activation of the next factor. Very small amounts of the initial factors are sufficient to trigger the cascade because of the catalytic nature of the process. These numerous steps yield a large amplification, thus ensuring a rapid and amplified response to trauma. A similar mechanism is involved in the dissolution of blood clots. A third important example of the coordinated action of serine proteases is the intestinal digestive enzymes. The apoptosis pathway is another important example of coordinated action of other types of proteases.

The cross-talk between kallikreins and the hypothesis that they are involved in a cascade enzymatic pathway are supported by strong, but mostly circumstantial evidence, as follows: many kallikreins are coexpressed in the same tissue (e.g., the adjacently localized kallikrein genes *KLK2, KLK3, KLK4*, and *KLK5* are all highly expressed in the prostate); some kallikreins have the ability to activate each other and the ability of other serine proteases to activate kallikreins (see following); the common patterns of steroid hormonal regulation; the parallel pattern of differential expression of many kallikreins in different malignancies (e.g., at least seven kallikrein genes are up-regulated in ovarian cancer, and at least seven kallikreins are down-regulated in breast cancer); and serine proteases commonly use other serine proteases as substrates.

Recent experiments have shown that hK3 can be activated by hK15 [161]. hK4 has also recently been shown to activate hK3 and does so much more efficiently compared to hK2 [106]. hK5 is predicted to be able to activate hK7 in the skin [103]. The activation of hK3 by hK2 is also possible. Although Takayama *et al.*, reported the ability of hK2 to activate hK3 [162], Denmeade *et al.* reported the opposite [105] and hypothesized that additional proteases may be required. It will be interesting to study all possible combinations of interactions among kallikreins, especially those with expression in the same tissues. Bhoola *et al.* have recently provided strong evidence of the involvement of a "kallikrein cascade" in initiating and maintaining systemic inflammatory responses and immune-modulated disorders [163].

Kallikreins might also be involved in cascade reactions involving nonkallikrein substrates. This is evident from the reported, but questionable, ability of hK3 to digest insulin-like growth factor–binding protein [164] and to inactivate parathyroid hormone-related protein [165]. Similar properties were reported for rodent kallikreins. There is also experimental evidence that hK2 and hK4 can activate the pro-form of another serine protease, the urokinase-type plasminogen activator [106, 125]. As mentioned above, other serine proteases, such as enterokinase and trypsin, are predicted to be able to activate many kallikreins. Furthermore, hK4 can degrade prostatic acid phosphatase in seminal plasma; hK7 can degrade the alpha chain of native human fibrinogen, and it is hypothesized that it is involved in an apoptotic-like mechanism that leads to skin desquamation [166]. A proposed model for the involvement of kallikreins in a cascade-like reaction and its association to the pathogenesis of ovarian cancer has recently been published [167].

13. Isoforms and Splice Variants of the Human Kallikreins

The mechanism of a single gene giving rise to greater than one mRNA transcript is referred to as differential splicing. This system is often tightly regulated in a cell-type or developmental stage–specific manner and increases genome complexity by generating different proteins from the same mRNA.

The presence of more than one mRNA form for the same gene is common among kallikreins. These variant mRNAs may result from alternative splicing, a retained intronic segment, or use of an alternative transcription initiation site. To date, there are at least 49 documented splice variants of the 15 kallikrein genes (Table 4), and more are currently being investigated (our unpublished data). Some of these variants may hold significant clinical value. Slawin *et al.* reported the prognostic significance of a splice-variant-specific RT-PCR assay for *KLK2* in detecting prostate cancer metastasis [168]. Nakamura *et al.* reported differential expression of the brain and prostate-types of *KLK11* among benign, hyperplastic, and malignant prostate cancer cell lines [169]. A novel ovarian cancer–specific variant of hK5 has been recently reported [170], and another *KLK5* transcript with a short 5′-untranslated region and a novel *KLK7* transcript with a long 3′-untranslated region were highly expressed in the ovarian cancer cell lines OVCAR-3 and PEO1, respectively, but were expressed at very low levels in normal ovarian epithelial cells. Both Western blot and immunohistochemistry analyses have shown that these two enzymes are secreted from ovarian carcinoma cells. Thus, the short *KLK5* and long *KLK7* transcripts may be useful as tumor markers for epithelial-derived serous carcinomas [170].

Some of these alternatively spliced forms were also found to be tissue specific. A 1.5-kb transcript of *KLK14* was only found in the prostate, and another 1.9 kb transcript was found only in skeletal muscle [171]. Several splice variants of *KLK13* were found to be testis specific [172]. Type 2 neuropsin (*KLK8*) is preferentially expressed in the hippocampus of the human adult brain [173], and a new splice variant of *KLK4* was isolated from prostatic tissue [174]. A *KLK6* splice variant (GenBank accession no. AY279381) was strongly expressed in adult brain compared to fetal brain. Some of these splice variants were found to be translated [175, 176].

Evolutionarily conserved sequences ensure that the 3′ and 5′ splice sites are correctly cleaved and the two ends properly joined. These consensus sequences contain invariant dinucleotides at each end—GT (donor site) and AG (acceptor site)—and are associated with a more flexible sequence, AG:**GT**(A/G)AGT....C**AG**:G. However, exceptions to the tightly regulated splice sites can arise. An alternative GT-GC intron may exist, but unlike a possible AT-AC intron boundary, it will still be processed by the same splicing pathway as the conventional GT-AG introns. The GT-GC boundary is present in some kallikrein splice variants including the 5′-untranslated region of *KLK5* splice variant 2 (GenBank accession no. AY279381) and intron 3 for *KLK10* transcript variant 2 (GenBank accession no. 145888).

14. Kallikreins in Normal Physiology

Little is known about the physiological functions of kallikreins in normal tissues. However, accumulating evidence indicates that kallikreins might have diverse functions, depending on the tissue and circumstances of expression. hK1 exerts its biological activity mainly through the release of lysyl-bradykinin (kallidin). It cleaves low–molecular weight kininogen to produce vasoactive kinin peptides. Intact kinin binds to bradykinin B_2 receptor in target tissues and exerts a broad spectrum of biological effects including blood pressure reduction via vasodilatation, smooth muscle relaxation or contraction, pain induction, and mediation of the inflammatory response [177]. Low renal synthesis and urinary excretion of tissue kallikrein have been repeatedly linked to hypertension in animals and humans [178]. It has also been reported that tissue kallikrein cleaves kininogen substrate to produce vasoactive kinin peptides that have been implicated in the proliferation of vascular smooth muscle cells. Abnormalities of the

TABLE 4
REPORTED SPLICE VARIANTS OF THE HUMAN KALLIKREIN GENES

Gene and splice variant name	Splice variant description	Reference
KLK1		
Intron containing kallikrein mRNA	Deletion of exon 2 + alternative first exon	[257]
ND[a]	104 bp deletion at beginning of exon 5	[258]
KLK2		
GK-10A	Additional 37 bp at the end of exon 4 + early 3′ UTR[b]	[150]
AF188745*	13-bp deletion at beginning of exon 4 + 37-bp extension at end of exon 4 + deletion of exon 5 + early 3′ UTR[b]	[259]
AF188747*	37-bp extension at end of exon 4 + early 3′ UTR[b]	[259]
AF336106*	Extension of exon 1 + deletion of exons 2, 3, 4, and 5 + early 3′ UTR[b]	[1]
KLK3		
PSA-RP1/PA 525	Alternative acceptor site for exon 5 starts 442 bp upstream of classical[c] exon 5 → 2 variations of PSA-RP1 with different 3 UTRs[b]	[175]
PSA-RP2	Extension from exon 3–4 + deletion of coding exon 4 and 5 + early 3′ UTR[b]	[2]
PSA-RP3	129-bp deletion in beginning of exon 3	[176]
PSA-RP4	123-bp deletion in middle of exon 3	[2]
PSA-RP5	386-bp extension at end of exon 4 + deletion of coding exon 5 + early 3′ UTR[b]	[175]
PSA-LM	269-bp extension of coding region at end of exon 1 + early 3′ UTR[b] → 2 variations of PSA-LM with different 3′ UTRs[b]	[1,2]
PA 424	Joining of exons 3 and 4 + 105-bp deletion at end of exon 4 + deletion of exon 5 + early 3′ UTR[b]	[45, 260]
KLK4		
ND[a]	12-bp extension at end of exon 2 + early 3′ UTR[b]	[174, 261]
KLK4-S	Deletion of exon 1 and 4 + exon 2 (alternative first exon) starts 61 bp downstream beginning of classical[c] exon 2 + early 3′ UTR[b]	[147]
ND[a]	Additional sequence in 5′ UTR[b]	[41]

KLK4-L	Exon 3 extended to exon 4 + exon 2 (alternative first exon) starts 61 bp downstream beginning of classical[c] exon 2 + early 3′ UTR[b]	[261]
ND[a]	Alternative first exon starts 61 bp downstream beginning of classical[c] exon 2	[261]
KLK5		
Kallikrein 5 splice variant 1	203-bp deletion at end of noncoding exon 1 (5′ UTR[b]] + 45 bp upstream extension in beginning of noncoding exon 1	AY279380*
Kallikrein 5 splice variant 2	135-bp deletion in middle of noncoding exon 1 (5′ UTR[b]] + 45 bp upstream extension in beginning of noncoding exon 1	AY279381*
Ovarian cancer klk5–long	40 bp upstream extension at beginning of noncoding exon 1	[170]
Ovarian cancer klk5–short	40 bp upstream extension at beginning of noncoding exon 1 + 204-bp deletion at end of noncoding exon 1 (5′ UTR[b]]	[170]
KLK6		
Kallikrein 6 splice variant 1	Deletion of coding exon 2 + early 3′ UTR[b]	AY279383*
KLK7		
KLK7 long	144 bp upstream extension to noncoding exon 1	[170]
KLK7 transcript variant 2	Noncoding exon 1 starts 36 bp downstream from the end of classical[c] noncoding exon 1	[152]
KLK8		
Neuropsin type 2	135 bp upstream extension of second coding exon	[173]
Neuropsin type 3	Deletion of coding exons 2 and 3	[233]
Neuropsin type 4	Deletion of exons 2, 3, and 4 + early 3′ UTR[b]	[233]
TADG14	Additional 491 bp in 5′ UTR[b]	[262]
KLK9		
KLK9 splice variant-1	106-bp deletion at the end of exon 3 + late 3′ UTR[b]	AF135026
KLK10		
KLK10 transcript variant 2	Noncoding exon 1 starts 122 bp downstream end of classical[c] noncoding exon 1 + second noncoding exon starts 3 bp downstream beginning of classical[c]	[239]
KLK11		
Brain-type TLSP	Prostate-type variant is located 282 bp downstream end of brain–type specific noncoding exon 1	[263]
KLK12		
KLK12 related protein-1	Split last coding exon with an intervening intron of 129 bp + late 3′ UTR[b]	[156]
KLK12 related protein-2	Deletion of coding exon 3 + early 3′ UTR[b]	[156]

(*continues*)

TABLE 4 (*Continued*)

Gene and splice variant name	Splice variant description	Reference
KLK13		
Short KLK-L4 variant	211-bp deletion at end of coding exon 3 + early 3′ UTR[b]	[90]
Long form	Additional exon between exons 1 and 2 + early 3′ UTR[b]	[172]
L4E-S	Alternative first exon + 211-bp deletion at end of exon + early 3′ UTR[b]	[172]
L4E-L	Alternative first exon + 80-bp extension at end of exon 3	[172]
L4D-S	Additional exon between exons 1 and 2 + 211-bp deletion at end of exon 3 + early 3′ UTR[b]	[172]
L4D-L	Additional exon between exons 1 and 2 + 80-bp extension at end of coding exon 3 + early 3′ UTR[b]	[172]
Kallikrein 13 splicing variant 2	211-bp deletion at end of coding exon 3	[90]
Kallikrein 13 splicing variant 3	Deletion of coding exons 2 and 3	[90]
KLK14		
ND[a]	1.5-kb transcript (not characterized)	[171]
ND[a]	1.9-kb transcript (not characterized)	[171]
KLK15		
KLK15 splice variant 1	118-bp deletion at end of exon 3 + early 3′ UTR[b]	[51]
KLK15 splice variant 2	Deletion of exon 4 + early 3′ UTR[b]	[51]
KLK15 splice variant 3	118 bp deletion at end of exon 3 + deletion of exon 4	

*Accession numbers in GenBank (variant name may not be determined, or submission may be unpublished).

[a]ND: not defined.

[b]Untranslated region (noncoding).

[c]"Classical" refers to the published report of the gene that follows the definition of a kallikrein gene, as described in the text.

tissue kallikrein–kinin system have been implicated in the pathogenesis of hypertension and cardiovascular and renal disorders [179].

An hK1 knockout mouse has recently been generated and found to be unable to generate significant levels of kinins in most tissues and develop cardiovascular abnormalities early in adulthood despite normal blood pressure [178].

However, the diverse expression pattern of hK1 has led to the suggestion that the functional role of this enzyme may be specific to different cell types [177]. Apart from its kininogenase activity, tissue kallikrein has been implicated in the processing of growth factors and peptide hormones in light of its presence in pituitary, pancreas, and other tissues. As summarized by Bhoola *et al.* [177], hK1 has been shown to cleave proinsulin, low-density lipoprotein, prorenin, angiotensinogen, vasoactive intestinal peptide, procollagenase, and the precursor of atrial natriuretic factor.

Seminal plasma hK2 was found to be able to cleave seminogelin I and seminogelin II, but at different cleavage sites and with lower efficiency than hK3 [180]. Because the amount of hK2 in seminal plasma is much lower than hK3 (1%–5%), the contribution of hK2 in the process of seminal clot liquefaction is expected to be relatively small [31].

Because hK3 is present at very high levels in seminal plasma, most studies focused on its biological activity within this fluid. Lilja has shown that hK3 rapidly hydrolyzes seminogelin I and seminogelin II, as well as fibronectin, resulting in liquefaction of the seminal clot after ejaculation [181]. Several other potential substrates for hK3 have been identified, including insulin-like growth factor–binding protein 3, TGF-β, parathyroid hormone-related peptide, and plasminogen [182]. The physiological relevance of these findings is still not clear.

The mouse and porcine orthologues of hK4 were originally designated enamel matrix serine protease because of their predicted role in normal teeth development [41]. The human *KLK4* gene, however, was shown to be highly expressed in the prostate, pointing to the possibility that it has a different function in humans. hK7 and, more recently, hK5 were found to be highly expressed in the skin, and it is believed that they are involved in skin keratinization and desquamation [183]. hK6, hK8, and hK11 are highly expressed in the central nervous system, where they are thought to play a role in neural plasticity.

Another possible mechanism for kallikrein action in physiology and pathobiology is the activation of proteinase-activated receptors (PARs). PAR is a recently described family of G-protein-coupled receptors with seven transmembrane domains that are stimulated by cleavage of their N-termini by a serine protease rather than by ligand-receptor interaction [184–186]. Four PARs have been identified so far, of which PAR1, PAR3, and PAR4 are activated by thrombin, whereas PAR2 is activated by trypsin or mast cell tryptase. Activation of these receptors elicits different responses in several tissues. In addition, they switch-on cell signaling pathways (e.g., the MAP-kinase pathway) leading to cell growth and division. Because most kallikreins have trypsin-like activity, they might also be involved in such mechanisms. The possible cleavage of PARs by kallikreins is an exciting new avenue of investigation, but no published data exist on this issue.

15. Association of Kallikreins with Human Diseases

Some kallikrein genes have been associated with the pathogenesis of human diseases. The *KLK1* gene is involved in inflammation [8], hypertension [187], renal nephritis, and diabetic renal disease [188], The relationships between hK5 and hK7 and skin diseases, including pathological keratinization and psoriasis, have already been reported [189, 190]. Much research is now focusing on the relation of kallikreins to diseases of the central nervous system (CNS) and skin, as well as to malignancy, as discussed below.

15.1. Kallikreins in CNS Diseases

Many kallikreins seem to play important physiological roles in the CNS. In mouse, neuropsin appears to have an important role in neural plasticity, and the amount of neuropsin mRNA is related to memory retention after a chemically induced ischemic insult [191]. The human neuropsin gene (hK8) was first isolated from the hippocampus. Recent reports describe the association of hK8 expression with diseases of the CNS, including epilepsy [192, 193]. In addition, an 11.5-fold increase in *KLK8* mRNA levels in Alzheimer's disease (AD) hippocampus compared to controls was recently reported [194]. The same study showed that *KLK1, KLK4, KLK5, KLK6, KLK7, KLK8, KLK10, KLK11, KLK13*, and *KLK14* are expressed in both the cerebral cortex and hippocampus, whereas *KLK9* is expressed in the cortex but not the hippocampus [194]. Another kallikrein, *KLK6*, was shown to have amyloidogenic activity in the AD brain [107, 195], indicating that it may play a role in AD development (107). *KLK6* was also found to be localized in perivascular cells and microglial cells in human AD brain [107]. Scarisbrick *et al.* have shown that hK6 is abundantly expressed by inflammatory cells at sites of CNS inflammation and demyelination in animal models of multiple sclerosis and in human lesions at autopsy, prompting the researchers to postulate that hK6 in inflammatory CNS lesions may promote demyelination [196, 197]. *KLK11*, another newly discovered kallikrein, was isolated from brain hippocampus cDNA and is thought to play a role in brain plasticity [198]. Similarly, *KLK5* and *KLK14* are also expressed at high levels in the brain [153, 199] and might have roles in normal and aberrant brain physiology. A recent report showed significant alterations of hK6, hK7, and hK10 concentrations in CSF of patients with AD and frontotemporal dementia [200]. For a more detailed discussion about the role of kallikreins in the CNS, we refer the reader to our recent review [201].

15.2. Kallikreins in Skin Diseases

The epidermis forms the external surface of the skin and is composed of differentiated keratinocytes that form four layers: the stratum basale, the stratum spinosum, the stratum granulosum, and the stratum corneum, where keratinocytes have been transformed into corneocytes. The stratum corneum functions as the protective, virtually water-impermeable skin barrier against external insults including desiccation and the entry of noxious chemicals and microbes. To maintain this barrier, old corneocytes are continuously desquamated from the stratum corneum by both stratum corneum trypsin-like and stratum corneum chymotrypsin-like enzymes [202]. The expression of hK5 and hK7 [203] and of several kallikrein mRNAs [204] in the upper epidermis (the stratum granulosum or the stratum corneum), indicate that kallikreins may be the stratum corneum serine proteases responsible for the desquamation of corneocytes [205]. A quantitative ELISA assay showed high concentrations of hK7 [206] and hK8 [78] in skin tissue extracts. In addition, given that kallikrein proteins or mRNAs are also expressed in skin appendages such as sweat glands, sebaceous glands, and hair follicles [203, 205, 207], kallikrein activities may be related to the maturation and secretion of sebum and sweat and to hair growth [205].

In addition, kallikrein expression may also be involved in the pathogenesis of several skin diseases. In psoriasis, an inflammatory skin disease, various kallikrein mRNAs were shown to be up-regulated in the upper epidermis [204] and associated with the conversion of the inactive hK7 precursor to active hK7 in the psoriatic lesion [189]. In ichthyoses and squamoproliferative disorders, hK7 expression was found to be low [208]. Transgenic mice overexpressing hK7 showed increased epidermal thickness, hyperkeratosis, and a dermal inflammation with pruritus [209] and expression of MHC II antigen [210]. These data indicated that hK7 may lead to skin changes that contribute to development of inflammatory skin diseases [210]. hK8-deficient mice showed a prolonged recovery of ultraviolet B–irradiated skin, indicating that hK8 might be involved in the process of differentiation [211]. Psoriasis vulgaris, seborrheic keratosis, lichen planus, and squamous cell carcinoma also display a high density of *KLK8* mRNA [207].

In Netherton syndrome, a severe inherited skin disorder, the molecular defect is thought to involve mutations of the serine protease inhibitor

TABLE 5

PROGNOSTIC VALUE OF KALLIKREIN GENES/PROTEINS IN OVARIAN CANCER

Kallikrein gene/protein and sample used	Method	Clinical applications	Reference
KLK4			
mRNA from normal and cancerous ovarian tissues	RT-PCR[a]	*Unfavorable* prognosis: • Overexpressed in patients with late-stage disease, higher-grade tumors, and no response to chemotherapy • Associated with shorter DFS[b] and OS[c] • Independent indicator of poor prognosis in patients with low-grade tumors	[226]
mRNA from normal, benign, and cancerous ovarian tissues and late-stage serous ovarian cancer cell lines	SQ-RT-PCR,[d] Southern blot	*Unfavorable* prognosis: • Overexpressed in late-stage serous ovarian tumor tissues and cell lines	[264]
KLK5			
mRNA from normal and cancerous ovarian tissues	RT-PCR	*Unfavorable* prognosis: • Overexpressed in patients with late-stage disease and higher-grade tumors • Associated with shorter DFS and OS • Independent indicator of poor prognosis in patients with low-grade tumors	[227]
KLK5/hK5			
mRNA and extracts from normal, benign and cancerous ovarian tissues and late-stage serous ovarian cancer cell lines	SQ-RT-PCR, Southern, Northern, and Western blots and immunohistochemistry	*Unfavorable* prognosis: • Overexpressed in ovarian tumor tissues and cell lines mainly of late stage and serous histotype	[170]

hK5			
Ovarian cancer cytosols	Immunoassay	*Unfavorable* prognosis: • Overexpressed in patients with late stage disease and higher grade tumors • Associated with shorter DFS and OS • Independent indicator of poor prognosis in patients with high-grade tumors and optimal debulking success	Our unpublished data
KLK6/hK6			
mRNA and extracts from normal, benign, and cancerous ovarian tissues	SQ-RT-PCR, Northern blot, immunohistochemistry	*Unfavorable* prognosis: • Overexpressed in ovarian cancer tissues	[228]
hK6			
Ovarian cancer cytosols	Immunoassay	*Unfavorable* prognosis: • Overexpressed in patients with late-stage disease and serous tumors • Associated with shorter DFS and OS • Independent indicator of poor prognosis in patients with low-grade tumors and optimal debulking success	[229]
Serum from normal women, women with benign disease and ovarian cancer	Immunoassay	*Unfavorable* prognosis: • Serum hK6 levels elevated in cancer vs. normal • Higher serum hK6 I patients with late-stage disease, higher grade, serous tumors, suboptimal debulking, and a poor response to chemotherapy • Indicator of decreased DFS and OS	[217]
KLK7			
mRNA and extracts from normal and cancerous ovarian tissues	SQ-RT-PCR, Northern and Western blots and immunohistochemistry	*Unfavorable* prognosis: • Overexpressed in ovarian cancer tissues	[265]

(*continues*)

TABLE 5 (*Continued*)

Kallikrein gene/protein and sample used	Method	Clinical applications	Reference
KLK7/hK7			
mRNA from normal, benign, and cancerous ovarian tissues and late-stage serous ovarian cancer cell lines	SQ-RT-PCR, Southern, Northern, and Western blots and immunohistochemistry	*Unfavorable* prognosis: • Overexpressed in ovarian tumor tissues and cell lines mainly of late stage and serous histotype	[170]
mRNA from cancerous ovarian tissue	Q-RT-PCR[e]	*Unfavorable* prognosis: • Overexpressed in higher-grade tumors • Associated with shorter DFS • Independent indicator of poor prognosis in patients with low-grade tumors and optimal debulking success	[230]
KLK8			
mRNA from ovarian cancer tissues	RT-PCR	*Favorable* prognosis: • Overexpressed in lower-grade tumors • Associated with longer DFS and OS • Independent indicator of longer DFS	[233]
KLK9			
mRNA from ovarian cancer tissues	Q-RT-PCR	*Favorable* prognosis: • Overexpressed in patients with early-stage disease and optimal debulking success • Associated with longer DFS and OS • Independent indicator of prolonged DFS in patients with low-grade tumors and optimal debulking success	[234]

hK10			
Normal, benign, and cancerous ovarian cytosols	Immunoassay	*Unfavorable* prognosis: • Overexpressed in cancer patients with late-stage disease, serous tumors, and suboptimal debulking success • Associated with shorter DFS and OS • Independent indicator of DFS and OS in patients with late-stage tumors	[91]
Serum from normal women, women with benign disease and ovarian cancer	Immunoassay	*Unfavorable* prognosis: • Serum hK10 levels elevated in cancer vs. normal • Higher serum hK10 in patients with late-stage disease, higher-grade, serous tumors, suboptimal debulking, and a poor response to chemotherapy • Indicator of decreased DFS and OS • Independent indicator of OS	[218]
hK11			
Ovarian cancer cytosols	Immunoassay	*Favorable* prognosis: • Overexpressed in patients with early-stage disease, pre-/perimenopausal status, and who responded to chemotherapy • Associated with longer DFS and OS • Independent indicator of OS • Independent indicator of DFS and OS in patients with low-grade tumors	[235]
hK13			
Ovarian cancer cytosols	Immunoassay	*Favorable* prognosis: • Overexpressed in early-stage disease, and patients with no residual tumor after surgery and optimal debulking success • Independent indicator of longer DFS and OS	[266]

(continues)

TABLE 5 (*Continued*)

Kallikrein gene/protein and sample used	Method	Clinical applications	Reference
KLK14			
mRNA from normal, benign and cancerous ovarian tissues	Q-RT-PCR	*Favorable* prognosis: • Stepwise decrease in the amount *KLK14* mRNA (normal > benign > cancerous tissues) • Overexpressed in patients with early-stage disease and optimal debulking success, who responded to chemotherapy • Independent indicator of longer DFS and OS	[157]
Serum and tissue from ovarian cancer patients	ELISA	Diagnosis: • Elevated serum levels in 65% of ovarian cancer patients vs. normal • Higher levels in 40% of ovarian cancer tissues compared to normal	[258]
KLK15			
mRNA from benign and cancerous ovarian tissues	Q-RT-PCR	*Unfavorable* prognosis: • Higher levels in cancerous tissues • Independent indicator of decreased DFS and OS	[232]

[a] Reverse transcriptase-polymerase chain reaction.
[b] Disease-free survival.
[c] Overall survival.
[d] Semiquantitative reverse transcriptase polymerase chain reaction (RT-PCR).
[e] Quantitative RT-PCR.

Kazal-type 5 (SPINK5), also known as LEKT1 [212–214]. It is likely that inactivation of the inhibitor leads to overactivity of hK5, hK7, and possibly other kallikreins in skin. The same general mechanism may apply to psoriasis (our unpublished data). Clearly, a possible kallikrein cascade in skin, including active enzymes and inhibitors, needs to be further investigated.

16. Kallikreins and Cancer

16.1. Overview

The association of kallikreins with cancer is well established. PSA (hK3) and, more recently, human glandular kallikrein (hK2) are useful biomarkers for prostate cancer. A more detailed discussion about hK2 and hK3 as cancer biomarkers can be found elsewhere [31]. In addition to its established role in prostate cancer diagnosis and monitoring, recent reports indicate that hK3 can be useful as a marker for breast cancer prognosis [215].

With the identification and characterization of all members of the kallikrein gene family, accumulating evidence indicates that other kallikreins might be also related to hormonal (e.g., breast, prostate, testicular, and ovarian cancers) and other malignancies. *KLK6* and *KLK10* were originally isolated by differential display from breast cancer libraries [216].

At the protein level, recent reports demonstrate that kallikrein proteins can be useful serum biomarkers for diagnosis and prognosis of cancer. In addition to hK3 and hK2, hK6 and hK10 are emerging diagnostic markers for ovarian cancer [217, 218]. Also, hK11 was shown to be a potential marker for ovarian and prostate cancer [219]. A synthetic hK1 inhibitor was recently found to suppress cancer cell invasiveness in human breast cancer cell lines [220]. hK1 was immunolocalized in the giant cells of squamous cell carcinoma of the esophagus and gastric carcinoma [163].

In Tables 5–8, we summarize published data on the analysis of kallikrein genes and proteins in tumor tissue extracts and serum of cancer patients for the purpose of disease diagnosis, monitoring, and prognosis. As discussed below, some kallikreins are very promising new cancer biomarkers.

The differential regulation of certain kallikreins in more than one type of cancer has been repeatedly reported. The phenotype of cells and tissues, benign or malignant, ultimately depends on which proteins, and at what level, are expressed at any time. Ultimately, the pathological classification of human neoplasia, which is now based on histological features, will be replaced with a biological portrait of different tumors, including hundreds or thousands of differentially expressed genes [221].

TABLE 6

PROGNOSTIC/PREDICTIVE VALUE OF KALLIKREIN GENES/PROTEINS IN BREAST CANCER

Kallikrein gene/protein and sample used	Method	Clinical applications	Reference
hK3/PSA			
Breast cancer cytosols	Immunoassay	*Favorable* prognosis: • Overexpressed in younger patients with early-stage disease, small ER[a]-positive, low S-phase, low-cellularity diploid tumors • Associated with a longer DFS[b] and OS[c] • Independent indicator of increased DFS for all patients as well as node-positive, ER-negative	[267, 268]
Breast cancer cytosols	Immunoassay	*Favorable* prognosis: • Overexpressed in younger, pre-/perimenopausal patients with smaller, steroid hormone receptor–positive tumors • Associated with a longer OS Predictive value: • Higher hK3 levels associated with a poor response to tamoxifen therapy	[269]
KLK5			
mRNA from breast cancer tissues	Q-RT-PCR[d]	*Unfavorable* prognosis: • Overexpressed in pre-/perimenopausal, node-positive patients with ER-negative tumors • Independently associated with decreased DFS and OS • Independent indicator of shorter DFS and OS in node-positive patients with large tumors • Associated with shorter DFS in patients with low-grade tumors	[93]
KLK9			
mRNA from breast cancer tissues	Q-RT-PCR	*Favorable* prognosis: • Overexpressed in patients with early-stage disease and small tumors • Independently associated with increased DFS and OS • Independent indicator of prolonged DFS and OS in patients with ER and PR[e]-negative tumors	[238]

hK10			
Breast cancer cytosols	Immunoassay	*Predictive value:* • Higher hK10 levels independently associated with a poor response to tamoxifen therapy	[239]
KLK13			
mRNA from breast cancer tissues	Q-RT-PCR	*Favorable* prognosis: • Overexpressed in older, estrogen receptor–positive patients • Associated with a prolonged DFS and OS • Independent indicator of longer DFS and OS in node-, ER-, and PR-positive patients with low-grade tumors	[89]
KLK14			
mRNA from breast cancer tissues	Q-RT-PCR	*Unfavorable* prognosis: • Overexpressed in patients with advanced-stage disease • Independent indicator of shorter DFS[b] and OS[c] • Independent indicator of shorter DFS and OS in patients with a tumor size $\leq$2 cm and positive nodal, ER,[d] and PR[e] status	[94]
Serum from breast cancer patients	ELISA	Diagnosis: • Overexpressed in patients with advanced-stage disease • Elevated serum hK14 levels in 40% breast cancer patients vs. normal	[270]
KLK15			
mRNA from breast cancer tissues	Q-RT-PCR	*Favorable* prognosis: • Overexpressed in node-negative patients • Independently associated with a longer DFS and OS • Independent indicator of longer DFS and OS in patients with lower-grade, ER- and PR-negative tumors	[155]

[a] Estrogen receptor.
[b] Disease-free survival.
[c] Overall survival.
[d] Quantitative reverse transcriptase polymerase chain reaction.
[e] Progesterone receptor.

TABLE 7

PROGNOSTIC VALUE OF KALLIKREIN GENES/PROTEINS IN PROSTATE CANCER

Kallikrein gene/protein and sample used	Method	Clinical applications	Reference
KLK5			
mRNA from matched normal and prostate cancer tissues	Q-RT-PCR[a]	*Favorable* prognosis: • Down-regulated in cancer vs. normal prostate tissues • Higher levels associated with low grade tumors and low Gleason score	[242]
KLK11			
mRNA from matched normal and prostate cancer tissues	Q-RT-PCR	*Favorable* prognosis: • Overexpressed in cancer vs. normal prostate tissues • Prostate-specific splice variant associated with early-stage disease, lower tumor grade, and Gleason score	[271]
KLK14			
mRNA from matched normal and prostate cancer tissues	Q-RT-PCR	*Unfavorable* prognosis: • Overexpressed in cancer vs. normal prostate tissues • Overexpressed in patients with late-stage disease, high-grade tumors, and higher Gleason score	[272]
KLK15			
mRNA from matched normal and prostate cancer tissues	Q-RT-PCR	*Unfavorable* prognosis: • Overexpressed in cancer vs. normal prostate tissues • Overexpressed in patients with late-stage disease, high-grade tumors, and higher Gleason score	[273]

[a] Quantitative reverse transcriptase polymerase chain reaction.

TABLE 8

PROGNOSTIC VALUE OF KALLIKREIN GENES/PROTEINS IN TESTICULAR CANCER

Kallikrein gene/protein and sample used	Method	Clinical applications	Reference
KLK5			
mRNA from matched normal and testicular cancer tissues	Q-RT-PCR[a]	*Favorable* prognosis: • Down-regulated in cancer vs. normal testicular tissues • Overexpressed in smaller, early stage, nonseminomas	[243]
KLK10			
mRNA from matched normal and testicular cancer tissues	RT-PCR	• Down-regulated in cancer vs. normal testicular tissues	[244]
KLK13			
mRNA from matched normal and testicular cancer tissues	RT-PCR	Testicular cancer–specific splice variants	[171]
KLK14			
mRNA from matched normal and malignant testicular tissues	RT-PCR	*Favorable* prognosis: • Down-regulated in cancer vs. normal testicular tissues	[199]

[a] Quantitative reverse transcriptase polymerase chain reaction (RT-PCR).

overexpression indicates an unfavorable prognosis for prostate cancer patients [51].

16.2.4. *Testicular Cancer*

We have recently published a review documenting the apparent relationship between kallikreins and testicular cancer [92]. The *KLK5* gene has potential as a favorable prognostic marker for testicular cancer patients [243]. Furthermore, the differential expression of *KLK10* and *KLK14* and *KLK13* splice variants in testicular cancer tissues have been recently reported [172, 199, 244]. The potential diagnostic or prognostic roles of kallikreins in testicular cancer are summarized in Table 8. Further studies are clearly warranted.

16.2.5. *Lung, Pancreatic, and Colon Cancers and Leukemias*

A microarray study has identified at least one kallikrein gene being overexpressed in lung carcinoma (*KLK11*), particularly in neuroendocrine tumors [245]. Because many kallikreins are coexpressed in normal lung tissue [43], we hypothesize that multiple kallikreins, in addition to *KLK11*, may also be deregulated in lung cancer.

Expression of multiple kallikrein genes was reported in endocrine and exocrine pancreas by immunohistochemistry [81–83]. More recently, our *in silico* analysis, using two independent databases of the Cancer Genome Anatomy Project, provided evidence that some kallikreins are differentially regulated in pancreatic cancer [246]. In particular, *KLK6* and *KLK10* were significantly upregulated. This finding is in accord with recent data from microarray analysis [247].

We have also recently provided evidence indicating the overexpression of three kallikreins (*KLK7, KLK8*, and *KLK10*) and downregulation of another kallikrein (*KLK1*) in colon cancer [246]. A recent report showed also a downregulation of the KLK10 gene in acute lymphoblastic leukemia [143].

16.2.6. *Brain Tumors*

The possible involvement of kallikreins in brain tumors has been examined recently. A recent report showed expression of the hK6 protein by glioblastoma cells implanted intracranially in nude mice. Moreover, hK6 expression was shown to colocalize with the expression of an invasion-associated matricellular protein called SPARC [248]. Given the high level of expression of some kallikreins in the brain, it is logical to speculate a possible involvement of kallikrein in brain tumors. Ongoing studies are now being conducted to evaluate the possible functional importance of kallikreins in brain tumor invasiveness [248].

16.3. How are Kallikreins Involved in Cancer?

Several mechanisms can be proposed by which kallikreins can be involved in the pathogenesis of endocrine-related malignancies. Proteolytic enzymes are thought to be involved in tumor progression because of their role in extracellular matrix degradation. Many studies have shown that a variety of proteolytic enzymes are overproduced either by the cancer cells themselves or by the surrounding stromal cells, and that their overexpression is associated with unfavorable clinical prognosis [249].

Breast, prostate, testicular, and ovarian cancers are all considered "hormonal" malignancies. Sex hormones are known to affect the initiation or progression of these malignancies. However, all kallikreins are under sex steroid hormonal regulation. Taken together, kallikreins may represent downstream targets by which hormones affect the initiation or progression of such tumors.

Experimental evidence indicates that hK2 and hK4 can activate the proform of another serine protease, the urokinase-type plasminogen activator [106, 162]. Urokinase activates plasmin from its inactive form (plasminogen), which is ubiquitously located in the extracellular space, leading to degradation of the extracellular matrix proteins. This might provide some clues about the role of kallikreins in cancer progression and could explain the differential expression of several kallikreins in tumors. Plasminogen can also activate precursor forms of collagenases, thus promoting the degeneration of the collagen in the basement membrane surrounding the capillaries and lymph nodes. Another kallikrein, hK7, can degrade the alpha chain of human fibrinogen, and it is hypothesized to be involved in an apoptotic-like mechanism that leads to desquamation of the skin [166]. The involvement in growth and apoptotic activities is also reported for hK3, which can digest insulin-like growth factor-binding protein 3 [250] and parathyroid hormone-related protein [251]. Similar findings were observed for some rodent kallikreins [252].

Bhoola *et al.* [163] have recently provided strong evidence indicating the presence of hK1 activity in the chemotactically attracted inflammatory cells of esophageal and renal cancers, indicating a role for kallikreins in these malignancies. Modulation of angiogenic activity is another possible mechanism for kallikrein involvement in cancer. The kinin family of vasoactive peptides, liberated by hK1 action, is believed to regulate the angiogenic process [253]. It was recently reported that immunolabeling of hK1 was intense in the angiogenic endothelial cells derived from mature corpora lutea. Immunoreactivity was lower in nonangiogenic endothelial cells, and least in angiogenic endothelial cultures of the regressing corpus luteum [253]. In addition, hK3 was reported to have antiangiogenic activities [254].

The elevation of serum concentration of kallikreins in cancer might be a result of the increased vasculature (angiogenesis), the destruction of the glandular architecture of the tissues involved, and the subsequent leakage of these proteins into the general circulation. It is possible that the concentration of kallikreins may also be increased in serum because of gene overexpression.

17. Therapeutic Applications

It is possible that some kallikreins may become valuable therapeutic targets when the biological pathways that are involved are delineated. For example, the enzymatic activity of these serine proteases may initiate (e.g., tumor invasion, activation of hormones, growth factors, other enzymes, receptors or cytokines, amyloid formation) or terminate (e.g., inhibition of angiogenesis, inactivation of growth factors, hormones, enzymes, cytokines, or receptors) biological events. Once known, these events could be manipulated, for therapeutic purposes, by specific enzyme inhibitors or activators. Another potential therapeutic approach is the cell-specific activation of therapeutic agents [255]. Preliminary reports show potential success by using the PSA promoter to express molecules in a tissue-specific fashion [256]. A third possible therapeutic approach involves immunotherapy or development of cancer vaccines. With our increasing knowledge of the hormonal regulation of kallikreins, hormonal activation (or repression) of kallikrein activity could be investigated in the future.

18. Future Directions

The kallikrein locus in humans has now been well characterized and confirmed by independent analyses of various investigators. The structure of these simple serine protease genes has now been fully elucidated. It is thus now appropriate to shift the research focus from characterization of the gene structure to study of the functional aspects of kallikreins in humans and other species. The completion of the mouse, rat, and other genome sequences will aid in comparative genomic analysis of the human and other animal kallikrein loci, toward establishing phylogenetic and functional relationships. The large number of kallikrein genes and proteins and their close localization point to an important proteolytic system that has not yet been recognized. We do not, therefore, know whether the tight clustering of these genes is related to their physiological functions or whether it represents a functional redundancy.

With very few exceptions, knockout animal models for kallikreins have not yet been developed. It will be very interesting to establish the phenotype of mice lacking not one but many kallikreins, or even the whole-mouse kallikrein locus. This will provide insights into the function of these genes in the mouse and further delineate the degree of redundancy of these genes. Mutational analysis of human kallikreins has not been performed in detail, and we do not yet know whether the functional inactivation of any of these genes in humans leads to recognizable diseases.

The localization of these genes next to each other and their parallel expression in many tissues strongly indicate that these genes are regulated by a locus-control region. This proposal merits further investigation. However, epigenetic changes seem to play a role in kallikrein gene silencing. Furthermore, kallikrein genes are regulated by steroid hormones and vitamin D in certain tissues. A better understanding of their mode of regulation will be important in the future.

It appears that small groups of kallikreins may represent enzymatic cascade pathways in certain tissues. For example, it is very likely that at least three kallikreins, hK2, hK3, and hK11, which are present in seminal plasma at relatively very large concentrations, may coordinately act as a cascade enzymatic pathway, involved in semen liquefaction or other activities. In contrast, another group of kallikreins, including hK5 and hK7 and possibly many others, seem to be involved in skin desquamation. Similar cascade pathways may be operating in the breast, testis and other tissues.

Some avenues for future kallikrein research include, first, their continued investigation as promising novel biomarkers for diagnosis, prognosis, and monitoring of many diseases, particularly cancer. Although most of the current reports describe single kallikreins as potential biomarkers, in the future, it may be possible to combine multiple kallikreins with other tumor markers in multiparametric panels. However, the recognition that kallikrein genes give rise to a very large number of splice variants (more than 70) offers new avenues of investigation regarding the applicability of these molecules as cancer biomarkers. It is possible that splice variants of some of these genes may be even more promising cancer biomarkers than the classical forms of the enzymes.

Second, the full utilization of the kallikrein gene family in various aspects of human physiology and pathobiology will necessitate the delineation of their physiological functions. Future investigations should include the examination of their enzymatic specificity and their regulation by specific or nonspecific tissue or circulating inhibitors. We are currently using synthetic peptide substrates, combinatorial substrate libraries, macromolecular protein substrates, and phage display technology to delineate the physiological function of these enzymes.

[25] Chapdelaine P, Gauthier E, Ho-Kim MA, Bissonnette L, Tremblay RR, Dube JY. Characterization and expression of the prostatic arginine esterase gene, a canine glandular kallikrein. DNA Cell Biol 1991; 10:49–59.

[26] Fiedler F, Betz G, Hinz H, Lottspeich F, Raidoo DM, Bhoola KD. Not more than three tissue kallikreins identified from organs of the guinea pig. Biol Chem 1999; 380:63–73.

[27] Giusti EP, Sampaio CA, Michelacci YM, Stella RC, Oliveira L, Prado ES. Horse urinary kallikrein, I. Complete purification and characterization. Biol Chem Hoppe Seyler 1988; 369:387–396.

[28] Carvalho AL, Sanz L, Barettino D, Romero A, Calvete JJ, Romao MJ. Crystal structure of a prostate kallikrein isolated from stallion seminal plasma: A homologue of human PSA. J Mol Biol 2002; 322:325–337.

[29] Karr JF, Kantor JA, Hand PH, Eggensperger DL, Schlom J. The presence of prostate-specific antigen-related genes in primates and the expression of recombinant human prostate-specific antigen in a transfected murine cell line. Cancer Res 1995; 55:2455–2462.

[30] Fujimori H, Levison PR, Schachter M. Purification and partial characterization of cat pancreatic and urinary kallikreins—comparison with other cat tissue kallikreins and related proteases. Adv Exp Med Biol 1986; 198(Pt A):219–228.

[31] Rittenhouse HG, Finlay JA, Mikolajczyk SD, Partin AW. Human Kallikrein 2 (hK2) and prostate-specific antigen (PSA): Two closely related, but distinct, kallikreins in the prostate. Crit Rev Clin Lab Sci 1998; 35:275–368.

[32] Riegman PH, Vlietstra RJ, Suurmeijer L, Cleutjens CB, Trapman J. Characterization of the human kallikrein locus. Genomics 1992; 14:6–11.

[33] Riegman PH, Vlietstra RJ, van der Korput JA, Romijn JC, Trapman J. Characterization of the prostate-specific antigen gene: A novel human kallikrein-like gene. Biochem Biophys Res Commun 1989; 159:95–102.

[34] Yousef GM, Chang A, Scorilas A, Diamandis EP. Genomic organization of the human kallikrein gene family on chromosome 19q13.3–q13.4. Biochem Biophys Res Commun 2000; 276:125–133.

[34a] Yousef GM, Borgono CA, Michael IP, Diamandis EP. Cloning of a Kallikrein pseudogene. Clin Biochem 2004; 37:961–967.

[35] Gan L, Lee I, Smith R, et al. Sequencing and expression analysis of the serine protease gene cluster located in chromosome 19q13 region. Gene 2000; 257:119–130.

[36] Yousef GM, Diamandis M, Jung K, Diamandis EP. Molecular cloning of a novel human acid phosphatase gene (ACPT) that is highly expressed in the testis. Genomics 2001; 74:385–395.

[37] Yousef GM, Borgono C, Michael IP, et al. Molecular cloning of a new gene which is differentially expressed in breast and prostate cancers. Tumour Biol 2004; 25:122–133.

[38] Yousef GM, Ordon M, Foussias G, Diamandis EP. Genomic organization of the Siglec gene locus on chromosome 19q13.4 and cloning of two new siglec pseudogenes. Gene 2002; 286:259–270.

[39] Clements J, Hooper J, Dong Y, Harvey T. The expanded human kallikrein (KLK) gene family: Genomic organisation, tissue-specific expression and potential functions. Biol Chem 2001; 382:5–14.

[40] Yousef GM, Bharaj BS, Yu H, Poulopoulos J, Diamandis EP. Sequence analysis of the human kallikrein gene locus identifies a unique polymorphic minisatellite element. Biochem Biophys Res Commun 2001; 285:1321–1329.

[41] Hu JC, Zhang C, Sun X, et al. Characterization of the mouse and human PRSS17 genes, their relationship to other serine proteases, and the expression of PRSS17 in developing mouse incisors. Gene 2000; 251:1–8.

[42] Diamandis EP, Yousef GM. Human tissue kallikreins: A family of new cancer biomarkers. Clin Chem 2002; 48:1198–1205.
[43] Yousef GM, Diamandis EP. The new human tissue kallikrein gene family: Structure, function, and association to disease. Endocr Rev 2001; 22:184–204.
[44] Yousef GM, Diamandis EP. Human kallikreins: Common structural features, sequence analysis and evolution. Curr Genomic 2003; 4:147–165.
[45] Riegman PH, Vlietstra RJ, Klaassen P, et al. The prostate-specific antigen gene and the human glandular kallikrein-1 gene are tandemly located on chromosome 19. FEB Lett 1989; 247:123–126.
[46] Schedlich LJ, Bennetts BH, Morris BJ. Primary structure of a human glandular kallikrein gene. DNA 1987; 6:429–437.
[47] Lundwall A. Characterization of the gene for prostate-specific antigen, a human glandular kallikrein. Biochem Biophys Res Commun 1989; 161:1151–1159.
[48] Bernett MJ, Blaber SI, Scarisbrick IA, Dhanarajan P, Thompson SM, Blaber M. Crystal structure and biochemical characterization of human kallikrein 6 reveals that a trypsin-like kallikrein is expressed in the central nervous system. J Biol Chem 2002; 277:24562–24570.
[49] Gomis-Ruth FX, Bayes A, Sotiropoulou G, et al. The structure of human prokallikrein 6 reveals a novel activation mechanism for the kallikrein family. J Biol Chem 2002; 277:27273–27281.
[50] Katz BA, Liu B, Barnes M, Springman EB. Crystal structure of recombinant human tissue kallikrein at 2.0 A resolution. Protein Sci 1998; 7:875–885.
[51] Yousef GM, Scorilas A, Jung K, Ashworth LK, Diamandis EP. Molecular cloning of the human kallikrein 15 gene (KLK15). Up-regulation in prostate cancer. J Biol Chem 2001; 276:53–61.
[52] Bharaj B, Scorilas A, Giai M, Diamandis EP. TA repeat polymorphism of the 5alpha-reductase gene and breast cancer. Cancer Epidemiol Biomarker Prev 2000; 9:387–393.
[53] Baffa R, Moreno JG, Monne M, Veronese ML, Gomella LG. A comparative analysis of prostate-specific antigen gene sequence in benign and malignant prostate tissue. Urology 1996; 47:795–800.
[54] Majumdar S, Diamandis EP. The promoter and the enhancer region of the KLK 3 (prostate specific antigen) gene is frequently mutated in breast tumours and in breast carcinoma cell lines. Br J Cancer 1999; 79:1594–1602.
[55] Angermann A, Bergmann C, Appelhans H. Cloning and expression of human salivary-gland kallikrein in *Escherichia coli*. Biochem J 1989; 262:787–793.
[56] Baker AR, Shine J. Human kidney kallikrein: cDNA cloning and sequence analysis. DNA 1985; 4:445–450.
[57] Chan H, Springman EB, Clark JM. Expression and characterization of human tissue kallikrein variants. Protein Expr Purif 1998; 12:361–370.
[58] Bharaj BB, Luo LY, Jung K, Stephan C, Diamandis EP. Identification of single nucleotide polymorphisms in the human kallikrein 10 (KLK10) gene and their association with prostate, breast, testicular, and ovarian cancers. Prostate 2002; 51:35–41.
[59] Bharaj B, Scorilas A, Diamandis EP, et al. Breast cancer prognostic significance of a single nucleotide polymorphism in the proximal androgen response element of the prostate specific antigen gene promoter. Breast Cancer Res Treat 2000; 61:111–119.
[60] Rawlings ND, Barrett AJ. Evolutionary families of peptidases. Biochem J 1993; 290:205–218.
[61] Barrett AJ, Rawlings ND. Families and clans of serine peptidases. Arch Biochem Biophys 1995; 318:247–250.

[62] Yousef GM, Kopolovic AD, Elliott MB, Diamandis EP. Genomic overview of serine proteases. Biochem Biophys Res Commun 2003; 305:28–36.
[63] Yousef GM, Elliott MB, Kopolovic AD, Serry E, Diamandis EP. Sequence and evolutionary analysis of the human trypsin subfamily of serine peptidases. Biochim Biophys Acta 2004; 1698:77–86.
[64] Lopez-Otin C, Overall CM. Protease degradomics: A new challenge for proteomics. Nat Rev Mol Cell Biol 2002; 3:509–519.
[65] Puente XS, Sanchez LM, Overall CM, Lopez-Otin C. Human and mouse proteases: A comparative genomic approach. Nat Rev Genet 2003; 4:544–558.
[66] Schultz RM, Liebman MN. Structure-function relationship in protein families. In: Devlin TM, editor. Textbook of Biochemistry with Clinical Correlations, fourth Ed. 1–2–116. 1997.
[67] Horl WH. Proteinases: potential role in health and disease. In: Sandler M, Smith HJ, editors. Design of Enzyme Inhibitors as Drugs. Oxford: Oxford University Press, 1989: 573–581.
[68] Henderson BR, Tansey WP, Phillips SM, Ramshaw IA, Kefford RF. Transcriptional and posttranscriptional activation of urokinase plasminogen activator gene expression in metastatic tumor cells. Cancer Res 1992; 52:2489–2496.
[69] Froelich CJ, Zhang X, Turbov J, Hudig D, Winkler U, Hanna WL. Human granzyme B degrades aggrecan proteoglycan in matrix synthesized by chondrocytes. J Immunol. 1993; 151:7161–7171.
[70] Yousef GM, Diamandis EP. Expanded human tissue kallikrein family—a novel panel of cancer biomarkers. Tumour Biol 2002; 23:185–192.
[71] Southan C. A genomic perspective on human proteases as drug targets. Dru Discov Toda 2001; 6:681–688.
[72] Dayhoff MO. Atlas of protein sequence and structure. Natl Biomed Res Found 1978; 5:79–81.
[73] Perona JJ, Craik CS. Structural basis of substrate specificity in the serine proteases. Protein Sci 1995; 4:337–360.
[74] Greer J. Comparative modeling methods: Application to the family of the mammalian serine proteases. Proteins 1990; 7:317–334.
[75] Krem MM, Di Cera E. Molecular markers of serine protease evolution. EMB J 2001; 20:3036–3045.
[76] Stryer L. Biochemistry, 4th ed. New York: W. H. Freeman, 1995.
[77] Diamandis EP, Yousef GM, Soosaipillai AR, et al. Immunofluorometric assay of human kallikrein 6 (zyme/protease M/neurosin) and preliminary clinical applications. Clin Biochem 2000; 33:369–375.
[78] Kishi T, Grass L, Soosaipillai A, Shimizu-Okabe C, Diamandis EP. Human kallikrein 8: Immunoassay development and identification in tissue extracts and biological fluids. Clin Chem 2003; 49:87–96.
[79] Yousef, G. M., Polymeris, M. E., Grass, L., et al. Human kallikrein 5: A potential novel serum biomarker for breast and ovarian cancer. Cancer Res 2003: 63:3958–3965.
[80] Petraki CD, Karavana VN, Skoufogiannis PT, et al. The spectrum of human kallikrein 6 (zyme/protease M/neurosin) expression in human tissues as assessed by immunohistochemistry. J Histochem Cytochem 2001; 49:1431–1441.
[81] Petraki CD, Karavana VN, Luo L-Y, Diamandis EP. Human kallikrein 10 expression in normal tissues by immmunohistochemistry. J Histochem Cytochem 2002; 50:1247–1261.
[82] Petraki CD, Karavana VN, Diamandis EP. Human kallikrein 13 expression in normal tissues. An immunohistochemical study. J Histochem Cytochem 2003; 51:493–501.

[83] Petraki CD, Karavana VN, Revelos KI, Luo LY, Diamandis EP. Immunohistochemical localization of human kallikreins 6 and 10 in pancreatic islets. Histochem J 2002; 34:313–322.

[84] Petraki CD, Gregorakis AK, Papanastasiou PA, Karavana VN, Luo LY, Diamandis EP. Immunohistochemical localization of human kallikreins 6, 10 and 13 in benign and malignant prostatic tissues. Prostat CancerProstatic Dis 2003; 6:223–227.

[85] Petraki CD, Gregorakis AK, Vaslamatzis MM, Papanastassiou P, Yousef GM, Diamandis EP. The immunohistochemical expression of human kallikreins 5, 6, 10 and 11 in renal cell carcinoma. Correlation with several prognostic factors. Proc AACR 2003; 44:1077, abstract 5412.

[86] McCormack RT, Rittenhouse HG, Finlay JA, et al. Molecular forms of prostate-specific antigen and the human kallikrein gene family: A new era. Urology 1995; 45:729–744.

[87] Chu TM. Prostate-specific antigen and early detection of prostate cancer. Tumour Biol 1997; 18:123–134.

[88] Stenman UH. New ultrasensitive assays facilitate studies on the role of human glandular kallikrein (hK2) as a marker for prostatic disease [editorial; comment]. Clin Chem 1999; 45:753–754.

[89] Chang A, Yousef GM, Scorilas A, et al. Human kallikrein gene 13 (KLK13) expression by quantitative RT-PCR: An independent indicator of favourable prognosis in breast cancer. Br J Cancer 2002; 86:1457–1464.

[90] Yousef GM, Chang A, Diamandis EP. Identification and characterization of KLK-L4, a new kallikrein-like gene that appears to be down-regulated in breast cancer tissues. J Biol Chem 2000; 275:11891–11898.

[91] Luo LY, Katsaros D, Scorilas A, et al. Prognostic value of human kallikrein 10 expression in epithelial ovarian carcinoma. Clin Cancer Res 2001; 7:2372–2379.

[92] Luo LY, Yousef G, Diamandis EP. Human tissue kallikreins and testicular cancer. Apmis 2003; 111:225–232; discussion 232–233.

[93] Yousef GM, Scorilas A, Kyriakopoulou LG, et al. Human kallikrein gene 5 (KLK5) expression by quantitative PCR: An independent indicator of poor prognosis in breast cancer. Clin Chem 2002; 48:1241–1250.

[94] Yousef GM, Borgono CA, Scorilas A, et al. Quantitative analysis of human kallikrein gene 14 expression in breast tumours indicates association with poor prognosis. Br J Cancer 2002; 87:1287–1293.

[95] Xi Z, Klokk TI, Korkmaz K, et al. Kallikrein 4 is a predominantly nuclear protein and is overexpressed in prostate cancer. Cancer Res 2004; 64:2365–2370.

[96] Cleutjens KB, van Eekelen CC, van der Korput HA, Brinkmann AO, Trapman J. Two androgen response regions cooperate in steroid hormone regulated activity of the prostate-specific antigen promoter. J Biol Chem 1996; 271:6379–6388.

[97] Cleutjens KB, van der Korput HA, van Eekelen CC, van Rooij HC, Faber PW, Trapman J. An androgen response element in a far upstream enhancer region is essential for high, androgen-regulated activity of the prostate-specific antigen promoter. Mol Endocrinol 1997; 11:148–161.

[98] Brookes DE, Zandvliet D, Watt F, Russell PJ, Molloy PL. Relative activity and specificity of promoters from prostate-expressed genes. Prostate 1998; 35:18–26.

[99] Huang W, Shostak Y, Tarr P, Sawyers C, Carey M. Cooperative assembly of androgen receptor into a nucleoprotein complex that regulates the prostate-specific antigen enhancer. J Biol Chem 1999; 274:25756–25768.

[100] Murtha P, Tindall DJ, Young CY. Androgen induction of a human prostate-specific kallikrein, hKLK2: Characterization of an androgen response element in the 5′ promoter region of the gene. Biochemistry 1993; 32:6459–6464.

[101] Yu DC, Sakamoto GT, Henderson DR. Identification of the transcriptional regulatory sequences of human kallikrein 2 and their use in the construction of calydon virus 764, an attenuated replication competent adenovirus for prostate cancer therapy. Cancer Res 1999; 59:1498–1504.

[102] Khan AR, James MN. Molecular mechanisms for the conversion of zymogens to active proteolytic enzymes. Protein Sci 1998; 7:815–836.

[103] Brattsand M, Egelrud T. Purification, molecular cloning, and expression of a human stratum corneum trypsin-like serine protease with possible function in desquamation. J Biol Chem 1999; 274:30033–30040.

[104] Hansson L, Stromqvist M, Backman A, Wallbrandt P, Carlstein A, Egelrud T. Cloning, expression, and characterization of stratum corneum chymotryptic enzyme. A skin-specific human serine proteinase. J Biol Chem 1994; 269:19420–19426.

[105] Denmeade SR, Lovgren J, Khan SR, Lilja H, Isaacs JT. Activation of latent protease function of pro-hK2, but not pro-PSA, involves autoprocessing. Prostate 2001; 48:122–126.

[106] Takayama TK, McMullen BA, Nelson PS, Matsumura M, Fujikawa K. Characterization of hK4 (Prostase), a prostate-specific serine protease: Activation of the precursor of prostate specific antigen (pro-PSA) and single-chain urokinase-type plasminogen activator and degradation of prostatic acid phosphatase. Biochemistry 2001; 40:15341–15348.

[107] Little SP, Dixon EP, Norris F, et al. Zyme, a novel and potentially amyloidogenic enzyme cDNA isolated from Alzheimer's disease brain. J Biol Chem 1997; 272:25135–25142.

[108] Laskowski M, Qasim MA. What can the structures of enzyme-inhibitor complexes tell us about the structures of enzyme substrate complexes? Biochim Biophys Acta 2000; 1477:324–337.

[109] Sottrup-Jensen L, Sand O, Kristensen L, Fey GH. The alpha-macroglobulin bait region. Sequence diversity and localization of cleavage sites for proteinases in five mammalian alpha-macroglobulins. J Biol Chem 1989; 264:15781–15789.

[110] Becker C, Lilja H. Individual prostate-specific antigen (PSA) forms as prostate tumor markers. Clin Chim Acta 1997; 257:117–132.

[111] Christensson A, Lilja H. Complex formation between protein C inhibitor and prostate-specific antigen *in vitro* and in human semen. Eur J Biochem 1994; 220:45–53.

[112] Stenman UH, Leinonen J, Alfthan H, Rannikko S, Tuhkanen K, Alfthan O. A complex between prostate-specific antigen and alpha 1-antichymotrypsin is the major form of prostate-specific antigen in serum of patients with prostatic cancer: Assay of the complex improves clinical sensitivity for cancer. Cancer Res 1991; 51:222–226.

[113] Lilja H, Christensson A, Dahlen U, et al. Prostate-specific antigen in serum occurs predominantly in complex with alpha 1-antichymotrypsin. Clin Chem 1991; 37:1618–1625.

[114] Christensson A, Bjork T, Nilsson O, et al. Serum prostate specific antigen complexed to alpha 1-antichymotrypsin as an indicator of prostate cancer. J Urol 1993; 150:100–105.

[115] Mikolajczyk SD, Millar LS, Marker KM, et al. Ala217 is important for the catalytic function and autoactivation of prostate-specific human kallikrein 2. Eur J Biochem 1997; 246:440–446.

[116] Grauer LS, Finlay JA, Mikolajczyk SD, Pusateri KD, Wolfert RL. Detection of human glandular kallikrein, hK2, as its precursor form and in complex with protease inhibitors in prostate carcinoma serum. J Androl 1998; 19:407–411.

[117] Black MH, Magklara A, Obiezu CV, Melegos DN, Diamandis EP. Development of an ultrasensitive immunoassay for human glandular kallikrein with no cross-reactivity from prostate-specific antigen [see comments]. Clin Chem 1999; 45:790–799.

[118] Heeb MJ, Espana F. alpha2-macroglobulin and C1-inactivator are plasma inhibitors of human glandular kallikrein. Blood Cell Mol Dis 1998; 24:412–419.
[119] Cao Y, Lundwall A, Gadaleanu V, Lilja H, Bjartell A. Anti-thrombin is expressed in the benign prostatic epithelium and in prostate cancer and is capable of forming complexes with prostate-specific antigen and human glandular kallikrein 2. Am J Pathol 2002; 161:2053–2063.
[120] Stephan C, Jung K, Diamandis EP, Rittenhouse HG, Lein M, Loening SA. Prostate-specific antigen, its molecular forms, and other kallikrein markers for detection of prostate cancer. Urology 2002; 59:2–8.
[121] Yousef GM, Kapadia C, Polymeris ME, et al. The human kallikrein protein 5 (hK5) is enzymatically active, glycosylated and forms complexes with two protease inhibitors in ovarian cancer fluids. Biochim Biophys Acta 2003; 1628:88–96.
[122] Hutchinson S, Luo LY, Yousef GM, Soosaipillai A, Diamandis EP. Purification of human kallikrein 6 from biological fluids and identification of its complex with alpha(1)-antichymotrypsin. Clin Chem 2003; 49:746–751.
[123] Kapadia C, Yousef GM, Mellati AA, Magklara A, Wasney GA, Diamandis EP. Complex formation between human kallikrein 13 and serum protease inhibitors. Clin Chim Acta 2004; 339:157–167.
[124] Christensson A, Laurell CB, Lilja H. Enzymatic activity of prostate-specific antigen and its reactions with extracellular serine proteinase inhibitors. Eur J Biochem. 1990; 194:755–763.
[125] Lovgren J, Rajakoski K, Karp M, Lundwall A, Lilja H. Activation of the zymogen form of prostate-specific antigen by human glandular kallikrein 2. Biochem Biophys Res Commun 1997; 238:549–555.
[126] Yousef GM, Polymeris ME, Yacoub GM, et al. Parallel overexpression of seven kallikrein genes in ovarian cancer. Cancer Res 2003; 63:2223–2227.
[127] Boutanaev AM, Kalmykova AI, Shevelyov YY, Nurminsky DI. Large clusters of co-expressed genes in the *Drosophila* genome. Nature 2002; 420:666–669.
[128] Li Q, Harju S, Peterson KR. Locus control regions: Coming of age at a decade plus. Trends Genet 1999; 15:403–408.
[129] Kramer JA, McCarrey JR, Djakiew D, Krawetz SA. Human spermatogenesis as a model to examine gene potentiation. Mol Reprod Dev 2000; 56:254–258.
[130] Vermaak D, Wolffe AP. Chromatin and chromosomal controls in development. Dev Genet 1998; 22:1–6.
[131] Cedar H. DNA methylation and gene activity. Cell 1988; 53:3–4.
[132] Davie JR, Hendzel MJ. Multiple functions of dynamic histone acetylation. J Cell Biochem 1994; 55:98–105.
[133] Eden S, Hashimshony T, Keshet I, Cedar H, Thorne AW. DNA methylation models histone acetylation. Nature 1998; 394:842.
[134] Stallmach A, Wittig BM, Kremp K, et al. Downregulation of CD44v6 in colorectal carcinomas is associated with hypermethylation of the CD44 promoter region. Exp Mol Pathol 2003; 74:262–266.
[135] Herman JG, Merlo A, Mao L, et al. Inactivation of the CDKN2/p16/MTS1 gene is frequently associated with aberrant DNA methylation in all common human cancers. Cancer Res 1995; 55:4525–4530.
[136] Issa JP, Ottaviano YL, Celano P, Hamilton SR, Davidson NE, Baylin SB. Methylation of the oestrogen receptor CpG island links ageing and neoplasia in human colon. Nat Genet 1994; 7:536–540.

[137] Nakayama S, Sasaki A, Mese H, Alcalde RE, Tsuji T, Matsumura T. The E-cadherin gene is silenced by CpG methylation in human oral squamous cell carcinomas. Int J Cancer 2001; 93:667–673.
[138] Esteller M, Hamilton SR, Burger PC, Baylin SB, Herman JG. Inactivation of the DNA repair gene O6-methylguanine-DNA methyltransferase by promoter hypermethylation is a common event in primary human neoplasia. Cancer Res 1999; 59:793–797.
[139] Dammann R, Schagdarsurengin U, Liu L, et al. Frequent RASSF1A promoter hypermethylation and K-ras mutations in pancreatic carcinoma. Oncogene 2003; 22:3806–3812.
[140] Ahuja N, Mohan AL, Li Q, et al. Association between CpG island methylation and microsatellite instability in colorectal cancer. Cancer Res 1997; 57:3370–3374.
[141] Cameron EE, Bachman KE, Myohanen S, Herman JG, Baylin SB. Synergy of demethylation and histone deacetylase inhibition in the re-expression of genes silenced in cancer. Nat Genet 1999; 21:103–107.
[142] Li B, Goyal J, Dhar S, et al. CpG methylation as a basis for breast tumor-specific loss of NES1/kallikrein 10 expression. Cancer Res 2001; 61:8014–8021.
[143] Roman-Gomez J, Jimenez-Velasco A, Agirre X, et al. The normal epithelial cell-specific 1 (NES1) gene, a candidate tumor suppressor gene on chromosome 19q13.3–4, is downregulated by hypermethylation in acute lymphoblastic leukemia. Leukemia 2004; 18:362–365.
[144] Liu XL, Wazer DE, Watanabe K, Band V. Identification of a novel serine protease-like gene, the expression of which is down-regulated during breast cancer progression. Cancer Res 1996; 56:3371–3379.
[145] Goyal J, Smith KM, Cowan JM, Wazer DE, Lee SW, Band V. The role for NES1 serine protease as a novel tumor suppressor. Cancer Res 1998; 58:4782–4786.
[146] Trapman J, Cleutjens KB. Androgen-regulated gene expression in prostate cancer. Semin Cancer Biol 1997; 8:29–36.
[147] Myers SA, Clements JA. Kallikrein 4 (KLK4), a new member of the human kallikrein gene family is up-regulated by estrogen and progesterone in the human endometrial cancer cell line, KLE. J Clin Endocrinol Metab 2001; 86:2323–2326.
[148] Nelson PS, Gan L, Ferguson C, et al. Molecular cloning and characterization of prostase, an androgen-regulated serine protease with prostate-restricted expression. Proc Natl Acad Sci USA 1999; 96:3114–3119.
[149] Riegman PH, Vlietstra RJ, van der Korput JA, Brinkmann AO, Trapman J. The promoter of the prostate-specific antigen gene contains a functional androgen responsive element. Mol Endocrinol 1991; 5:1921–1930.
[150] Riegman PH, Vlietstra RJ, van der Korput HA, Romijn JC, Trapman J. Identification and androgen-regulated expression of two major human glandular kallikrein-1 (hGK-1) mRNA species. Mol Cell Endocrinol 1991; 76:181–190.
[151] Luo LY, Grass L, Diamandis EP. The normal epithelial cell-specific 1 (NES1) gene is up-regulated by steroid hormones in the breast carcinoma cell line BT-474. Anticancer Res 2000; 20:981–986.
[152] Yousef GM, Scorilas A, Magklara A, Soosaipillai A, Diamandis EP. The KLK7 (PRSS6) gene, encoding for the stratum corneum chymotryptic enzyme is a new member of the human kallikrein gene family-genomic characterization, mapping, tissue expression and hormonal regulation. Gene 2000; 254:119–128.
[153] Yousef GM, Diamandis EP. The new kallikrein-like gene, KLK-L2. Molecular characterization, mapping, tissue expression, and hormonal regulation. J Biol Chem 1999; 274:37511–37516.

[154] Yousef GM, Luo LY, Scherer SW, Sotiropoulou G, Diamandis EP. Molecular characterization of Zyme/Protease M/Neurosin (PRSS9), a hormonally regulated kallikrein-like serine protease. Genomics 1999; 62:251–259.

[155] Yousef GM, Scorilas A, Magklara A, et al. The androgen-regulated gene human kallikrein 15 (KLK15) is an independent and favourable prognostic marker for breast cancer. Br J Cancer 2002; 87:1294–1300.

[156] Yousef GM, Magklara A, Diamandis EP. KLK12 is a novel serine protease and a new member of the human kallikrein gene family-differential expression in breast cancer. Genomics 2000; 69:331–341.

[157] Yousef GM, Fracchioli S, Scorilas A, et al. Steroid hormone regulation and prognostic value of the human kallikrein gene 14 in ovarian cancer. Am J Clin Pathol 2003; 119:346–355.

[158] Palmer HG, Sanchez-Carbayo M, Ordonez-Moran P, Larriba MJ, Cordon-Cardo C, Munoz A. Genetic signatures of differentiation induced by 1alpha,25-dihydroxyvitamin D3 in human colon cancer cells. Cancer Res 2003; 63:7799–7806.

[159] Krem MM, Rose T, Di Cera E. Sequence determinants of function and evolution in serine proteases. Trends Cardiovasc Med 2000; 10:171–176.

[160] Arnason U, Janke A. Mitogenomic analyses of eutherian relationships. Cytogenet Genome Res 2002; 96:20–32.

[161] Takayama TK, Carter CA, Deng T. Activation of prostate-specific antigen precursor (pro-PSA) by prostin, a novel human prostatic serine protease identified by degenerate PCR. Biochemistry 2001; 40:1679–1687.

[162] Takayama TK, Fujikawa K, Davie EW. Characterization of the precursor of prostate-specific antigen. Activation by trypsin and by human glandular kallikrein. J Biol Chem 1997; 272:21582–21588.

[163] Bhoola K, Ramsaroop R, Plendl J, Cassim B, Dlamini Z, Naicker S. Kallikrein and kinin receptor expression in inflammation and cancer. Biol Chem 2001; 382:77–89.

[164] Cohen P, Graves HC, Peehl DM, Kamarei M, Giudice LC, Rosenfeld RG. Prostate-specific antigen (PSA) is an insulin-like growth factor binding protein-3 protease found in seminal plasma. J Clin Endocrinol Metab 1992; 75:1046–1053.

[165] Iwamura M, Hellman J, Cockett AT, Lilja H, Gershagen S. Alteration of the hormonal bioactivity of parathyroid hormone-related protein (PTHrP) as a result of limited proteolysis by prostate-specific antigen. Urology 1996; 48:317–325.

[166] Egelrud T. Stratum corneum chymotryptic enzyme. In: Barrett AJ, RND, Woessner JF, editors. Handbook of Proteolytic Enzymes, Vol. 2. London: Academic Press, 1998: 1556–1558.

[167] Yousef GM, Diamandis EP. Human tissue kallikreins: A new enzymatic cascade pathway? Biol Chem 2002; 383:1045–1057.

[168] Slawin KM, Shariat SF, Nguyen C, et al. Detection of metastatic prostate cancer using a splice variant-specific reverse transcriptase-polymerase chain reaction assay for human glandular kallikrein. Cancer Res 2000; 60:7142–7148.

[169] Nakamura T, Mitsui S, Okui A, et al. Alternative splicing isoforms of hippostasin (PRSS20/KLK11) in prostate cancer cell lines. Prostate 2001; 49:72–78.

[170] Dong Y, Kaushal A, Brattsand M, Nicklin J, Clements JA. Differential splicing of KLK5 and KLK7 in epithelial ovarian cancer produces novel variants with potential as cancer biomarkers. Clin Cancer Res 2003; 9:1710–1720.

[171] Hooper JD, Bui LT, Rae FK, et al. Identification and characterization of KLK14, a novel kallikrein serine protease gene located on human chromosome 19q13.4 and expressed in prostate and skeletal muscle. Genomics 2001; 73:117–122.

[172] Chang A, Yousef GM, Jung K, Meyts ER, Diamanids EP. Identification and molecular characterization of five novel kallikrein gene 13 (KLK13;KLK-L4) splice variants: Differential expression in human testis and testicular cancer. Anticancer Res 2001; 21:3147–3152.

[173] Mitsui S, Tsuruoka N, Yamashiro K, Nakazato H, Yamaguchi NA. Novel form of human neuropsin, a brain-related serine protease, is generated by alternative splicing and is expressed preferentially in human adult brain. Eur J Biochem 1999; 260:627–634.

[174] Obiezu CV, Diamandis EP. An alternatively spliced variant of KLK4 expressed in prostatic tissue. Clin Biochem 2000; 33:599–600.

[175] Heuze N, Olayat S, Gutman N, Zani ML, Courty Y. Molecular cloning and expression of an alternative hKLK3 transcript coding for a variant protein of prostate-specific antigen. Cancer Res 1999; 59:2820–2824.

[176] Tanaka T, Isono T, Yoshiki T, Yuasa T, Okada Y. A novel form of prostate-specific antigen transcript produced by alternative splicing. Cancer Res 2000; 60:56–59.

[177] Bhoola KD, Figueroa CD, Worthy K. Bioregulation of kinins: Kallikreins, kininogens, and kininases. Pharmacol Rev 1992; 44:1–80.

[178] Meneton P, Bloch-Faure M, Hagege AA, et al. Cardiovascular abnormalities with normal blood pressure in tissue kallikrein-deficient mice. Proc Natl Acad Sci USA 2001; 98:2634–2639.

[179] Sharma JN, Uma K, Noor AR, Rahman AR. Blood pressure regulation by the kallikrein-kinin system. Gen Pharmacol 1996; 27:55–63.

[180] Deperthes D, Frenette G, Brillard-Bourdet M, et al. Potential involvement of kallikrein hK2 in the hydrolysis of the human seminal vesicle proteins after ejaculation. J Androl 1996; 17:659–665.

[181] Lilja H. A kallikrein-like serine protease in prostatic fluid cleaves the predominant seminal vesicle protein. J Clin Invest 1985; 76:1899–1903.

[182] Heidtmann HH, Nettelbeck DM, Mingels A, Jager R, Welker HG, Kontermann RE. Generation of angiostatin-like fragments from plasminogen by prostate-specific antigen. Br J Cancer 1999; 81:1269–1273.

[183] Egelrud T, Lundstrom A. The dependence of detergent-induced cell dissociation in non-palmo-plantar stratum corneum on endogenous proteolysis. J Invest Dermatol 1990; 95:456–459.

[184] Macfarlane SR, Seatter MJ, Kanke T, Hunter GD, Plevin R. Proteinase-activated receptors. Pharmacol Rev 2001; 53:245–282.

[185] Noorbakhsh F, Vergnolle N, Hollenberg MD, Power C. Proteinase-activated receptors in the nervous system. Nat Rev Neurosci 2003; 4:981–990.

[186] Coughlin SR. Thrombin signalling and protease-activated receptors. Nature 2000; 407:258–264.

[187] Margolius HS, Horwitz D, Pisano JJ, Keiser HR. Urinary kallikrein excretion in hypertensive man. Relationships to sodium intake and sodium-retaining steroids. Circ Res 1974; 35:820–825.

[188] Jaffa AA, Chai KX, Chao J, Chao L, Mayfield RK. Effects of diabetes and insulin on expression of kallikrein and renin genes in the kidney. Kidney Int 1992; 41:789–795.

[189] Ekholm E, Egelrud T. Stratum corneum chymotryptic enzyme in psoriasis. Arch Dermatol Res 1999; 291:195–200.

[190] Sondell B, Dyberg P, Anneroth GK, Ostman PO, Egelrud T. Association between expression of stratum corneum chymotryptic enzyme and pathological keratinization in human oral mucosa. Acta Derm Venereol 1996; 76:177–181.

[191] Akita H, Matsuyama T, Iso H, Sugita M, Yoshida S. Effects of oxidative stress on the expression of limbic-specific protease neuropsin and avoidance learning in mice. Brain Res 1997; 769:86–96.

[192] Momota Y, Yoshida S, Ito J, et al. Blockade of neuropsin, a serine protease, ameliorates kindling epilepsy. Eur J Neurosci 1998; 10:760–764.

[193] Kishi T, Kato M, Shimizu T, et al. Crystal structure of neuropsin, a hippocampal protease involved in kindling epileptogenesis. J Biol Chem 1999; 274:4220–4224.

[194] Shimizu-Okabe C, Yousef GM, Diamandis EP, Yoshida S, Shiosaka S, Fahnestock M. Expression of the kallikrein gene family in normal and Alzheimer's disease brain. Neuroreport 2001; 12:2747–2751.

[195] Yamashiro K, Tsuruoka N, Kodama S, et al. Molecular cloning of a novel trypsin-like serine protease (neurosin) preferentially expressed in brain. Biochim Biophys Acta 1997; 1350:11–14.

[196] Scarisbrick IA, Blaber SI, Lucchinetti CF, Genain CP, Blaber M, Rodriguez M. Activity of a newly identified serine protease in CNS demyelination. Brain 2002; 125:1283–1296.

[197] Blaber SI, Ciric B, Christophi GP, et al. Targeting kallikrein 6 proteolysis attenuates CNS inflammatory disease. FASEB J 2004; 18:920–922.

[198] Yoshida S, Taniguchi M, Suemoto T, Oka T, He X, Shiosaka S. cDNA cloning and expression of a novel serine protease, TLSP. Biochim Biophys Acta 1998; 1399:225–228.

[199] Yousef GM, Magklara A, Chang A, Jung K, Katsaros D, Diamandis EP. Cloning of a new member of the human kallikrein gene family, KLK14, which is down-regulated in different malignancies. Cancer Res 2001; 61:3425–3431.

[200] Diamandis EP, Scorilas A, Kishi T, et al. Altered kallikrein 7 and 10 concentrations in cerebrospinal fluid of patients with Alzheimer's disease and frontotemporal dementia. Clin Biochem 2004; 37:230–237.

[201] Yousef GM, Kishi T, Diamandis EP. Role of kallikrein enzymes in the central nervous system. Clin Chim Acta 2003; 329:1–8.

[202] Suzuki Y, Koyama J, Moro O, et al. The role of two endogenous proteases of the stratum corneum in degradation of desmoglein-1 and their reduced activity in the skin of ichthyotic patients. Br J Dermatol 1996; 134:460–464.

[203] Ekholm IE, Brattsand M, Egelrud T. Stratum corneum tryptic enzyme in normal epidermis: A missing link in the desquamation process? J Invest Dermatol 2000; 114:56–63.

[204] Komatsu N, Takata M, Takehara K, Ohtsuki N. Kallikrein family: The expression, the localization and the role in skin tissue. J Dermatol Sci 2002; 139.

[205] Komatsu N, Takata M, Otsuki N, et al. Expression and localization of tissue kallikrein mRNAs in human epidermis and appendages. J Invest Dermatol 2003; 121:542–549.

[206] Kishi T, Soosaipillai A, Grass L, Little SP, Johnstone EM, Diamandis EP. Development of an immunofluorometric assay and quantification of human kallikrein 7 in tissue extracts and biological fluids. Clin Chem 2004; 50:709–716.

[207] Kuwae K, Matsumoto-Miyai K, Yoshida S, et al. Epidermal expression of serine protease, neuropsin (KLK8) in normal and pathological skin samples. Mol Pathol 2002; 55:235–241.

[208] Johnson B, Horn T, Sander C, Kohler S, Smoller BR. Expression of stratum corneum chymotryptic enzyme in ichthyoses and squamoproliferative processes. J Cutan Pathol 2003; 30:358–362.

[209] Hansson L, Backman A, Ny A, et al. Epidermal overexpression of stratum corneum chymotryptic enzyme in mice: A model for chronic itchy dermatitis. J Invest Dermatol 2002; 118:444–449.

[210] Ny A, Egelrud T. Transgenic mice over-expressing a serine protease in the skin: Evidence of interferon gamma-independent MHC II expression by epidermal keratinocytes. Acta Derm Venereol 2003; 83:322–327.

[211] Kirihara T, Matsumoto-Miyai K, Nakamura Y, Sadayama T, Yoshida S, Shiosaka S. Prolonged recovery of ultraviolet B-irradiated skin in neuropsin (KLK8)-deficient mice. Br J Dermatol 2003; 149:700–706.

[212] Chavanas S, Bodemer C, Rochat A, et al. Mutations in SPINK5, encoding a serine protease inhibitor, cause Netherton syndrome. Nat Genet 2000; 25:141–142.

[213] Mitsudo K, Jayakumar A, Henderson Y, et al. Inhibition of serine proteinases plasmin, trypsin, subtilisin A, cathepsin G, and elastase by LEKTI: A kinetic analysis. Biochemistry 2003; 42:3874–3881.

[214] Jayakumar A, Kang Y, Mitsudo K, et al. Expression of LEKTI domains 6-9′ in the baculovirus expression system: Recombinant LEKTI domains 6–9′ inhibit trypsin and subtilisin A. Protein Expr Purif 2004; 35:93–101.

[215] Black MH, Diamandis EP. The diagnostic and prognostic utility of prostate-specific antigen for diseases of the breast. Breast Cancer Res Treat 2000; 59:1–14.

[216] Anisowicz A, Sotiropoulou G, Stenman G, Mok SC, Sager R. A novel protease homolog differentially expressed in breast and ovarian cancer. Mol Med 1996; 2:624–636.

[217] Diamandis EP, Scorilas A, Fracchioli S, et al. Human kallikrein 6 (hK6): A new potential serum biomarker for diagnosis and prognosis of ovarian carcinoma. J Clin Oncol 2003; 21:1035–1043.

[218] Luo LY, Katsaros D, Scorilas A, et al. The serum concentration of human kallikrein 10 represents a novel biomarker for ovarian cancer diagnosis and prognosis. Cancer Res 2003; 63:807–811.

[219] Diamandis EP, Okui A, Mitsui S, et al. Human kallikrein 11: A new biomarker of prostate and ovarian carcinoma. Cancer Res 2002; 62:295–300.

[220] Wolf WC, Evans DM, Chao L, Chao J. A synthetic tissue kallikrein inhibitor suppresses cancer cell invasiveness. Am J Pathol 2001; 159:1797–1805.

[221] Alizadeh AA, Ross DT, Perou CM, van de Rijn M. Towards a novel classification of human malignancies based on gene expression patterns. J Pathol 2001; 195:41–52.

[222] Jemal A, Murray T, Samuels A, Ghafoor A, Ward E, Thun MJ. Cancer statistics, 2003. CA. Cancer J Clin 2003; 53:5–26.

[223] Schink JC. Current initial therapy of stage III and IV ovarian cancer: Challenges for managed care. Semin Oncol 1999; 26:2–7.

[224] Trope C. Prognostic factors in ovarian cancer. Cancer Treat Res 1998; 95:287–352.

[225] Yousef GM, Diamandis EP. Kallikreins, steroid hormones and ovarian cancer: Is there a link? Minerva Endocrinol 2002; 27:157–166.

[226] Obiezu CV, Scorilas A, Katsaros D, et al. Higher human kallikrein gene 4 (KLK4) expression indicates poor prognosis of ovarian cancer patients. Clin Cancer Res 2001; 7:2380–2386.

[227] Kim H, Scorilas A, Katsaros D, et al. Human kallikrein gene 5 (KLK5) expression is an indicator of poor prognosis in ovarian cancer. Br J Cancer 2001; 84:643–650.

[228] Tanimoto H, Underwood LJ, Shigemasa K, Parmley TH, O'Brien TJ. Increased expression of protease M in ovarian tumors. Tumour Biol 2001; 22:11–18.

[229] Hoffman BR, Katsaros D, Scorilas A, et al. Immunofluorometric quantitation and histochemical localisation of kallikrein 6 protein in ovarian cancer tissue: A new independent unfavourable prognostic biomarker. Br J Cancer 2002; 87:763–771.

[230] Kyriakopoulou LG, Yousef GM, Scorilas A, et al. Prognostic value of quantitatively assessed KLK7 expression in ovarian cancer. Clin Biochem 2003; 36:135–143.

[231] Luo L, Bunting P, Scorilas A, Diamandis EP. Human kallikrein 10: A novel tumor marker for ovarian carcinoma? Clin Chim Acta 2001; 306:111–118.
[232] Yousef GM, Scorilas A, Katsaros D, et al. Prognostic value of the human kallikrein gene 15 expression in ovarian cancer. J Clin Oncol 2003; 21:3119–3126.
[233] Magklara A, Scorilas A, Katsaros D, et al. The human KLK8 (neuropsin/ovasin) gene: Identification of two novel splice variants and its prognostic value in ovarian cancer. Clin Cancer Res 2001; 7:806–811.
[234] Yousef GM, Kyriakopoulou LG, Scorilas A, et al. Quantitative expression of the human kallikrein gene 9 (KLK9) in ovarian cancer: A new independent and favorable prognostic marker. Cancer Res 2001; 61:7811–7818.
[235] Borgono CA, Fracchioli S, Yousef GM, et al. Favorable prognostic value of tissue human kallikrein 11 (hK11) in patients with ovarian carcinoma. Int. J. Cancer 2003; 106:605–610.
[236] Parkin DM, Pisani P, Ferlay J. Estimates of the worldwide incidence of 25 major cancers in 1990. Int J Cancer 1999; 80:827–841.
[237] Bundred NJ. Prognostic and predictive factors in breast cancer. Cancer Treat Rev 2001; 27:137–142.
[238] Yousef GM, Scorilas A, Nakamura T, et al. The prognostic value of the human kallikrein gene 9 (KLK9) in breast cancer. Breast Cancer Res Treat 2003; 78:149–158.
[239] Luo LY, Diamandis EP, Look MP, Soosaipillai AP, Foekens JA. Higher expression of human kallikrein 10 in breast cancer tissue predicts tamoxifen resistance. Br J Cancer 2002; 86:1790–1796.
[240] Yousef GM, Yacoub GM, Polymeris ME, Popalis C, Soosaipillai A, Diamandis EP. Kallikrein gene downregulation in breast cancer. Br J Cancer 2004; 90:167–172.
[241] Magklara A, Scorilas A, Stephan C, et al. Decreased concentrations of prostate-specific antigen and human glandular kallikrein 2 in malignant versus nonmalignant prostatic tissue. Urology 2000; 56:527–532.
[242] Yousef GM, Scorilas A, Chang A, et al. Down-regulation of the human kallikrein gene 5 (KLK5) in prostate cancer tissues. Prostate 2002; 51:126–132.
[243] Yousef GM, Obiezu CV, Jung K, Stephan C, Scorilas A, Diamandis EP. Differential expression of Kallikrein gene 5 in cancerous and normal testicular tissues. Urology 2002; 60:714–718.
[244] Luo LY, Rajpert-De Meyts ER, Jung K, Diamandis EP. Expression of the normal epithelial cell-specific 1 (NES1; KLK10) candidate tumour suppressor gene in normal and malignant testicular tissue. Br J Cancer 2001; 85:220–224.
[245] Bhattacharjee A, Richards WG, Staunton J, et al. Classification of human lung carcinomas by mRNA expression profiling reveals distinct adenocarcinoma subclasses. Proc Natl Acad Sci USA 2001; 98:13790–13795.
[246] Yousef GM, Borgono CA, Popalis C, et al. *In-silico* analysis of kallikrein gene expression in pancreatic and colon cancers. Anticancer Res 2004; 24:43–51.
[247] Iacobuzio-Donahue CA, Ashfaq R, Maitra A, et al. Highly expressed genes in pancreatic ductal adenocarcinomas: A comprehensive characterization and comparison of the transcription profiles obtained from three major technologies. Cancer Res 2003; 63:8614–8622.
[248] Goble JM, James KM, Rodriguez M, James D, Scarisbrick-Paulsen IA, Uhm JH. Kallikrein 6, a degradative serine protease, is expressed by gliobalstoma cells *in vivo*. In: Proceedings of the American Association of Cancer Research, 2004. Orlando, Florida, abstract 1811.
[249] Sappino AP, Busso N, Belin D, Vassalli JD. Increase of urokinase-type plasminogen activator gene expression in human lung and breast carcinomas. Cancer Res 1987; 47:4043–4046.

[250] Sutkowski DM, Goode RL, Baniel J, et al. Growth regulation of prostatic stromal cells by prostate-specific antigen. J Natl Cancer Inst 1999; 91:1663–1669.

[251] Abate-Shen C, Shen MM. Molecular genetics of prostate cancer. Gene Dev 2000; 14:2410–2434.

[252] Matsui H, Takahashi T. Mouse testicular leydig cells express klk21, a tissue kallikrein that cleaves fibronectin and igf-binding protein-3. Endocrinology 2001; 142:4918–4929.

[253] Plendl J, Snyman C, Naidoo S, Sawant S, Mahabeer R, Bhoola KD. Expression of tissue kallikrein and kinin receptors in angiogenic microvascular endothelial cells. Biol Chem 2000; 381:1103–1115.

[254] Fortier AH, Nelson BJ, Grella DK, Holaday JW. Antiangiogenic activity of prostate-specific antigen. J Natl Cancer Inst 1999; 91:1635–1640.

[255] Denmeade SR, Nagy A, Gao J, Lilja H, Schally AV, Isaacs JT. Enzymatic activation of a doxorubicin-peptide prodrug by prostate-specific antigen. Cancer Res 1998; 58:2537–2540.

[256] Lee SJ, Kim HS, Yu R, et al. Novel prostate-specific promoter derived from PSA and PSMA enhancers. Mol Ther 2002; 6:415–421.

[257] Chen LM, Murray SR, Chai KX, Chao L, Chao J. Molecular cloning and characterization of a novel kallikrein transcript in colon and its distribution in human tissues. Braz J Med Biol Res 1994; 27:1829–1838.

[258] Rae F, Bulmer B, Nicol D, Clements J. The human tissue kallikreins (KLKs 1–3) and a novel KLK1 mRNA transcript are expressed in a renal cell carcinoma cDNA library. Immunopharmacology 1999; 45:83–88.

[259] Liu XF, Essand M, Vasmatzis G, Lee B, Pastan I. Identification of three new alternate human kallikrein 2 transcripts: Evidence of long transcript and alternative splicing. Biochem Biophys Res Commun 1999; 264:833–839.

[260] Riegman PH, Klaassen P, van der Korput JA, Romijn JC, Trapman J. Molecular cloning and characterization of novel prostate antigen cDNA's. Biochem Biophys Res Commun 1988; 155:181–188.

[261] Korkmaz KS, Korkmaz CG, Pretlow TG, Saatcioglu F. Distinctly different gene structure of KLK4/KLK-L1/Prostase/ARM1 compared with other members of the kallikrein family: Intracellular localization, alternative cDNA forms, and regulation by multiple hormones. DNA Cell Biol 2001; 20:435–445.

[262] Underwood LJ, Tanimoto H, Wang Y, Shigemasa K, Parmley TH, O'Brien TJ. Cloning of tumor-associated differentially expressed gene-14, a novel serine protease over-expressed by ovarian carcinoma. Cancer Res 1999; 59:4435–4439.

[263] Mitsui S, Okui A, Kominami K, Uemura H, Yamaguchi N. cDNA cloning and tissue-specific splicing variants of mouse hippostasin/TLSP (PRSS20). Biochim Biophys Acta 2000; 1494:206–210.

[264] Dong Y, Kaushal A, Bui L, et al. Human kallikrein 4 (KLK4) is highly expressed in serous ovarian carcinomas. Clin Cancer Res 2001; 7:2363–2371.

[265] Tanimoto H, Underwood LJ, Shigemasa K, et al. The stratum corneum chymotryptic enzyme that mediates shedding and desquamation of skin cells is highly overexpressed in ovarian tumor cells. Cancer 1999; 86:2074–2082.

[266] Scorilas A, Borgono CA, Harbeck N, et al. Human kallikrein 13 protein in ovarian cancer cytosols: A new favorable prognostic marker. J Clin Oncol 2004; 22:678–685.

[267] Yu H, Giai M, Diamandis EP, et al. Prostate-specific antigen is a new favorable prognostic indicator for women with breast cancer. Cancer Res 1995; 55:2104–2110.

[268] Yu H, Levesque MA, Clark GM, Diamandis EP. Prognostic value of prostate-specific antigen for women with breast cancer: A large United States cohort study. Clin Cancer Res 1998; 4:1489–1497.

[269] Foekens JA, Diamandis EP, Yu H, Look MP, et al. Expression of prostate-specific antigen (PSA) correlates with poor response to tamoxifen therapy in recurrent breast cancer. Br J Cancer 1999; 79:888–894.

[270] Borgono CA, Grass L, Soosaipillai A, et al. Human kallikrein 14: A new potential biomarker for ovarian and breast cancer. Cancer Res 2003; 63:9032–9041.

[271] Nakamura T, Stephan C, Scorilas A, Yousef GM, Jung K, Diamandis EP. Quantitative analysis of hippostasin/KLK11 gene expression in cancerous and noncancerous prostatic tissues. Urology 2003; 61:1042–1046.

[272] Yousef GM, Stephan C, Scorilas A, et al. Differential expression of the human kallikrein gene 14 (KLK14) in normal and cancerous prostatic tissues. Prostate 2003; 56:287–292.

[273] Stephan C, Yousef GM, Scorilas A, et al. Quantitative analysis of kallikrein 15 gene expression in prostate tissue. J Urol 2003; 169:361–364.

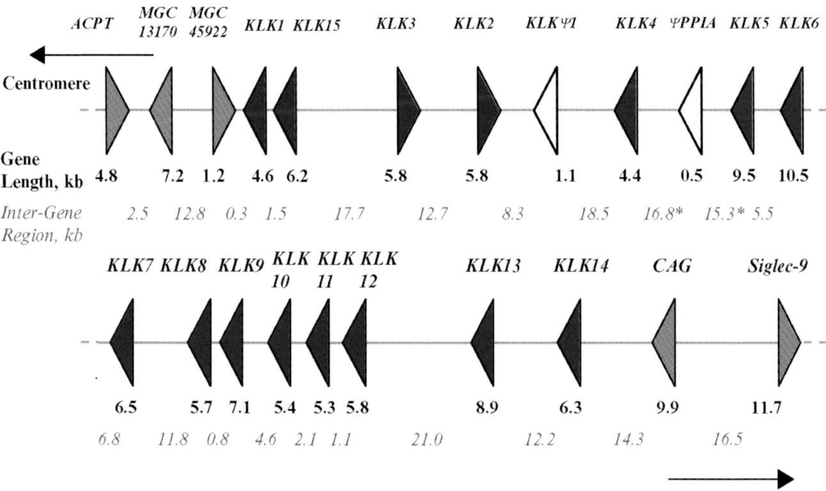

YOUSEF *ET AL.*, FIG. 1. Schematic representation of the human kallikrein gene locus on chromosome 19q13.4. Genes are represented by arrowheads indicating the direction of transcription. Kallikrein genes are shown in blue, and nonkallikrein genes and pseudogenes are presented by grey and white arrows, respectively. Official gene symbol is shown above each gene, and the approximate gene length is shown below each gene in kilobytes. The approximate intergenic regions are shown in red in Kb. CAG, cancer associated gene (GenBank accession AY279382); ACPT, testicular acid phosphatase (GenBank accession AF321918). The position of the PPIA pseudogene is provisional and the length of the intergenic regions (shown with asterisks) may change in the future. MGC, mammalian gene collection; these two genes have not as yet been characterized.

	KLK1	KLK2	KLK3	KLK4	KLK5	KLK6	KLK7	KLK8	KLK9	KLK10	KLK11	KLK12	KLK13	KLK14	KLK15
ADIPOSE															
ADRENAL GLAND															
BONE MARROW															
CEREBELLUM															
ADULT BRAIN															
CERVIX															
COLON															
ESOPHAGUS															
FALLOPIAN TUBE															
FETAL BRAIN															
FETAL LIVER															
HEART															
HIPPOCAMPUS															
KIDNEY															
LIVER															
LUNG															
MAMMARY GLAND															
OVARY															
PANCREAS															
PLACENTA															
PROSTATE															
PITUITARY															
SALIVARY GLAND															
SKELETAL MUSCLE															
SKIN															
SPINAL CORD															
SPLEEN															
SMALL INTESTINE															
STOMACH															
TESTIS															
THYMUS															
THYROID															
TONSIL															
TRACHEA															
UTERUS															
VAGINA															

SEMIQUANTITATIVE EXPRESSION SCORING SYSTEM					
VERY HIGH	HIGH	MODERATE	LOW	VERY LOW	NOTHING

YOUSEF *ET AL.*, FIG. 2. Expression map of human tissue kallikreins in a variety of tissues, as determined by reverse transcriptase polymerase chain reaction. The relative semiquantitative expression levels for each gene are indicated.

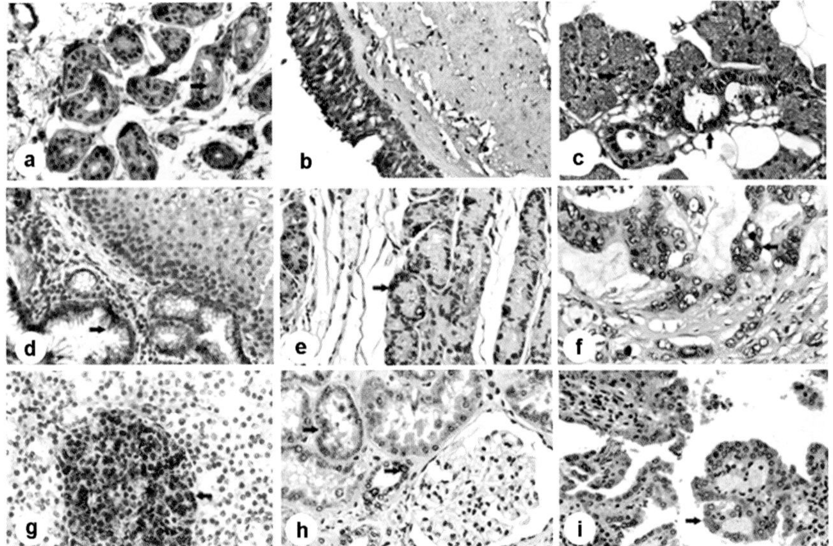

YOUSEF *ET AL.*, FIG. 3. Immunohistochemical expression of (a) hK7 by the epithelium of eccrine glands of the skin (monoclonal antibody, clone 73.2), (b) hK13 by the epithelium of the bronchus (monoclonal antibody, clone IIC1), (c) hK5 by the ductal epithelium of the parotid gland (polyclonal antibody), (d) hK7 by the esophageal glands (monoclonal antibody, clone 73.2), (e) hK13 by the gastric mucosa (monoclonal antibody, clone 2–17), (f) hK6 by the large intestine mucosa (polyclonal antibody), (g) hK10 in an islet of Langerhans in the pancreas (monoclonal antibody, clone 5D3), (h) hK11 by the epithelium of the urinary tubuli (monoclonal antibody), (i) hK11 by a papillary renal cell carcinoma (monoclonal antibody).

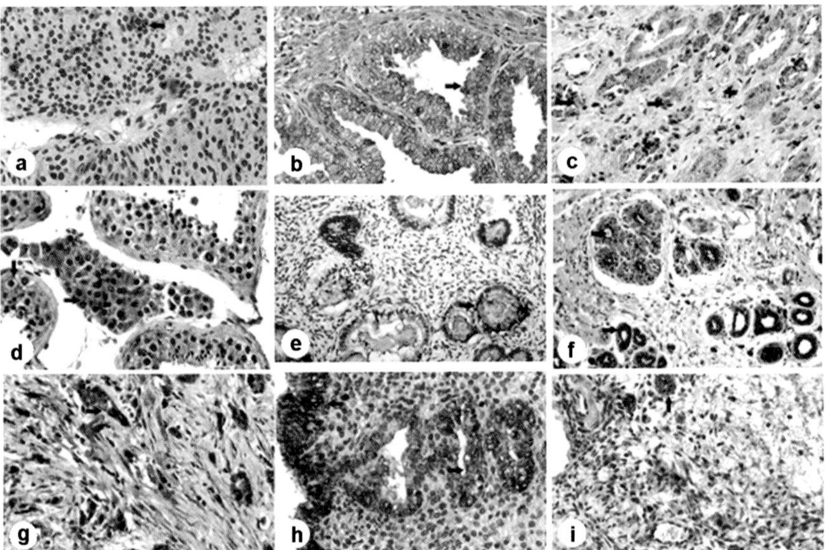

YOUSEF *ET AL.*, FIG. 4. (*continues*)

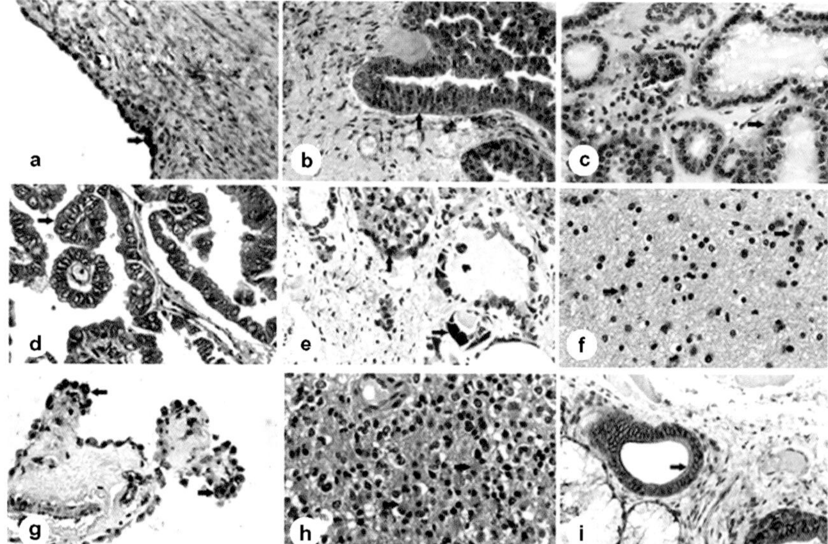

YOUSEF *ET AL.*, FIG. 5. Immunohistochemical expression of (a) hK14 by the ovarian surface epithelium (polyclonal antibody), (b) hK14 by a cystadenocarcinoma of the ovary (polyclonal antibody), (c) hK10 by hyperplastic follicles of the thyroid gland (monoclonal antibody, clone 5D3), (d) hK6 by a papillary thyroid carcinoma (polyclonal antibody), (e) hK13 by endocrine cells in the pituitary gland (monoclonal antibody, clone 2–17), (f) hK10 by glial cells in the brain (monoclonal antibody, clone 5D3), (g) hK13 by the choroid plexus epithelium (monoclonal antibody, clone 2–17), (h) hK5 by a glioma (monoclonal antibody, clone 6.10), (i) hK7 by the ducts of the submucosal glands of the tonsils (monoclonal antibody, clone 85.2).

YOUSEF *ET AL.*, FIG. 4. (*Continued*) Immunohistochemical expression of: (a) hK10 by a low-grade urothelial carcinoma (monoclonal antibody, clone 5D3), (b) hK11 by the secretory epithelium of the prostate gland (polyclonal antibody), (c) hK11 by a Gleason score 6 prostate carcinoma (polyclonal antibody), (d) hK14 by the spermatic epithelium and the stromal Leydig cells in the testis (polyclonal antibody), (e) hK10 by epithelial elements in a testicular immature teratoma (monoclonal antibody, clone 5D3), (f) hK14 by lobuloalveolar structures of the breast (polyclonal antibody), (g) hK14 by a ductal breast carcinoma, grade II (polyclonal antibody), (h) hK13 by the glandular epithelium of the endometrium (polyclonal antibody), (i) hK14 by luteinized stromal cells of the ovary (polyclonal antibody).

CHOLESTEROL AND LIPIDS IN DEPRESSION: STRESS, HYPOTHALAMO-PITUITARY-ADRENOCORTICAL AXIS, AND INFLAMMATION/IMMUNITY

Tiao-Lai Huang* and Jung-Fu Chen[†]

***Department of Psychiatry, Chang Gung Memorial Hospital, Kaohsiung 833, Taiwan, Republic of China**
[†]Department of Internal Medicine, Chang Gung Memorial Hospital, Kaohsiung 833, Taiwan, Republic of China

1. Introduction

There are many different ways to classify lipids. Here, we focus on lipid profiles in depression, including total cholesterol (TC), triglyceride (TG), high-density lipoprotein (HDL) cholesterol, low-density lipoprotein (LDL) cholesterol, very-low density lipoprotein (VLDL) cholesterol, and the ratios of TC/HDL and LDL/HDL (atherogenic index).

0065-2423/05 $35.00
DOI: 10.1016/S0065-2423(04)39003-7

Several studies have discussed the relationship between serum cholesterol and suicide, violence, anxiety disorders, depressive disorders, and schizophrenia [1–3]. Some of these papers suggested that low or lowering cholesterol levels could cause or worsen depressive symptoms and increase the risks of suicide and violence death. There are many reports that discussed the relationships between the lipid profiles, depression, and suicide from the viewpoints of decreased serotonergic transmission on suicide behavior [4, 5], lower serum cholesterol and serotonin levels [6, 7], serum cholesterol levels and polymorphism in the promoter region of the serotonin transporter gene for depression and suicide [8–10], low serum cholesterol and suicide risk [11, 12], and serotonergic receptor function [13, 14]. These studies supported the hypothesis that reduced cholesterol levels resulted in reduced central serotonin transmission.

Major depression is associated with altered changes in hypothalamo-pituitary-adrenocortical (HPA) axis activity and immune and endocrine systems. Because these systems can affect each other reciprocally, interactions between the lipid profiles and the neuroendoimmune systems in depression should be addressed [15–20]. Penttinen has suggested that low cholesterol concentration and suicidal behavior are connected with interleukin 2, which caused decreased serum cholesterol level and increased serum TG level [15]. In major depression, activation of the inflammatory response system and increased concentrations of proinflammatory cytokines, prostaglandin E_2, and negative immuno-regulatory cytokines in peripheral blood have been reported [16]; Myint and Kim suggested one neurodegeneration hypothesis of depression that involved cytokine–serotonin interaction through the enzyme indoleamine-2,3-dioxygenase [16]. Some authors also discussed brain-immune interactions [17], glucocorticoid receptors in major depression [18], physiopathology of depression in HPA axis [19], and neurobiological consequences of adverse early-life experiences [20].

In clinical practice, depressive symptoms were common in patients with physical illness, including cardiovascular disease, diabetes mellitus, end-stage renal disease, and women in pregnancy, following delivery or menopause. However, data that specifically addressed serum lipid profiles in patients with depressive disorders and physical illnesses were still scarce.

In this review we discuss the relationships between serum lipid profile levels, major depression, and suicide attempts, as well as the interactions between lipid profiles, stress, HPA axis, and inflammation/immunity in depressive disorders. The conclusion emphasizes the importance of integrated data between clinical phenotypes and molecular mechanisms in depressive disorders.

2. The Relationships Between Lipid Profile Levels, Depression, and Suicide Attempts

2.1. Depressive Symptoms or Depressive Disorders

Most of the data that examined depressive symptoms and lipid profile levels have been from the community studies performed in different countries by the self-related depression scales. Most of these studies were performed by psychiatrists according to specifically clinical criteria in psychiatric inpatients or outpatients or in patients admitted for general health screening.

2.1.1. *Depressive Symptoms and Lipid Profile Levels in the Community*

Lower cholesterol levels have been found in some patients with depressive symptoms [21, 22], but not in others [23–26]. Troisi *et al.* also found no significant association in the younger age group (<50 years of age) [27]. In contrast, in a subgroup of older women, serum cholesterol was negatively and significantly correlated to negative mood with the self-rated scales. Restricting analysis to the subjects in the highest quartile of the age distribution (>60 years of age) yielded a much stronger correlation between cholesterol and mood. This result was not, however, compatible with those reported previously [26]. McCallum *et al.* examined the relationship between low serum cholesterol and depressive symptoms in the elderly and found that low serum cholesterol was not associated with depressive symptoms in older men or women from the cross-sectional data in a community [26]. However, they did point out the significance of financial status, low self esteem, adequacy of practical help and emotional support, and recent widowhood in depression. Hu *et al.* found that hypocholesterolemia was not an independent risk factor for increased overall mortality in older men and women [28]. The association between low total cholesterol and high mortality is mainly confounded by common cardiovascular risk factors, rather than underlying inflammation or lack of adequate nutrition [28].

In addition, Nakao *et al.* have examined the effects of mood states on "persistent" versus "temporary" hypercholesterolemia in students entering a university [29]. They found that depressive mood appeared to relate to hypercholesterolemia when the university students were screened with tension–anxiety, depression, anger–hostility, vigor, fatigue, and confusion scales [29]. In children, after covariance adjustment for age, race, and sex, Glueck *et al.* pointed out that children having adjustment disorders with depression had much lower covariance-adjusted TC values than control schoolchildren, whereas those with disruptive behavior with oppositional defiant disorder had much higher adjusted TC value [30].

In addition to total cholesterol, Lindberg *et al.* tried to explore the relationship between other lipids and depressive symptoms [31]. They found that total cholesterol and LDL cholesterol values were lower in men, but serum triglyceride concentration was not. In women, however, the serum triglyceride value, but not the total cholesterol or LDL cholesterol, was lower in those who reported low mood, depression, or anxiety during the last 6 months [31]. Chen *et al.* examined the correlation between serum lipid levels and psychological distress [32]. They found that women with an HDL-C level lower than 35 mg/dL scored significantly higher on depression, interpersonal sensitivity, phobia, anxiety, somatization, and aggressive hostility, whereas subjects with a total cholesterol concentration lower than 160 mg/dL scored significantly higher on anxiety, aggressive hostility, phobia, and psychosis [32]. Huang *et al.* also investigated the correlation between serum lipid, lipoprotein concentrations, and anxious state, depressive state, or major depressive disorder [33]. A total of 207 patients admitted for general health screening were recruited in this study during a 1-year period. When the patients were not associated with systemic diseases ($n = 162$), the researchers found that HDL and the ratio of TC/HDL displayed significant differences among anxious state, depressive state, and normal groups in men after age adjustment. However, the ratios of TC/HDL and LDL/HDL showed significant differences between patients with major depressive disorder and normal controls in women [33].

2.1.2. *Depressive Disorders and Lipid Profile Levels in Psychiatric Inpatients*

The relationships between cholesterol, lipids, and depressive symptoms have been investigated but require further study. Lower cholesterol levels have been found in patients with major depression [34–38], but not in others [39].

Glueck *et al.* assessed hypocholesterolemia in 203 patients hospitalized with affective disorders (depression, bipolar disorder, and schizoaffective disorder), 1595 self-referred subjects in an urban supermarket screening, and 11,864 subjects in the National Health and Nutrition Examination Survey II (a national probability sample) [34]. Low plasma cholesterol concentration (<160 mg/dL) was much more common in patients with affective disorders than in those found in urban supermarket screening subjects or in the National Health and Nutrition Examination Survey II subjects. When paired with supermarket screening subjects by age and sex, patients with affective disorders had much lower TC, LDL, HDL, and higher TG concentrations. However, there was no evidence that low plasma cholesterol could cause or worsen affective disorders [34].

Apter *et al.* investigated the relationship between serum cholesterol levels and suicidal behaviors in adolescent psychiatric inpatients [40]. They found that serum cholesterol levels were significantly higher in adolescent patients who were currently suicidal than in nonsuicidal adolescents. Within the suicidal group, but not in the total inpatient group, serum cholesterol correlated negatively with the degree of suicidal behavior. No correlation between serum cholesterol levels and depression, violence, and impulsivity was detected. No significant differences were found in serum cholesterol levels between diagnoses or between suicidal and nonsuicidal patients within each diagnostic group [40].

In Taiwan, Huang *et al.* investigated the correlation between serum lipid, lipoprotein concentrations and major depressive disorder in patients admitted for general health screening [33]. They found that the ratios of TC/HDL and LDL/HDL showed significant differences between patients with major depressive disorder and normal controls in women. Huang and Chen also pointed out that no significant differences were found in lipid concentrations of TC, TG, HDL, VLDL, LDL, TC/HDL, and LDL/HDL between patients with dysthymia and normal controls [41].

2.1.3. *Depression in Patients with Physical Illness*

Depressive symptoms are common in patients with physical illness, including cardiovascular disease, diabetes mellitus, end-stage renal disease, and women in pregnancy, delivery, or menopause. The depressive symptoms in patients with physical illness include apathy, anorexia, sleep disorder, fatigue, and cognitive deficits [42].

2.1.3.1. *Depression, Cardiovascular Disease, and Cognitive Impairment.* The effect of depression on patients with heart disease has been mentioned and is strongly associated with increased morbidity and mortality [43–45]. Coelho *et al.* showed that there were significant differences according to gender regarding almost every psychometric dimension assessed [46]. After adjusting for the presence of different biomedical risk factors, significant decreasing mean behavior pattern scores were found with increasing age. Mean depression scores were significantly higher in women and in individuals with lower educational level [46].

In a previous study, van Doornen and van Blokland pointed out that type A behavior and a vital exhaustion/depression cluster appeared to be the most crucial elements of the psychological "coronary risk profile" [47]. The authors found that type A behavior was related to a stronger response of adrenaline and diastolic blood pressure to the stressor. Vital exhaustion was also positively correlated with the adrenaline reaction and, moreover, with cholesterol base level, stress-induced cholesterol change, and noradrenaline and cholesterol stress levels [47]. Recently, Rutledge *et al.* investigated associations between atherosclerosis risk factors (smoking behavior, serum cholesterol, hypertension, body mass index, and functional capacity) and psychological characteristics with suspected linkages to coronary disease (depression, hostility, and anger expression) in an exclusively female cohort [48]. High depression scores were associated with a nearly threefold risk of smoking after covariate adjustment, and women reporting higher depression symptoms were approximately four times more likely to describe themselves in the lowest category of functional capacity. High anger-out scores were associated with a fourfold or greater risk of low HDL concentration ($<$50 mg/dL) and high LDL concentration ($>$160 mg/dL). In conclusion, these results demonstrate consistent and clinically relevant relationships between psychosocial factors and atherosclerosis risk factors among women and may aid our understanding of the increased mortality risk among women reporting high levels of psychological distress [48].

Although cholesterol is a major cardiovascular risk factor, its association with stroke remains controversial. In Taiwan, Su investigated the relation between job strain status and cardiovascular risk factors (high serum total cholesterol, low serum HDL cholesterol, and high plasma fibrinogen) [49]. The author found that plasma fibrinogen is a possible intermediate factor linking occupational stress to elevated cardiovascular risk [49]. Engstrom *et al.* also explored whether the cholesterol-related incidence of stroke and myocardial infarction is modified by plasma markers of inflammation, using a large, population-based cohort with a long follow-up [50]. The researchers found that hypercholesterolemia is associated with high plasma levels of inflammation-sensitive plasma proteins (fibrinogen, α_1-antitrypsin, haptoglobin, ceruloplasmin, and orosomucoid). These proteins increase the cholesterol-related incidence of cardiovascular diseases [50]. In addition, C-reactive protein (CRP) is a prototypic marker of inflammation. Numerous prospective studies in healthy volunteers have confirmed that high-sensitivity CRP predicts cardiovascular events, and high-sensitivity CRP seems additive to an elevated total cholesterol level and a TC/HDL ratio in men and women in predicting risk [51].

Further, certain dietary risk factors for physical ill health are also risk factors for depression and cognitive impairment. For example, cognitive impairment is associated with atherosclerosis, type 2 diabetes, and hypertension, and findings from a broad range of studies show significant relationships between cognitive function and intakes of various nutrients, including long-chain polyunsaturated fatty acids, antioxidant vitamins, and folate and vitamin B12. Further support is provided by data on nutrient status and cognitive function [52–54].

Alzheimer disease (AD) is characterized by the presence of senile plaques, neurofibrillary tangles, and neuronal cell loss associated with membrane cholesterol release. 24S-hydroxycholesterol (24S-OH-Chol) is an enzymatically oxidized product of cholesterol mainly synthesized in the brain. Lutjohann *et al.* found that the concentration of 24S-OH-Chol in AD and non-AD demented patients was significantly higher than in healthy controls and in depressed patients [55]. However, there was not a significant difference in the concentrations of 24S-OH-Chol between depressed patients and healthy controls, or between AD and non-AD demented patients. The researchers speculated that 24S-OH-Chol plasma levels may potentially be used as an early biochemical marker for an altered cholesterol homeostasis in the central nervous system [55]. Even though it is known that apolipoprotein E is deeply involved in major age-related disorders such as atherosclerosis or AD [56], the control of cell-specific apolipoprotein E expression is still poorly understood. The response of KYN-2 cells to both cytokines and cholesterol differs from that found in astrocytoma cells [57]. Brahimi *et al.* suggested that blood variation of apolipoprotein E concentrations in AD does not reflect the lack of regulation taking place in the brain [57]. In addition, glial fibrillary acidic protein autoantibody level may be a late marker for neurodegeneration [58]. To date, serum α_1-antichymotrypsin concentration is the most convincing marker for CNS inflammation. Increased serum homocysteine concentrations have also been consistently reported in AD [58].

2.1.3.2. *Depression and Diabetes Mellitus.* Patients with chronic medical illness have a high prevalence of major depressive disorder [59]. Depression may be three times more prevalent in the diabetic population when compared with its occurrence in nondiabetic individuals [60]. In addition, microalbuminuria, hypertension, and hyperinsulinemia are another three independent risk factors for cardiac disease in non-insulin-dependent diabetes mellitus (NIDDM) [61]. Nosadini *et al.* showed that peripheral insulin resistance, hypertension, microalbuminuria, and lipid abnormalities are associated with NIDDM [61]. Further, Helkala *et al.* determined that cognitive and memory dysfunction are associated with NIDDM and explored the disease's relationship with depression, metabolic control, and serum lipids. The results showed that the NIDDM patients had impaired control of their learning processes [62]. Obviously, future research examining the causal relationship of depression to the onset on diabetes and the effect of depression on the natural course of diabetes is needed [60].

2.1.3.3. *Depression and End-Stage Renal Disease.* There are few data on the epidemiology, consequences, and treatment of depression in patients with renal disease, and the role of depression in these patients needs to be discussed in more detail [42, 63]. From a biochemical standpoint, hyperlipidemia is a common manifestation of the nephrotic syndrome, and serum lipid concentrations have been observed by others to be negatively correlated with serum protein concentration. Hyperlipidemia has been postulated to result from a coordinate increase in the synthesis of both albumin and lipoproteins, as well as from their decreased catabolism [64–67].

Kaysen *et al.* have shown that serum cholesterol concentration was dependent only on the renal clearance of albumin, and changes in serum cholesterol concentration were dependent only on changes in the renal clearance of albumin. Serum cholesterol concentration was completely independent of the rate of albumin synthesis [64]. Nihei *et al.* suggested that resveratrol, a polyphenolic compound, is a potent anti-glomerulonephritic food factor capable of suppressing proteinuria, hypoalbuminemia, and hyperlipidemia concurrently [65]. In addition, endothelin 1 (ET-1) is able to determine functional and structural renal alterations in diabetic patients. In a select group of type 2 normotensive diabetic patients with microalbuminuria, Bruno *et al.* showed that circulating ET-1 values were increased and correlated with albumin excretion rate [66]. These findings confirm that endothelial dysfunction, as expressed by ET-1 levels, occurs early in these patients, and it supports the hypothesis of a potential role for this peptide in development of microalbuminuria in diabetic nephropathy [66]. Using clinical data, Huang and Lee also showed that hemodialysis patients with major depression had lower serum albumin and higher ferritin levels than the hemodialysis patients without major depression [67].

2.1.3.4. *Depression and Metabolic Syndrome.* Abnormal serum albumin levels and lipid profiles have both been observed in patients with major depression, as well as cardiovascular disease, diabetes mellitus, and end-stage renal disease. Depressive symptoms are very common in patients with these chronic illnesses. Recent clinical data have shown that cardiovascular disease, diabetes mellitus, end-stage renal disease, and obesity are all related to metabolic syndromes [68–74], and especially insulin resistance [75, 76]. However, the data examining major depression without physical illness and insulin resistance are still scarce. In the future, the biological relationship between depression and physical illness needs to be more fully explored.

2.1.3.5. *Depression and Women in Pregnancy, Postpartum, or Menopause.* Lipids and lipoproteins are known to increase substantially during pregnancy and to decrease rapidly after delivery. The factors responsible for the changes have not been identified; however, they could be related to changes in one or more of the endocrine hormones [77–80]. During pregnancy, the total serum cholesterol concentration rises up to 45%, followed by a rapid fall after delivery. Schwertner *et al.* found that the increases in cholesterol during pregnancy and labor could be, in part, a result of the metabolic- and stress-related increases in cortisol [77]. The studies also indicate that both pregnancy and labor and delivery might be useful "natural" models for studying hormonal mechanisms involved in lipid and lipoprotein metabolism.

Mild depressive symptoms ("postpartum blues") are a common complication of the puerperium and affect 30%–85% of women in the early postpartum period. On the basis of these observations, it has been suggested that the sudden fall in cholesterol levels after delivery could serve as a "natural model" to test the suggested association between cholesterol and mood [77]. Troisi *et al.* expanded the database concerning the association between cholesterol levels and mood in the postpartum period [78]. They found significant relationships between serum cholesterol levels and mood symptoms in the postpartum period that were not present during late pregnancy. Lower postpartum levels of total cholesterol were associated with symptoms of anxiety, anger–hostility, and depression, and lower postpartum levels of HDL cholesterol were associated with symptoms of anxiety [78]. The study confirmed that the physiological fall in blood lipids in the postpartum period provided a useful model to test the relationship between serum cholesterol levels and mood [78].

Nasta *et al.* also tried to investigate the relationship between cholesterol and mood states in the initial puerperal period. Their results showed that reduced plasma cholesterol concentration was associated with major feelings of fatigue and depressed mood [79]. In addition, West *et al.* compared the effects of transdermal versus oral estrogens on the vascular resistance index, mean arterial pressure, serum lipid concentrations, norepinephrine, and left ventricular structure in 10 postmenopausal women. The results showed that oral and transdermal estrogen significantly decreased the vascular resistance index, mean arterial pressure, norepinephrine, and total and low-density lipoprotein cholesterol to a similar extent [80].

2.2. Suicide Attempt, Aggression, or Violence

Because depression is a major factor in most suicides, investigating the association between low serum cholesterol levels and suicide may be important. Some studies have shown an association between low cholesterol and increased risk of death resulting from injuries or suicide [81–85]. Other studies have shown no such association [40, 86–89].

Fawcett *et al.* proposed four hypothetical pathways leading to suicide in clinical depression: an acute pathway involving severe anxiety/agitation associated with high brain corticotrophin-releasing factor levels, trait baseline and reactivity hopelessness, severe anhedonia, and trait impulsiveness associated with low brain serotonin turnover, with low total cholesterol as a possible peripheral correlate [90]. Future possibilities and the applications of these findings are discussed.

2.2.1. *Depression, Cholesterol–Serotonin Theory, and Suicide*

Both suicidal behavior and impulsive aggression have been associated with low levels of brain serotonergic activity [91, 92]. Engelberg suggested that a reduction in serum cholesterol may decrease brain-cell-membrane cholesterol, lower lipid microviscosity, and decrease exposure of protein serotonin receptors on the membrane surface, thus resulting in a poorer uptake of serotonin from the blood and less serotonin entry into brain cells [4]. Other reports have discussed the relationships between cholesterol, serotonin, and depression [6, 93–96].

In clinical studies, Steegmans *et al.* found that middle-aged men with chronically low serum cholesterol levels ($\leq$4.5 mmol/L or 172 mg/dL) have a higher risk of having depressive symptoms, according to scores on the Beck Depression Inventory, when compared with a reference group of men with cholesterol levels between 6 and 7 mmol/L [97]. These data may be important in the ongoing debate on the putative association between low cholesterol levels and violent death in the future.

Epidemiological and clinical studies have described an association between lower serum cholesterol concentrations and increased suicide risk that is not entirely attributable to depression-related malnutrition and weight loss. Recent epidemiological studies with greater samples and longer follow-up periods, however, have even shown a positive correlation between cholesterol concentrations and suicide risk after controlling for potential confounding variables [97, 98]. A meta-analysis of earlier intervention trials indicated that cholesterol lowering could cause or worsen depressive symptoms and increase the risk of suicide. However, some large trials of statins (simvastatin, lovastatin, and pravastatin) did not show an increase of suicide mortality [98, 99]. Recently, it was hypothesized that a decreased consumption of polyunsaturated fatty acids, especially omega-3 fatty acids, may be a risk factor for depression and suicide [98].

2.2.2. *Depression, Poor Social Support, and Suicide*

Horsten *et al.* examined the inverse relationship between cholesterol levels and death from violent causes, including suicide, in a group of 300 middle-aged healthy Swedish women [100]. The authors also investigated the association between cholesterol and other psychosocial factors (social support, vital exhaustion, and stressful life events), which are known to be related to depression. The results showed that women with low serum cholesterol, defined as the lowest tenth of the cholesterol distribution ($\leq$4.7 mmol/L or 180 mg/dL), reported significantly more depressive symptoms. In addition, low cholesterol was found to be strongly associated with lack of social support [100]. The findings may constitute a possible mechanism for the association found between low cholesterol and increased mortality, particularly suicide.

3. The Interactions Between Stress, Lipid Profiles, Cortisone/HPA Axis, and Inflammation/Immunity

To our knowledge, there are many interactions between stress, lipid profiles, cortisone/HPA axis, and inflammation/immunity. In addition to lipid profiles and depression, we also must investigate the relationships between lipid profiles and stress, cortisone/HPA axis, and inflammation/immunity, as described below.

3.1. Psychological Factors: Lipid Profiles and Stress

Recent research on the causes of disease and aging has increasingly supported the importance of stress [101]. One theory of the relationship between stress and disease is based on the concept of homeostasis [101]. Many researchers held that full, normal function of the self-regulating or homeostatic power of the body maintains the balanced, integrated condition we recognize as health. Failures in this capacity, such as those produced by frequent stressful experiences, can result in disease or death [101]. Walton and Pugh, in a review article, discussed both the fundamental elements of these theories and the current neuroendocrine research supporting their validity and immediate relevance [101].

In an earlier study, Francis investigated the correlations between serum uric acid, cortisol, HDL cholesterol, LDL cholesterol, and psychometric indices of stress, including anxiety, hostility, and depression, in 20 students over a 2.5-month academic quarter. The students' serum cholesterol and the LDL cholesterol levels were significantly elevated above control levels. The ratio of HDL cholesterol/total cholesterol was significantly lower [102]. Agarwal *et al.* also estimated the serum cholesterol, triglycerides, and total lipids in 12 students exposed to a varying degree of examination stress. Serum cholesterol and triglycerides exhibited a rise proportional to the degree of examination stress, whereas other lipids exhibited an initial rise followed by a fall. The rise in serum cholesterol and triglycerides seems to be caused by stress-induced changes in hormonal levels and peripheral lipolysis, respectively [103].

In humans, stress can increase the risk of cardiovascular disease by altering lipoprotein metabolism, but what about animal studies? In 1988,

Prabhakaran *et al.* indicated that noise could be a potent stressor and cause disturbances in the biochemical parameters of the body. It is presumed that most of the effects are indirect, being manifested through the activation of autonomic nervous system that liberates catecholamines and HPA axis responsible for the liberation of corticosteroids [104]. In 1989, Hershock and Vogel found that in male Sprague-Dawley rats, acute immobilization stress could affect serum triglyceride and nonesterified fatty acid values and that these effects were diet and time dependent; however, the rats' total cholesterol levels were unaffected by stress [105]. In addition, in 1996 Brennan *et al.* pointed out that stressed animals had higher levels of the cholesterol parameters than did home cage controls [106].

Abd el Mohsen *et al.* studied the changes in sex hormones and lipid profiles in adult female albino rats subjected to treatment with nicotine (N), immobilization stress (S), or a combination of the two (N + S). The researchers found that the changes in the lipid pattern could be attributed to the alterations occurring in corticosterol and female sex hormones, caused by N, S, or a combination of the two [107]. Further, Ricart-Jane *et al.* studied the metabolic response to acute and chronic stress following a model of immobilization in rats and evaluated the resulting circulating lipoprotein levels [108]. Both acute and chronic stress decreased the plasmatic triacylglycerol concentration, as reflected by the reduction in the number of VLDL particles. This may be caused by an increase in the metabolism of triacylglycerol, as suggested by the slightly higher amounts of circulating LDL [108]. Chronic stress, but not acute stress, significantly increased both the number and the estimated size of circulating HDL, as shown by the plasma cholesterol concentration. In addition, acute stress did not have an additive effect over chronic stress on the lipoprotein parameters studied [108].

3.2. Biological Factors: Lipid Profiles, Cortisone/HPA Axis, and Inflammation/Immunity

3.2.1. *Cholesterol, Lipids, and Cortisone/HPA Axis*

In animal studies, Cassano and D'mello found that stress-induced activation of the HPA axis stimulated the release of both facilitatory and inhibitory components [109]. Stress can alter plasma corticosterone-binding globulin levels, and aminoglutethimide administration can cause accumulation of the corticosterone biosynthetic precursor, adrenal cholesterol. Their results indicated that stress-induced facilitation of the HPA axis is associated with the release of serotonin [109]. Djordjevic *et al.* also investigated the reaction of the HPA system to various stressors (fasting, crowding, cold, and heat) by measuring blood corticosterone concentration as well as the cholesterol content in the adrenals [110].

The HPA axis, the mediator of cortisol, also plays a central role in the homeostatic processes in human. Subjects with abdominal obesity show several signs of a perturbed regulation of the HPA axis. This is known to occur after chronic, submissive stress. In contrast, perceived environmental stress depends on personality characteristics [111]. Rosmond *et al.* found that men with cluster A personality disorders showed centralized body fat distribution independent of dexamethasone suppression. In contrast, men with impulsive (cluster B) and anxious (cluster C) personality disorders often have abdominal obesity in combination with a blunted dexamethasone suppression test, indicating a HPA axis disturbance [111].

Rosmond and Bjorntorp also addressed the potential effect of HPA axis activity on established anthropometric, metabolic, and hemodynamic risk factors for cardiovascular disease, type 2 diabetes mellitus, and stroke [112]. Strong and consistent correlations were found not only within but also between different clusters of risk factors, including lipid profiles [110]. The close association to HPA axis abnormality may explain the previously reported powerful risk indication of abdominal obesity for the diseases mentioned under environmental stress challenges [112].

3.2.2. *Cholesterol, Lipids, and Inflammation/Immunity*

In addition to those that discuss lipid profiles and HPA systems, there are many papers that examine the relationships between lipid profiles and inflammation/immunity, especially that which is apparent in studies of atherosclerosis.

Membrane lipids play an important role in cellular responses to exogenous signals. Lipid profile alterations lead to increased concentration of membrane cholesterol. Inflammation is accompanied by changes in the plasma concentrations of acute phase reactants including the CRPs in man and other species [113, 114]. In addition, serum amyloid A, as an apolipoprotein, is present on HDL only during inflammatory states [115, 116]. Acute inflammation results in a profound change in the apolipoprotein composition of HDL. Lindhorst *et al.* suggested that SAA plays a role in cholesterol metabolism during the course of acute inflammation [117].

Endotoxin, via cytokines, could induce marked changes in lipid metabolism [118], and HDL has been found to neutralize lipopolysaccharide (endotoxin) activity *in vitro* and, in animals, *in vivo* [116, 119, 120].

From the studies of atherosclerosis, we know that oxidative stress and inflammatory processes are of major importance because they stimulate oxidized LDL (Ox-LDL)–induced macrophage cholesterol accumulation and foam cell formation—the hallmark of early disease. The monocyte chemo-attractant protein 1, macrophage colony-stimulating factor, and extracellular matrix are also involved in the atherogenesis [121]. In contrast, apolipoprotein E plays a key protective role in atherosclerosis. Its capacity to safeguard against this disease can be attributed to at least three distinct functions. First, plasma apolipoprotein E maintains overall plasma cholesterol homeostasis by facilitating efficient hepatic uptake of lipoprotein remnants. Second, lesion apolipoprotein E in concert with apolipoprotein A-I facilitates cellular cholesterol efflux from macrophage foam cells within the intima of the lesion. Third, lesion apolipoprotein E directly modifies both macrophage- and T lymphocyte–mediated immune responses that contribute to this chronic inflammatory disease. In addition, HDL plays an initial role in reverse cholesterol transport by mediating cholesterol removal from cells [122, 123]. The adenosine triphosphate-binding cassette transporter 1 and endotoxin-induced endothelial cell adhesion molecules are also involved in the biology of atherosclerosis [124, 125].

Peroxisome proliferator–activated receptors are lipid-activated transcription factors that regulate lipid and lipoprotein metabolism, glucose homeostasis, and inflammation [126]. Experiments using animal models of atherosclerosis and clinical studies in humans strongly support an antiatherosclerotic role for the peroxisome proliferator–activated receptors alpha and gamma *in vivo*. Thus, peroxisome proliferator–activated receptors remain attractive therapeutic targets for the development of drugs used in the treatment of chronic inflammatory diseases such as atherosclerosis [126–128]. Klappacher and Glass, in their 2002 review article, pointed out growing evidence of the regulation of both inflammatory responses and cholesterol homeostasis in macrophages by peroxisome proliferator–activated receptor gamma ligands and addressed the ligands' overall effect as antiatherogenic agents [129]. Statins can exert antiinflammatory, antiatherosclerotic effects through an antiinflammatory action, independent of lowering cholesterol [130]. In 2003, Kleemann *et al.* addressed the question of whether the antiinflammatory activities of statins can reduce atherosclerosis beyond the reduction achieved by cholesterol lowering, *per se* [131].

In summary, biological markers identified from studies of atherosclerosis could potentially be applied to the pathophysiological investigation of major depression. For example, lower albumin and cholesterol levels were also noted in depressive patients during the acute phase [33, 38, 41, 67, 132], not just in patients with cardiovascular disease or end-stage renal disease. These data indicate that major depression, cardiovascular disease, and end-stage renal disease might all have common pathologies.

4. Clinical Phenotypes in Depression and Molecular Levels

In clinical studies, major depressive disorder has had four specifiers, including melancholic feature, atypical feature, catatonic feature, and postpartum onset. In the future, we should investigate the distributions of the four specifiers of depression among patients with physical illness and discuss which biological markers could link the depressive disorder and the physical illness.

4.1. Lipid Profile Levels in Patients with Major Depression with Melancholic Feature, Atypical Feature, Catatonic Feature, or Postpartum

Although there are papers that discuss the relationships between cholesterol, lipid profiles, and major depression [34–39], there are few data that discuss the association between lipid profiles and depressive disorders with different phenotypes. Huang and Chen investigated the correlation between serum lipid, lipoprotein concentration, and major depressive disorder in patients evaluated for general health screening [41]. They found that analysis of covariance after age adjustment revealed significant differences in patients with melancholic feature and patients with atypical feature in serum concentrations of TG and VLDL in men and HDL in women [41]. However, there are still no reports that discuss the relationships between lipid profiles and major depression with postpartum onset or catatonic feature. In the future, large sample numbers will be needed to clarify the clinical differences in this field.

4.2. Possible Molecular Pathway, Genetics, and Mechanism

In addition to the studies of clinical biological changes in lipid profile levels in patients with major depression, the mechanism of lipid metabolism should be noted and discussed [133]. In past studies, the main plasma lipid transport forms have been free fatty acids, triglycerides, and cholesteryl esters.

Free fatty acids, derived primarily from adipocyte triglycerides, are transported as a physical complex with plasma albumin. Triglycerides and cholesteryl esters are transported in the core of plasma lipoproteins [134]. Deliconstantinos observed the physical state of the Na^+/K^+-ATPase lipid microenvironment as it changed from a liquid-crystalline form to a gel phase [135]. The studies concerning the albumin-cholesterol complex, its behavior, and its role in the structure of biomembranes provided important new clues as to the role of this fascinating molecule in normal and pathological states [135].

In addition, apolipoprotein E is a plasma protein that serves as a ligand for LDL receptors and, through its interaction with some receptors, participates in the transport of cholesterol and other lipids among various cells of the body [136, 137]. Lecithin-cholesterol acyltransferase is the enzyme responsible for the formation of the bulk of cholesteryl ester in human plasma. The lecithin-cholesterol acyltransferase reaction occurs mainly on HDLs and requires an apolipoprotein as activator [138].

Macrophages and cytokines can also influence lipoprotein metabolism [139]. Grove *et al.* indicated that macrophages can secrete several proteins, including 27-oxygenated metabolites of cholesterol, that upregulate LDL receptors in HepG2 cells [140]. This mechanism was compared with the classical HDL-dependent reverse cholesterol transport. With albumin as extracellular acceptor, the major secreted product was 3-β-hydroxy-5-cholestenoic acid; with HDL as acceptor, 27-hydroxycholesterol was the major secreted product [140, 141].

Sterol carrier protein 2 has also been shown to be involved in the intracellular transport and metabolism of cholesterol. Hirai *et al.* (1994) suggested that sterol carrier protein 2 plays an important role during foam cell formation induced by acetylated LDL and may be an important step in atherosclerosis [142]. Lipoproteins can bind lipopolysaccharide and decrease the lipopolysaccharide-stimulated production of proinflammatory cytokines [142, 143]. In addition, lipoprotein entrapment by the extracellular matrix can lead to the progressive oxidation of LDL because of the action of lipoxygenases, reactive oxygen species, peroxynitrite, or myeloperoxidase [144, 145].

In past research, many biochemical material or genes were found to be involved in cholesterol homeostasis and atherosclerosis, including carriers of the 22/23EK allele [146], LDL receptor allele [147, 148], cholesterol-7α-hydroxylase (CYP7A1) gene [149], ET-1 level [66], acyl-CoA:cholesterol acyl transferase activity [150], signal transducer and activator of transcription (STAT transcription factors signal) [151], sterol regulatory element binding protein 1 [152], ATP-binding cassette transporter A1 [153], and T helper cells [154]. Finally, Vainio and Ikonen, in their 2003 review article, summarized key aspects of intracellular cholesterol processing in macrophages, including mechanisms of lipoprotein cholesterol uptake, fate of the internalized cholesterol, and mechanisms implicated in cholesterol efflux. The importance of inflammatory cues, the cellular compartmentalization of cholesterol homeostatic responses, and the increasing information on the transcriptional control of cholesterol balance are also discussed [155].

5. Conclusion

Depression is a physical and psychological disorder that affects every aspect of human physiology. Stewart and Atlas have discussed the chaos between depression and metabolic syndrome [71]. As we know, some biological markers in cardiovascular disease are also used in the study of renal disease and diabetes and could similarly be applied to the study of major depression. As discussed above, these markers may be especially relevant to those involved in lipid metabolism. In addition the influence of antipsychotics and antidepressants on the human body should be considered [71]. Identifying the important links between physical illness, depression, and biological markers will be very important in the future, as many diseases have overlapping pathophysiology aspects (i.e., information obtained from one disease can potentially be applied to other diseases). Perhaps better methods will be developed to help resolve these difficult problems [156]. However, the exact relationships between lipid metabolism, HPA axis, immune abnormalities in major depression, and medical illness are still unknown. More studies are clearly needed to specifically address the relationship between major depression and medical illness.

References

[1] Boston PF, Dursun SM, Reveley MA. Cholesterol and mental disorder. Br J Psychiatry 1996; 169:682–689.

[2] Golomb BA. Cholesterol and violence: Is there a connection? Ann Int Med 1998; 128:478–487.

[3] Muldoon MF, Manuck SB, Mendelsohn AB, Kaplan JR, Belle S. Cholesterol reduction and non-illness mortality: Meta-analysis of randomized clinical trials. BMJ 2001; 322:11–15.

[4] Engelberg H. Low serum cholesterol and suicide. Lancet 1992; 339:727–729.

[5] Hawton K, Cowen P, Owens D, Bond A, Elliott M. Low serum cholesterol and suicide. Br J Psychiatry 1993; 162:818–825.

[6] Steegmans PH, Fekkes D, Hoes AW, Bak AA, van der Does E, Grobbee DE. Low serum cholesterol concentration and serotonin metabolism in men. BMJ 1996; 312:221.

[7] Alvarez JC, Cremniter D, Gluck N, et al. Low serum cholesterol in violent but not in non-violent suicide attempters. Psychiatry Res 2000; 95:103–108.

[8] Comings DE, MacMurray JP, Gonzalez N, Ferry L, Peters W. Association of the serotonin transporter gene with serum cholesterol levels and heart disease. Mol Genet Metab 1999; 67:248–253.

[9] Collier DA, Stober G, Li T, et al. A novel functional polymorphism within the promoter of the serotonin transporter gene: Possible role in susceptibility to affective disorders. Mol Psychiatry 1996; 1:453–460.

[10] Du L, Faludi G, Palkovits M, et al. Frequency of long allele in serotonin transporter gene is increased in depressed suicide victims. Biol Psychiatry 1999; 46:196–201.

[11] Bocchetta A, Chillotti C, Carboni G, Oi A, Ponti M, Del Zompo M. Association of personal and familial suicide risk with low serum cholesterol concentration in male lithium patients. Acta Psychiatr Scand 2001; 104:37–41.

[12] Kim YK, Myint AM. Clinical application of low serum cholesterol as an indicator for suicide risk in major depression. J Affect Disord 2004; 81:161–166.

[13] Terao T, Yoshimura R, Ohmori O, et al. Effects of serum cholesterol levels of meta-chlorophenylpiperazine-evoked neuroendocrine responses in healthy subjects. Biol Psychiatry 1997; 41:974–978.
[14] Sarchiapone M, Camardese G, Roy A, et al. Cholesterol and serotonin indices in depressed and suicidal patients. J Affect Disord 2001; 62:217–219.
[15] Penttinen J. Hypothesis: Low cholesterol, suicide and interleukin-2. Am J Epidemiol 1996; 141:716–718.
[16] Myint Am, Kim YK. Cytokine-serotonin interaction through IDO: A neurodegeneration of depression. Med Hypotheses 2003; 61:519–525.
[17] Anisman H, Merali Z. Cytokines, stress and depressive illness: Brain-immune interactions. Ann Med 2003; 35:2–11.
[18] Pariante CM, Miller AH. Glucocorticoid receptors in major depression: Relevance to pathophysiology and treatment. Biol Psychiatry 2001; 49:391–404.
[19] Barden N. Implication of the hypothalamic-pituitary-adrenal axis in the physiopathology of depression. J Psychiatry Neurosci 2004; 29:185–193.
[20] Nemeroff CB. Neurobiological consequences of childhood trauma. J Clin Psychiatry 2004; 65(Suppl. 1):18–28.
[21] Morgan RE, Palinkas LA, Barrett-Connor EL, Wingard DL. Plasma cholesterol and depressive symptoms in older men. Lancet 1993; 341:75–79.
[22] Terao T, Iwata N, Kanazawa K, et al. Low serum cholesterol levels and depressive state in human dock visitors. Acta Psychiatr Scand 2000; 101:231–234.
[23] Swartz CM. Albumin decrement in depression and cholesterol decrement in mania. J Affect Disord 1990; 19:173–176.
[24] Bajwa WK, Asnis GM, Sanderson WC, Irfan A, van Praag HM. High cholesterol levels in patients with panic disorder. Am J Psychiatry 1992; 149:376–378.
[25] Maes M, Delanghe J, Meltzer HY, et al. Lower degree of esterification of serum cholesterol in depression: Relevance for depression and suicide research. Acta Psychiatr Scand 1994; 90:252–258.
[26] McCallum J, Simons L, Simons J, Friedlander Y. Low serum cholesterol is not associated with depression in the elderly: Data from an Australian community study. Aus NZ J Med 1994; 24:561–564.
[27] Troisi A, Scucchi S, San Martino L, Montera P, d'Amore A, Moles A. Age specificity of the relationship between serum cholesterol and mood in obese women. Physiol Behav 2001; 72:409–413.
[28] Hu P, Seeman TE, Harris TB, Reuben DB. Does inflammation or undernutrition explain the low cholesterol-mortality association in high-functioning older persons? MacArthur studies of successful aging. J Am Geriatr Soc 2003; 51:80–84.
[29] Nakao M, Ando K, Nomura S, et al. Depressive mood accompanies hypercholesterolemia in young Japanese adults. Jpn Heart J 2001; 42:739–748.
[30] Glueck CJ, Kuller FE, Hamer T, et al. Hypocholesterolemia, hypertriglyceridemia, suicide, and suicide ideation in children hospitalized for psychiatric diseases. Pediatr Res 1994; 35:602–610.
[31] Lindberg G, Larsson G, Setterlind S, Rastam L. Serum lipids and mood in working men and women in Sweden. J Epidemiol Community Health 1994; 48:360–363.
[32] Chen CC, Lu FH, Wu JS, Chang CJ. Correlation between serum lipid concentrations and psychological distress. Psychiatry Res 2001; 93:93–102.
[33] Huang TL, Wu SC, Chiang YS, Chen JF. Correlation between serum lipid, lipoprotein concentrations and anxious state, depressive state or major depressive disorder. Psychiatry Res 2003; 118:147–153.

[34] Glueck CJ, Tieger M, Kunkel R, Hamer T, Tracy T, Speirs J. Hypocholesterolemia and affective disorders. Am J Med 1994; 308:218–225.
[35] Olusi SO, Fido AA. Serum lipid concentrations in patients with major depressive disorder. Biol Psychiatry 1996; 40:1128–1131.
[36] Maes M, Smith R, Christophe, A, et al. Lower serum high-density lipoprotein cholesterol (HDL-C) in major depression and in depressed men with serious suicidal attempts: Relationship with immune-inflammatory markers. Acta Psychiatr Scand 1997; 95:212–221.
[37] Partonen T, Haukka J, Virtamo J, Taylor PR, Lonnqvist J. Association of low serum total cholesterol with major depression and suicide. Br J Psychiatry 1999; 175:259–262.
[38] Huang TL. Serum cholesterol levels in mood disorders associated with physical violence or suicide attempts in Taiwanese. Chang Gung Med J 2001; 24:563–568.
[39] Brown SL, Salive ME, Harris TB, Simonsick EM, Guralnik JM, Kohout FJ. Low cholesterol concentrations and severe depressive symptoms in elder people. BMJ 1994; 308:1328–1332.
[40] Apter A, Laufer N, Bar-Sever M, Har-Even D, Ofek H, Weizman A. Serum cholesterol, suicidal tendencies, impulsivity, aggression, and depression in adolescent psychiatric inpatients. Biol Psychiatry 1999; 46:532–541.
[41] Huang TL, Chen JF. Lipid and lipoprotein levels in depressive disorders with melancholic feature or atypical feature and dysthymia. Psychiatry Clin Neurosci 2004; 58:295–299.
[42] Kimmel PL, Weihs K, Peterson RA. Survival in hemodialysis patients: The role of depression. J Am Soc Nephrol 1993; 3:12–27.
[43] Frasure-Smith N, Lesperance F, Talajic M. Depression following myocardial infarction: Impact of 6-month survival. JAMA 1993; 270:1819–1825.
[44] Barefoot JC, Schroll M. Symptoms of depression, acute myocardial infarction, and total mortality in a community sample. Circulation 1996; 93:1976–1980.
[45] Penninx BWJH, Beekman ATF, Honig A, et al. Depression and cardiac mortality: Results from a community-based longitudinal study. Arch Gen Psychiatry 2001; 58:221–227.
[46] Coelho R, Ramos E, Prata J, Barros H. Psychosocial indexes and cardiovascular risk factors in a community sample. Psychother Psychosom 2000; 69:261–274.
[47] van Doornen LJ, van Blokland RW. The relation of type A behavior and vital exhaustion with physiological reactions to real life stress. J Psychosom Res 1989; 33:715–725.
[48] Rutledge T, Reis SE, Olson M, et al. Psychosocial variables are associated with atherosclerosis risk factors among women with chest pain: The WISE study. Psychosom Med 2001; 63:282–288.
[49] Su CT. Association between job strain status and cardiovascular risk in a population of Taiwanese white-collar workers. Jpn Circ J 2001; 65:509–513.
[50] Engstrom G, Lind P, Hedblad B, Stavenow L, Janzon L, Lindgarde F. Effects of cholesterol and inflammation-sensitive plasma proteins on incidence of myocardial infarction and stroke in men. Circulation 2002; 105:2632–2637.
[51] Jialal I, Devaraj S. Inflammation and atherosclerosis: The value of the high-sensitivity C-reactive protein assay as a risk marker. Am J Clin Pathol 2001; 116(Suppl.): S108–S115.
[52] Yetiv JZ. Clinical applications of fish oils. JAMA 1988; 260:665–670.

[53] Shibata H, Kumagai S, Watanabe S, Suzuki T. Relationship of serum cholesterols and vitamin E to depressive status in the elderly. J Epidemiol 1999; 9:261–267.

[54] Rogers PJ. A healthy body, a healthy mind: Long-term impact of diet on mood and cognitive function. Proc Nutr Soc 2001; 60:135–143.

[55] Lutjohann D, Papassotiropoulos A, Bjorkhem I, et al. Plasma 24S-hydroxycholesterol (cerebrosterol) is increased in Alzheimer and vascular demented patients. J Lipid Res 2000; 41:195–198.

[56] Curtiss LK, Boisvert WA. Apolipoprotein E and atherosclerosis. Curr Opin Lipidol 2000; 11:243–251.

[57] Brahimi F, Bertrand P, Starck M, Galteau MM, Siest G. Control of apolipoprotein E secretion in the human hepatoma cell line KYN-2. Cell Biochem Funct 2001; 19:51–58.

[58] Teunissen CE, de Vente J, Steinbusch HW, De Bruijn C. Biochemical markers related to Alzheimer's dementia in serum and cerebrospinal fluid. Neurobiol Aging 2002; 23:485–508.

[59] Katon WJ. Clinical and health services relationships between major depression, depressive symptoms, and general medical illness. Biol Psychiatry 2003; 54:216–226.

[60] Harris MD. Psychosocial aspects of diabetes with an emphasis on depression. Curr Diab Rep 2003; 3:49–55.

[61] Nosadini R, Manzato E, Solini A, et al. Peripheral, rather than hepatic, insulin resistance and atherogenic lipoprotein phenotype predict cardiovascular complications in NIDDM. Eur J Clin Invest 1994; 24:258–266.

[62] Helkala EL, Niskanen L, Viinamaki H, Partanen J, Uusitupa M. Short-term and long-term memory in elderly patients with NIDDM. Diabetes Care 1995; 18:681–685.

[63] Kimmel PL, Patel SS, Peterson RA. Depression in African-American patients with kidney disease. J Nati Med Assoc 2002; 94(Suppl.):S92–S103.

[64] Kaysen GA, Gambertoglio J, Felts J, Hutchison FN. Albumin synthesis, albuminuria and hyperlipemia in nephrotic patients. Kid Int 1987; 31:1368–1376.

[65] Nihei T, Miura Y, Yagasaki K. Inhibitory effect of resveratrol on proteinuria, hypoalbuminemia and hyperlipidemia in nephritic rats. Life Sci 2001; 68:2845–2852.

[66] Bruno CM, Meli S, Marcinno M, Ierna D, Sciacca C, Neri S. Plasma endothelin-1 levels and albumin excretion rate in normotensive, microalbuminuric type 2 diabetic patients. J Biol Regul Homeost Agents 2002; 16:114–117.

[67] Huang TL, Lee CT. Serum albumin and ferritin levels in chronic hemodialysis patients with or without major depression. International Congress of Biological Psychiatry, February 9–13, 2004. Sydney, abstract 88.

[68] Executive Summary of the Third Report of the National Cholesterol Education Program (NCEP) Expert Panel on Detection, Evaluation, and Treatment of High Blood Cholesterol in Adults (Adult Treatment Panel III). JAMA 2001; 285:2486–2497.

[69] Isomaa B, Almgren P, Tuomi T, et al. Cardiovascular morbidity and mortality associated with the metabolic syndrome. Diabetes Care 2001; 24:683–689.

[70] Chiba M, Suzuki S, Hinokio Y, et al. Tyrosine hydroxylase gene microsatellite polymorphism associated with insulin resistance in depressive disorder. Metabolism 2000; 49:1145–1149.

[71] Stewart TD, Atlas SA. Syndrome X, depression, and chaos: Relevance to medical practice. Conn Med 2000; 64:343–345.

[72] Haffner SM, Valdez RA, Hazuda HP, Mitchell BD, Morales PA, Stern MP. Prospective analysis of the insulin-resistance syndrome (syndrome X). Diabetes 1992; 41:715–722.

[73] Festa A, D'Agostino R, Jr, Howard G, Mykkanen L, Tracy RP, Haffner SM. Chronic subclinical inflammation as part of the insulin resistance syndrome: The Insulin Resistance Atherosclerosis Study (IRAS). Circulation 2000; 102:42–47.
[74] Lawlor DA, Smith GD, Ebrahim S. British Women's Heart and Health Study. Association of insulin resistance with depression: Cross sectional findings from the British Women's Heart and Health Study. BMJ 2003; 327:1383–1384.
[75] Hotamisligil GS, Shargill NS, Spiegelman BM. Adipose expression of tumor necrosis factor-alpha: Direct role in obesity-linked insulin resistance. Science 1993; 259:87–91.
[76] Ramasubbu R. Insulin resistance: A metabolic link between depressive disorder and atherosclerotic vascular diseases. Med Hypotheses 2002; 59:537–551.
[77] Schwertner HA, Torres L, Jackson WG, Maldonado HA, Whitson JD, Troxler RG. Cortisol and the hypercholesterolemia of pregnancy and labor. Atherosclerosis 1987; 67:237–244.
[78] Troisi A, Moles A, Panepuccia L, Lo Russo D, Palla G, Scucchi S. Serum cholesterol levels and mood symptoms in the postpartum period. Psychiatry Res 2002; 109:213–219.
[79] Nasta MT, Grussu P, Quatraro RM, Cerutti R, Grella PV. Cholesterol and mood states at 3 days after delivery. J Psychosom Res 2002; 52:61–63.
[80] West SG, Hinderliter AL, Wells EC, Girdler SS, Light KC. Transdermal estrogen reduces vascular resistance and serum cholesterol in postmenopausal women. Am J Obstet Gynecol 2001; 184:926–933.
[81] Muldoon MF, Manuck SB, Matthews KA. Lowering cholesterol concentrations and mortality: A quantitative review of primary prevention trials. BMJ 1990; 301:309–314.
[82] Lindberg G, Rastam L, Gullberg B, Ekulund GA. Low serum cholesterol concentration and short term mortality from injuries in men and women. BMJ 1992; 305:277–279.
[83] Neaton JD, Blackburn H, Jacobs D, et al. Serum cholesterol level and mortality findings for men screened in the multiple risk factor intervention trial. Arch Int Med 1992; 152:1490–1500.
[84] Zureik M, Courbon D, Ducimetiere P. Serum cholesterol concentration and death from suicide in men. BMJ 1996; 313:649–651.
[85] Kunugi H, Takei N, Aoki H, Nanko S. Low serum cholesterol in suicide attempter. Biol Psychiatry 1997; 41:196–200.
[86] Pekkanen J, Nissinen A, Punsar S, Karvonen MJ. Serum cholesterol and risk of accidental or violent death in a 25-year follow-up: The Finnish cohorts of the Seven Countries Study. Arch Int Med 1989; 149:1589–1591.
[87] Jacobs D, Blackburn H, Higgins M, et al. Report of the conference on low blood cholesterol: Mortality associations. Circulation 1992; 86:1046–1060.
[88] Vartiainen E, Puska P, Pekkanen J, Tuomilehto J, Lonnqvist J, Ehnholm C. Serum cholesterol and accidental or other violent deaths. BMJ 1994; 309:445–447.
[89] Tanskanen A, Vartiainen E, Tuomilehto J, Viinamaki H, Lehtonen J, Puska P. High serum cholesterol and risk of suicide. Am J Psychiatry 2000; 157:648–650.
[90] Fawcett J, Busch KA, Jacobs D, Kravitz HM, Fogg L. Suicide: A four-pathway clinical-biochemical model. Ann NY Acad Sci 1997; 836:288–301.
[91] Coccaro EF. Central serotonin and impulsive aggression. Br J Psychiatry 1989; 155(Suppl. 8):52–62.
[92] Mann JJ, Marzuk PM, Arango V, McBride PA, Leon AC, Tierney H. Neurochemical studies of violent and nonviolent suicide. Psychopharmacol Bull 1989; 25:407–413.

[93] Ringo DL, Lindley SE, Faull KF, Faustman WO. Cholesterol and serotonin: Seeking a possible link between blood cholesterol and CSF 5–HIAA. Biol Psychiatry 1994; 35:957–959.

[94] Terao T, Nakamura J, Yoshimura R, et al. Relationship between serum cholesterol levels and meta-chlorophenypiperazine-induced cortisol responses in healthy men and women. Psychiatry Res 2000; 96:167–173.

[95] Papakostas GI, Petersen T, Mischoulon D, et al. Serum cholesterol and serotonergic function in major depressive disorder. Psychiatry Res 2003; 118:137–145.

[96] Buydens-Branchey L, Branchey M, Hudson J, Fergeson P. Low HDL cholesterol, aggression and altered central serotonergic activity. Psychiatry Res 2000; 102:153–162.

[97] Steegmans PH, Hoes AW, Bak AA, van der Does E, Grobbee DE. Higher prevalence of depressive symptoms in middle-aged men with low serum cholesterol levels. Psychosom Med 2000; 62:205–211.

[98] Brunner J, Parhofer KG, Schwandt P, Bronisch T. Cholesterol, essential fatty acids, and suicide. Pharmacopsychiatry 2002; 35:1–5.

[99] Manfredini R, Caracciolo S, Salmi R, Boari B, Tomelli A, Gallerani M. The association of low serum cholesterol with depression and suicidal behaviours: New hypotheses for the missing link. J Int Med Res 2000; 28:247–257.

[100] Horsten M, Wamala SP, Vingerhoets A, Orth-Gomer K. Depressive symptoms, social support, and lipid profile in healthy middle-aged women. Psychosom Med 1997; 59:521–528.

[101] Walton KG, Pugh ND. Stress, steroids, and "ojas": Neuroendocrine mechanisms and current promise of ancient approaches to disease prevention. Indian J Physiol Pharmacol 1995; 39:3–36.

[102] Francis KT. Psychologic correlates of serum indicators of stress in man: A longitudinal study. Psychosom Med 1979; 41:617–628.

[103] Agarwal V, Gupta B, Singhal U, Bajpai SK. Examination stress: Changes in serum cholesterol, triglycerides and total lipids. J Physiol Pharmacol 1997; 41:404–408.

[104] Prabhakaran K, Suthanthirarajan N, Namasivayam A. Biochemical changes in acute noise stress in rats. Indian J Physiol Pharmacol 1988; 32:100–104.

[105] Hershock D, Vogel WH. The effects of immobilization stress on serum triglycerides, nonesterified fatty acids, and total cholesterol in male rats after dietary modifications. Life Sci 1989; 45:157–165.

[106] Brennan FX, Jr, Fleshner M, Watkins LR, Maier SF. Macrophage stimulation reduces the cholesterol levels of stressed and unstressed rats. Life Sci 1996; 58:1771–1776.

[107] Abd el Mohsen MM, Fahim AT, Motawi TM, Ismail NA. Nicotine and stress: Effect on sex hormones and lipid profile in female rats. Pharmacol Res 1997; 35:121–128.

[108] Ricart-Jane D, Rodriguez-Sureda V, Benavides A, Peinado-Onsurbe J, Lopez-Tejero MD, Llobera M. Immobilization stress alters intermediate metabolism and circulating lipoproteins in the rat. Metabolism 2002; 51:925–931.

[109] Cassano WJ, Jr, D'mello AP. Acute stress-induced facilitation of the hypothalamic-pituitary-adrenal axis: Evidence for the roles of stressor duration and serotonin. Neuroendocrinology 2001; 74:167–177.

[110] Djordjevic J, Cvijic G, Davidovic V. Different activation of ACTH and corticosterone release in response to various stressors in rats. Physiol Res 2003; 52:67–72.

[111] Rosmond R, Eriksson E, Bjorntorp P. Personality disorders in relation to anthropometric, endocrine and metabolic factors. J Endocrinol Invest 1999; 22:279–288.

[112] Rosmond R, Bjorntorp P. The hypothalamic-pituitary-adrenal axis activity as a predictor of cardiovascular disease, type 2 diabetes and stroke. J Int Med 2000; 247:188–197.

[113] Cabana VG, Gewurz H, Siegel JN. Inflammation-induced changes in rabbit CRP and plasma lipoproteins. J Immunol 1983; 130:1736–1742.

[114] Erlinger TP, Miller ER, 3rd, Charleston J, Appel LJ. Inflammation modifies the effects of a reduced-fat low-cholesterol diet on lipids: Results from the DASH-sodium trial. Circulation 2003; 108:150–154.

[115] Kisilevsky R. Serum amyloid A (SAA), a protein without a function: Some suggestions with reference to cholesterol metabolism. Med Hypotheses 1991; 35:337–341.

[116] van Leeuwen HJ, Heezius EC, Dallinga GM, van Strijp JA, Verhoef J, van Kessel KP. Lipoprotein metabolism in patients with severe sepsis. Crit Care Med 2003; 31:1359–1366.

[117] Lindhorst E, Young D, Bagshaw W, Hyland M, Kisilevsky R. Acute inflammation, acute phase serum amyloid A and cholesterol metabolism in the mouse. Biochim Biophys Acta 1997; 1339:143–154.

[118] Hardardottir I, Grunfeld C, Feingold KR. Effects of endotoxin and cytokines on lipid metabolism. Curr Opin Lipidol 1994; 5:207–215.

[119] Pajkrt D, Doran JE, Koster F, et al. Anti-inflammatory effects of reconstituted high-density lipoprotein during human endotoxemia. J Exp Med 1996; 184:1601–1608.

[120] Feingold KR, Hardardottir I, Grunfeld C. Beneficial effects of cytokine induced hyperlipidemia. Z. Ernahrungswiss 1998; 37(Suppl. 1):66–74.

[121] Kaplan M, Aviram M. Oxidized low-density lipoprotein: Atherogenic and pro-inflammatory characteristics during macrophage foam cell formation. An inhibitory role for nutritional antioxidants and serum paraoxonase. Clin Chem Lab Med 1999; 37:777–787.

[122] Khovidhunkit W, Shigenaga JK, Moser AH, Feingold KR, Grunfeld C. Cholesterol efflux by acute-phase high density lipoprotein: Role of lecithin: Cholesterol acyltransferase. J Lipid Res 2001; 42:967–975.

[123] Mallat Z, Gojova A, Brun V, et al. Induction of a regulatory T cell type 1 response reduces the development of atherosclerosis in apolipoprotein E-knockout mice. Circulation 2003; 108:1232–1237.

[124] Fazio S, Linton MF. The inflamed plaque: Cytokine production and cellular cholesterol balance in the vessel wall. Am J Cardiol 2001; 88:12E–15E.

[125] Cerwinka WH, Cheema MH, Granger DN. Hypercholesterolemia alters endotoxin-induced endothelial cell adhesion molecule expression. Shock 2001; 16:44–50.

[126] Duval C, Chinetti G, Trottein F, Fruchart JC, Staels B. The role of PPARs in atherosclerosis. Trends Mol Med 2002; 8:422–430.

[127] Diep QN, Amiri F, Touyz RM, et al. PPARalpha activator effects on Ang II-induced vascular oxidative stress and inflammation. Hypertension 2002; 40:866–871.

[128] Lee CH, Chawla A, Urbiztondo N, Liao D, Boisvert WA, Evans RM. Transcriptional repression of atherogenic inflammation: Modulation by PPARdelta. Science 2003; 302:453–457.

[129] Klappacher GW, Glass CK. Roles of peroxisome proliferator-activated receptor gamma in lipid homeostasis and inflammatory responses of macrophages. Curr Opin Lipidol 2002; 13:305–312.

[130] Mason JC. Statins and their role in vascular protection. Clin Sci 2003; 105:251–266.

[131] Kleemann R, Princen HM, Emeis JJ, et al. Rosuvastatin reduces atherosclerosis development beyond and independent of its plasma cholesterol-lowering effect in APOE*3-Leiden transgenic mice: Evidence for anti-inflammatory effects of rosuvastatin. Circulation 2003; 108:1368–1374.

[132] Huang TL. Lower serum albumin levels in patients with mood disorders. Chang Gung Med J 2002; 25:509–513.

[133] Gurr MI, Harwood JL, Frayn KN, Lipid biochemistry, fifth Ed. Oxford: Blackwell Science 2002: 170–214.
[134] Spector AA. Plasma lipid transport. Clin Physiol Biochem 1984; 2:123–134.
[135] Deliconstantinos G. Effect of rat serum albumin-cholesterol on the physical properties of biomembranes. Biochem Int 1987; 15:467–474.
[136] Mahley RW. Apolipoprotein E: Cholesterol transport protein with expanding role in cell biology. Science 1988; 240:622–630.
[137] Kaul D. Molecular link between cholesterol, cytokines and atherosclerosis. Mol Cell Biochem 2001; 219:65–71.
[138] Steinmetz A, Hocke G, Saile R, Puchois P, Fruchart JC. Influence of serum amyloid A on cholesterol esterification in human plasma. Biochim Biophys Acta 1989; 1006:173–178.
[139] Etingin OR, Hajjar DP. Evidence for cytokine regulation of cholesterol metabolism in herpesvirus-infected arterial cells by the lipoxygenase pathway. J Lipid Res 1990; 31:299–305.
[140] Grove RI, Mazzucco C, Allegretto N, et al. Macrophage-derived factors increase low-density lipoprotein uptake and receptor number in cultured human liver cells. J Lipid Res 1991; 32:1889–1897.
[141] Babiker A, Andersson O, Lund E, et al. Elimination of cholesterol in macrophages and endothelial cells by the sterol 27-hydroxylase mechanism. Comparison with high-density lipoprotein-mediated reverse cholesterol transport. J Biol Chem 1997; 272:26253–26261.
[142] Hirai A, Kino T, Tokinaga K, Tahara K, Tamura Y, Yoshida S. Regulation of sterol carrier protein 2 (SCP2) gene expression in rat peritoneal macrophages during foam cell formation. A key role for free cholesterol content. J Clin Invest 1994; 94:2215–2223.
[143] Netea MG, Demacker PN, Kullberg BJ, et al. Low-density lipoprotein receptor-deficient mice are protected against lethal endotoxemia and severe gram-negative infections. J Clin Invest 1996; 97:1366–1372.
[144] Pomerantz KB, Nicholson AC, Hajjar DP. Signal transduction in atherosclerosis: Second messengers and regulation of cellular cholesterol trafficking. Adv Exp Med Biol 1995; 369:49–64.
[145] Hajjar DP, Haberland ME. Lipoprotein trafficking in vascular cells. Molecular Trojan horses and cellular saboteurs. J Biol Chem 1997; 272:22975–22978.
[146] van Rossum EF, Koper JW, Huizenga NA, et al. A polymorphism in the glucocorticoid receptor gene, which decreases sensitivity to glucocorticoids *in vivo*, is associated with low insulin and cholesterol levels. Diabetes 2002; 51:3128–3134.
[147] Zabalawi M, Bhat S, Loughlin T, et al. Induction of fatal inflammation in LDL receptor and ApoA-I double-knockout mice fed dietary fat and cholesterol. Am J Pathol 2003; 163:1201–1213.
[148] Major AS, Fazio S, Linton MF. B-lymphocyte deficiency increases atherosclerosis in LDL receptor-null mice. Arterioscler Thromb Vasc Biol 2002; 22:1892–1898.
[149] Davis RA, Miyake JH, Hui TY, Spann NJ. Regulation of cholesterol-7alpha-hydroxylase: BAREly missing a SHP. J Lipid Res 2002; 43:533–543.
[150] Tam SP, Flexman A, Hulme J, Kisilevsky R. Promoting export of macrophage cholesterol: The physiological role of a major acute-phase protein, serum amyloid. A J Lipid Res 2002; 43:1410–1420.
[151] Sehgal PB, Guo GG, Shah M, Kumar V, Patel K. Cytokine signaling: STATS in plasma membrane rafts. J Biol Chem 2002; 277:12067–12074.

[152] Diomede L, Albani D, Bianchi M, Salmona M. Endotoxin regulates the maturation of sterol regulatory element binding protein-1 through the induction of cytokines. Eur Cytokine Netw 2001; 12:625–630.

[153] Santamarina-Fojo S, Remaley AT, Neufeld EB, Brewer HB, Jr.. Regulation and intracellular trafficking of the ABCA1 transporter. J Lipid Res 2001; 42:1339–1345.

[154] Huber SA, Sakkinen P, David C, Newell MK, Tracy RP. T helper-cell phenotype regulates atherosclerosis in mice under conditions of mild hypercholesterolemia. Circulation 2001; 103:2610–2616.

[155] Vainio S, Ikonen E. Macrophage cholesterol transport: A critical player in foam cell formation. Ann Med 2003; 35:146–155.

[156] Grant PJ. The genetics of atherothrombotic disorders: A clinician's view. J Thromb Haemost 2003; 1:1381–1390.

ADVANCES ON BIOLOGICAL MARKERS IN EARLY DIAGNOSIS OF ALZHEIMER DISEASE

Alessandro Padovani,* Barbara Borroni,*
and Monica Di Luca†

*Department of Neurological Sciences,
University of Brescia, 25100 Brescia, Italy
†Institute of Pharmacological Sciences, University of Milan,
20133 Milan, Italy

1. Introduction

Alzheimer disease (AD) represents the most common neurodegenerative disease worldwide, accounting for 60%–70% of cases of progressive cognitive impairment in elderly patients. The prevalence of dementia of the Alzheimer type (DAT) doubles every 5 years after the age of 60 years, increasing from a prevalence of 1% among those 60–64 years old up to 40% of those aged 85 years and older [1, 2]. The disease is more common among women than men

0065-2423/05 $35.00
DOI: 10.1016/S0065-2423(04)39004-9

by a ratio of 1.2 to 1.5. Dementia patients have a substantially shortened life expectancy; the average survival is 8 years from diagnosis. The characteristic findings at the microscopic level are degeneration of the neurons and their synapses, together with extensive amounts of senile plaques and neurofibrillary tangles [3].

Extensive exploration of possible risk factors for AD has been somewhat disappointing so far. Age, dementia in a close family member, and apolipoprotein E (APOE) ε4 allele are the only confirmed risk factors for the disease [4]. Female sex, herpes infection, low serum levels of folate and vitamin B12, elevated plasma and total homocysteine levels, low lipid plasma concentrations, a history of head trauma, and the protective effect of hormonal replacement therapy are all factors that likely bimodally interact with APOE genotype to modify relative risk [5]. Several other possible risk factors for AD such as exposure to anesthetic agents, diabetes mellitus, and the protective effects of antiinflammatory drugs, smoking, and alcohol are being reevaluated, using improving methodologies [5]. Having more education and an active lifestyle have been associated with low rates of AD, which could be at least partly attributable to compensatory strategies that delay detection of the disease.

The introduction of acetylcholine esterase inhibitors as symptomatic treatment has highlighted the importance of diagnostic markers for AD [6, 7]. The awareness in the population of the availability of drug treatment has also made patients seek medical advice at an earlier stage of the disease. This has increased the diagnostic challenge for physicians, because the characteristic clinical picture of AD with slowly progressive memory disturbances combined with parietal lobe symptoms is rather subtle in the early stage and can be barely noticeable. Accordingly, there is no clinical method to accurately identify AD in the very early phase and to determine which at-risk cases will progress to DAT, except for having a very long clinical follow-up period [8]. During the so-called preclinical phase of AD, the neuronal degeneration proceeds and the amount of plaques and tangles increase, and at a certain threshold, the first symptoms—most often including impairment of episodic memory—appear. This preclinical period probably starts 20–30 years before the first clinical symptoms appear [9]. According to current diagnostic criteria, AD cannot be diagnosed clinically before the disease has progressed so far that dementia is present. This means that the symptoms must be severe enough to significantly interfere with work and social activities or relations.

Thus, new diagnostic tools to aid the diagnosis of early AD and to identify incipient AD in at-risk cases would be of great importance. Such diagnostic markers would be of even higher significance if new drugs, with the promise of disease-arresting effects, prove to have clinical effect. Such drugs will probably be more effective in the earlier stages of the disease, before neurodegeneration is too severe and widespread.

Our goal with this review concerns early recognition of AD, focusing on the use of biological markers in identifying, among individuals at risk, those who will progress to DAT. In particular we will review the contribution of biological markers in early diagnosis of AD. "Early diagnosis" refers to the capability to diagnose AD at a very early stage, before symptoms and clinical signs have reached the stage at which a diagnosis of clinically probable DAT can be made according to currently recommended criteria. Accordingly, in this review AD is conceptualized as a chronic degenerative disease clinically characterized by a long presymptomatic phase followed by a prodromal transitional stage, hereby termed mild cognitive impairment (MCI) which is subsided by the ultimate stage of dementia, more properly called DAT [10]. The goal of research in this area is to develop highly specific and sensitive tools capable of identifying, among at-risk subjects either in the preclinical or in the prodromal stage, those who will eventually progress to DAT.

Before beginning this review, a brief account of some aspects of AD instrumental to understand the topic of this review are given.

2. Biological Aspects of AD Pathogenesis

2.1. Neuropathological Hallmarks

Increases in the understanding of the biology of AD during the last decades have been substantial and have led to potential treatment strategies that are just beginning to undergo clinical evaluation. In particular, molecular biology and biochemistry studies have led to a better understanding of the process of AD main pathological hallmark formation (i.e., plaques and tangles). Animal models of lesion formation have been developed, though it is still unclear how the lesions relate to each other, why some neurons are more vulnerable than others, and which factors determine individual susceptibility to neuronal lesions.

In AD, the main cause of dementia is assumed to result from the progressive loss of synapse and the neuronal degeneration [11]. The neuropathologic hallmarks of neurofibrillary tangles (NFTs) and senile plaques were originally described by Alois Alzheimer, a German psychiatrist, at the beginning of the last century. Both the senile plaques and the NFTs, although not individually unique to AD, have a characteristic spreading and density in this disease [12].

The plaque is an extracellular lesion composed of a core of an amyloid peptide of 40–42 amino acids designated as Abeta (Aβ). This peptide is in turn derived from an amyloid precursor protein (APP), the gene for which is located on chromosome 21. APP is a transmembrane protein and contains a large extracellular region, a transmembrane helix, and a short cytoplasmic tail, with the Aβ motif extending from the exterior to halfway through the cell membrane. Different isoforms of APP are generated by alternative splicing. The three major isoforms are constituted of 770, 751, and 695 amino acid residues. APP isoforms are expressed in several cellular systems such as muscle, epithelial, and circulating cells. Supporting the finding of ubiquitous expression of APP isoforms, large N-terminal fragments—products of secretase activity—are found in cerebrospinal fluid, blood, and urine [13].

The other lesion of AD is the neurofibrillary tangle—an intraneuronal lesion of highly phosphorylated and aggregated tau [14, 15]. Tau protein is a normal and essential component of neurons, where it stabilizes the microtubule cytoskeleton that is essential for axonal transport. It is the incorporation of excess phosphate groups (i.e., hyperphosphorylation), which leads to the formation of paired helical filaments, or tau [16]. The process of NFT accumulation begins with granular neuronal deposits of abnormally phosphorylated tau, which becomes progressively more fibrillar, followed by the formation of filamentous intracellular inclusions that ultimately fill the neuron. The neuron eventually dies, leaving insoluble filaments in the neuropil. In AD this cytoskeleton is lost and the aggregation of tau is an early event in pathogenesis that correlates well with cognitive impairment. It is still unclear why tau aggregates, but it might be caused by the loss of normal tau function resulting from an imbalance between protein kinases and phosphatases, which may also be influenced by the deposition of fibrillar Aβ.

Although the etiology of AD is still largely unknown, increasing evidence indicates that the conversion of the Aβ peptide to amyloid is central to the pathogenesis of AD [17, 18]. The amyloid hypothesis states that neuronal dysfunction and death, neurofibrillary degeneration, microglia activation and the full manifestations of Alzheimer pathology are initiated by Aβ deposition [19].

In particular, conditions of Aβ overproduction or impaired cerebral Aβ clearance mechanisms result in elevated Aβ levels that promote Aβ aggregation, oligomerization, and fibrillogenesis [20]. Deposition of Aβ in an insoluble beta-pleated conformation (amyloid) occurs as plaques in the neuropil and as amyloid angiopathy in the cerebral vasculature. The deposition of amyloid initiates a cascade of injurious events, including free radical production, glial activation, and direct neuronal damage. Kinase activation in response to amyloid-induced injury may increase the amount of hyperphosphorylated tau and lead to destabilization of microtubule and altered cellular transport.

Both immunohistochemical and biochemical studies indicate that amyloid deposits occur very early in the AD process, preceding the appearance of neocortical neurofibrillary tangles. It should be noted that tau pathology occurs in the form of at least a few tangles, which are almost always restricted to the entorhinal cortex and hippocampus, in many older, nondemented individuals [21].

An interaction between Aβ and tau has been proposed wherein the appearance of Aβ deposits accelerates the formation of tangles. Increasing evidence indicates that Aβ42, the 42-amino acid Aβ isoform, is the pathogenetic Aβ species, though both the shorter and more abundant Aβ40, as well as Aβ42, are constitutively produced both *in vitro* and *in vivo*. Aβ42, however, aggregates more readily than Aβ40 *in vitro* and is consistently more abundant in the brain.

2.2. Genetic Risk Factors

The most compelling evidence for the amyloid hypothesis comes from the recognition that the known mutations for autosomal AD in the APP (chromosome 21), presenilin 1 (chromosome 14), and presenilin 2 (chromosome 1) genes are all associated with Aβ overproduction or with increased relative amounts of Aβ1-42 [22–26]. These mutations cause early-onset AD but are rare, accounting for less than 1% of all AD cases, whereas there is increasing evidence that much more common sporadic forms of AD are associated with several susceptibility genes, which are believed to influence age at onset in early-onset AD individuals with autosomic mutations [27]. Among these, a relevant role has been identified for ε4 allele of APOE [28–30]. Individuals with AD who have the ε4 allele show more amyloid deposits, consistent with the observation that APOE-deficient animals show virtually no Aβ deposits in the brain [31]. The precise mechanism by which APOE ε4 increases AD risk is not clear. APOE is produced predominantly in astrocytes and is carried by the low-density lipoprotein receptors into neurons. The three variants of APOE vary in their affinity for beta-amyloid, and the APOE ε4 variant increases the deposition of fibrillar beta-amyloid, resulting in the beta-pleated sheets that stain with birefringent stains for amyloid. Though early clinical studies reported that APOE ε4 allele was specific in up to 90% of cases with dementia, subsequent population studies showed lower rates with a decreased gene dosage effect with advancing age [32–34]. In particular, lifetime risk of AD for an individual without the ε4 allele is approximately 9%; the lifetime risk of AD for an individual carrying at least one ε4 allele is 29%. Other susceptibility loci for AD include chromosome 12, which is

(typically memory) or an overall mild cognitive decline that is greater than would be expected for an individual's age or education but that is insufficient to interfere with social and occupational functioning. Patients fulfilling these criteria have a 5- to 10-fold greater risk of developing DAT within 3–5 years. In fact, as a group, MCI patients have a higher incidence of dementia than cognitively intact persons of similar age. In studies on different populations with mean age of 66–81 years, the conversion rate of those with MCI to DAT ranged between 6% and 25% per year (or 70% over 4 years), as opposed to about 0.2%–2.3% per year in cognitively intact persons of similar age [54]. This indicates that the MCI concept captures a high proportion of patients with AD at the prodromal stage, as confirmed by pathological evidence [55, 56]. Several observations, however, indicate that not all MCI patients deteriorate over time, thus suggesting that MCI is a heterogenous condition [57, 58]. A report from an international group of experts on MCI emphasized that nonamnestic presentations of MCI exist and might progress to dementing illnesses other than DAT [53]. Moreover, some MCI patients do not have prodromal dementia of any type but, instead, have reversible impairment; over time, these individuals may revert to normal, whereas others may have stable impairment [59].

Although methodological issues including characteristics of the population studied, MCI criteria, and settings might account for the contrasting conversion rates across studies, these observations caution against regarding all MCI patients *tout court* as affected by incipient AD. In fact, subjects with amnestic MCI usually progress to DAT at a high rate, and most demonstrate neurofibrillary pathologic conditions in the medial temporal lobes [60]. Though the adoption of these strict operational criteria might be of help in increasing the likelihood of early recognition of AD, predictivity over time is still relatively low, whereas narrowing the definition of MCI would impede the identification of AD patients with atypical clinical presentation. The obvious consequence is that the clinical diagnosis of MCI is doubtfully helpful, as many MCI patients do have AD but many do not. In fact, available data seem to indicate that the likelihood of a subject with MCI developing DAT in the long term is around 50%.

Pathological and clinical data have accumulated in the last few years, showing that some biological indicators of AD might be used to distinguish those MCI patients who will progress from those who will not. Among biological indicators, there is convergent evidence about the usefulness of hippocampal atrophy measures (because of early plaque and tangle deposition in the medial temporal lobes) and functional defects in the temporo-parietal and posterior cingulate cortex (because of deafferentation from medial temporal damage) [9].

4. Biological Markers and Early Diagnosis of AD

The search of relevant biomarkers of AD in living patients has been an active part of clinical research for the last few decades. The assumption of such an enterprise is that a biomarker provides at least an indirect link to the disease process or, ideally, directly relates to the primary mechanism of the disease. It is in this context that, along with the progress in the understanding of the pathogenetic mechanisms of AD, there has been an increasing interest in discovering both central and peripheral markers that may be closely linked to the pathophysiology of the disease. With the advent of symptomatic treatment and the prospect of stabilizing therapies for AD, the race for reliable disease and progression markers has picked up speed, and standardized criteria have been developed. According to a recent consensus report [61], an ideal biomarker for AD should fulfill the following criteria: detect a fundamental feature of the neuropathology; be validated in neuropathologically confirmed cases; have sensitivity higher than 85% for detecting AD and have a specificity of more than 75% for distinguishing AD from other causes of dementia, preferably established by two independent studies appropriately powered; be precise, reliable, and inexpensive; and be convenient to use and not harmful to the patient. A well-characterized biological marker that fulfills these requirements would have several advantages. An ideal biological marker would identify AD cases at a very early stage of the disease, before the cognitive symptoms are found in neuropsychological tests, and before there is clear-cut degeneration in brain-imaging studies. Further, biomarkers in AD could serve several purposes according to different perspectives. First, biomarkers indirectly might help in understanding and confirming pathogenetic mechanisms. Diagnostic markers would aid in the recognition of AD, thus improving reliability of epidemiologic investigations and enabling patients to benefit from available symptomatic drug therapies. Prognostic markers that serve as indicators of AD severity could be used as surrogate endpoints for monitoring progression and for assessing therapeutic response, possibly reducing the time, effort, and costs associated with drug trials that currently require large samples to be followed for long periods to measure clinical outcomes. Antecedent markers could identify those individuals with AD in the predementia stage or at risk of developing symptomatic AD and might predict those patients who might benefit most from the development of disease-modifying or preventive interventions. One biomarker is unlikely to serve all these purposes, thus implying that different biomarkers would eventually be used either in combination in the same stage or in different stages of the disease.

With regard to AD, antemortem biochemical markers have been sought for years in different tissues and cells, such as CSF, plasma, erythrocytes [62], lymphocytes [63], urine [64, 65], hair [66, 67], and skin [68, 69].

In fact, despite the attempts to identify biochemical markers related to AD have led to promising results, the field is far from being satisfied, as the diagnostic value of most hitherto proposed biochemical markers has been limited by considerable overlap between AD patients, patients with other dementing illnesses, and normal subjects.

One of the major difficulties in identifying potential biomarkers related to AD is likely a result of the fact that specific proteins involved in the pathogenetic mechanisms are low abundant in CSF and serum, while a small number of proteins such as albumin, transferrin, and immunoglobulins provide a significant background that makes it difficult to identify low-abundance proteins.

More important, some biochemical markers were not at all designed to assess a given biochemical metabolite as a diagnostic test, as in some cases they were designed to study pathophysiological processes involved in neurodegeneration and to assess the possible involvement of the given metabolite in AD.

Clinical findings on a wide range of measures of inflammation, cholesterol metabolism, and oxidative stress appear to be altered in AD, thus confirming the multiplicity of pathophysiological implicated processes in AD. In most cases, however, these markers have been assessed in small samples of patients that are often lacking neurologic controls and showed insufficient discriminatory power, in particular for early AD phase, as they likely refer to processes involved in neurodegeneration rather than in AD, per se. Nevertheless, although it is still possible that combining markers from several disease mechanisms may eventually be useful in the development of a "diagnostic" test, the importance of these biomarkers may be of greater value for predicting risk for disease or follow-up and for investigating the contributing role of concomitant biological processes and their treatment.

In the last decade, along with a better definition of the main pathogenetic characteristics of AD, there have been consistent findings with regard to selected CSF and "peripheral" biomarkers related to the "key players" of AD pathology (i.e., degeneration of neurons and synapses, disturbance in the APP metabolism and its subsequent deposition in senile plaques, and the hyperphosphorylation of tau with subsequent formation of neurofibrillary tangles), which have provided robust accuracy values reaching high specificity and sensitivity levels, even in early AD, and opened a new era in the diagnostic approach to AD. The following sections will review the main findings on these biomarkers, focusing on their diagnostic and predictive value for early AD.

5. CSF Biomarkers

5.1. APP AND Aβ IN CSF

Several studies have found that CSF levels of sAPPα are slightly decreased in AD compared to controls, with too much overlap for diagnostic utility, whereas levels of sAPPβ do not appear to differ [70].

Initial studies on Aβ in CSF used methods for total Aβ and showed that, in spite of a slight decrease, there was large overlap between DAT patients and controls. Similarly, studies focusing on Aβ-40 failed to show a significant difference between DAT and normals [71–73].

At variance, the majority of studies, to date involving more than 1000 patients with clinical DAT and 500 healthy individuals, showed that A$\tilde{\beta}$42 concentrations are decreased in CSF of DAT patients to about 50% of control levels [74, 75]. The sensitivity of A$\tilde{\beta}$42 reported in most reports varies from 55% to 100% (mean 86%), whereas specificity to distinguish patients with DAT from controls varies from 67% to 100% (mean 89%). Accuracy data between DAT and other dementias are relatively limited, and decreased CSF $\alpha\beta$ 42 levels are found in patients with vascular dementia, frontotemporal dementia, Creutzfeld-Jakob disease, amiotrophic lateral sclerosis, and multiple systemic atrophy, thus indicating a limited value in the differential diagnosis of dementia. However, levels of A$\tilde{\beta}$42 seem to be unrelated to age, and a cut-off level of >500 pg/mL has been suggested to discriminate AD from normal aging [71].

Reduced CSF levels of Aβ-42 have also been found in early AD [76] and in cases with MCI [77], indicating that decreased A$\tilde{\beta}$42 levels might identify the early stage of AD and be detectable before clinical symptoms of dementia become overt. Few studies have evaluated the performance of Aβ-42 in MCI cases, focusing on the magnitude of the index test to predict conversion to DAT within a given interval, and several studies have shown that there is a difference among MCI patients between those who convert to DAT and those who do not and that low levels of Aβ-42 significantly predict conversion to DAT in MCI patients with high sensitivity and specificity [78, 79]. Further, a recent study has showed reduced CSF levels of Aβ42 in nondemented individuals who developed DAT during a 3-year follow-up, with an odds ratio for conversion of 8.2 for individuals in the lower 50th percentile of CSF A$\tilde{\beta}$42, whereas none of those in the highest 33rd percentile developed dementia [80]. These findings support the view that disturbance of the metabolism of Aβ-42 is present very early in the disease process of AD, even before subtle cognitive impairment or MCI become apparent, and that the measurement of CSF levels of Aβ-42 holds the potential to be a reliable and predictive diagnostic test.

5.2. Total Tau and Phospho-Tau in CSF

As previously described, one of the major neuropathological hallmarks of AD are NFTs. NFTs are composed of paired helical filaments, with the principal protein subunit of paired helical filaments being abnormally hyperphosphorylated tau (p-tau). Physiologically, tau protein is located in the neuronal axons and in the cytoskeleton. There are six different tau isoforms. Total tau (t-tau) and truncated forms of monomeric and phosphorylated tau are released and can be found in the CSF [71, 80].

The t-tau, a general marker of neuronal destruction, has been intensely studied in more than 2000 DAT patients and 1000 age-matched elderly controls over the last 10 years. The most consistent finding is a significant increase of CSF levels of t-tau protein in DAT (about 300% compared to normal controls) [81].

Specificity levels were between 65% and 86%, and sensitivity was between 40% and 86%. In mild dementia, the potential of CSF t-tau protein to discriminate between DAT and normal aging is also high, with a mean sensitivity of 75% and specificity of 85%. The usefulness of t-tau is, however, limited by the relatively low power in differentiating DAT from other dementing illnesses [82, 83]. In fact, elevated total tau has been observed in vascular dementia, fronto-temporal dementia, and Lewy body dementia [84]. According to these findings, t-tau protein does not seem to provide sufficient power for discriminating DAT. Nevertheless, there is increasing evidence that t-tau might be of help for the prediction of DAT in MCI subjects. In fact, t-tau is not only increased in patients suffering MCI but is particularly increased in those who converted to DAT during follow-up [78, 79]. Different studies have reported that t-tau predicts conversion to DAT with a high sensitivity (range 88%–90%) and specificity (range 48%–90%) [77, 79].

More recently, immunoassays to specifically detect phosphorylated tau (p-tau) at different epitopes (threonine 231, serine 199, serine 396, serine 404, and threonine 181) have been developed. Evidence from recent studies indicates that all these biomarkers with minimal variation for different p-tau epitopes are more sensitive and specifically related to AD. In fact, CSF p-tau231 distinguished between DAT patients and subjects with other neurological disorders with a sensitivity of 85% and a specificity of 97% [85]. In DAT versus fronto-temporal dementia, p-tau231 raised sensitivity levels compared to t-tau from 57.7% to 90.2% at a specificity level of 92.3% for both markers. Similar discriminative values have been obtained by using either p-tau199 or p-tau181. According to a recent study comparing DAT patients to controls including patients with vascular dementia and neurological patients, by using p-tau396/404, which is highly sensitive to hyperphosphorylated tau, elevated p-tau was reported to be significantly associated with DAT, and the ratio of p-tau to total tau confirmed the clinical diagnosis in 96% of the cases, thus indicating the usefulness of this marker in the differential diagnosis [85]. Few studies have investigated p-tau and a significant correlation between baseline p-tau levels and subsequent cognitive decline in MCI subjects [75].

5.3. Combination of CSF $A\tilde{\beta}42$ and CSF Tau

Because diagnostic accuracy of CSF tau and $A\tilde{\beta}42$ alone is not clear-cut in the differentiation of AD from other clinically relevant dementing disorders, a combination of both markers has been proposed to increase accuracy power. The potential of the combination of these CSF biomarkers has been evaluated in several studies comparing DAT with normal controls and other demented groups such as fronto-temporal dementia, Lewy body dementia, and vascular dementia [74, 83, 87]. These studies have reported slightly higher sensitivity and specificity values (85%–89% and 80%–90%, respectively) for the combination than for t-tau and $A\tilde{\beta}42$ alone in DAT patients versus normal controls. However, other studies, contrasting DAT with other dementia, showed that the combined assays yielded rather low discriminative power, indicating that the combination may not sufficiently improve differential diagnosis between DAT and other clinically relevant dementias. On the contrary, a recent study, comparing patients with early-onset AD, normal controls, and patients with fronto-temporal dementia, by measuring CSF $A\tilde{\beta}42$, total tau, and p-tau181, showed that the combination of $A\tilde{\beta}42$ and p-tau181 allowed DAT patients to be distinguished from fronto-temporal dementia patients with a sensitivity of 72% and a specificity of 93%, thus supporting the claim that this specific combination might eventually be helpful in clinical practice [88, 89].

6. Peripheral Biomarkers

6.1. $A\beta$ in Plasma

In the last decade, several attempts have been made to identify peripheral markers by using plasma, serum or circulating cells. In particular, numerous investigators have tried to assess plasma $A\beta$-40 and $A\tilde{\beta}42$ levels, following findings on $A\tilde{\beta}42$ levels in CSF. Up to now, several cross-sectional studies and two longitudinal studies investigated plasma $A\beta$ measures in DAT patients and controls. Most studies have shown that plasma $A\beta$-40 and $A\tilde{\beta}42$ levels are not different in AD and control groups, thus minimizing their diagnostic usefulness [90, 91]. However, recent longitudinal studies showed that high plasma A$\tilde{\beta}$42 levels were a risk factor for developing DAT. In fact, in a study on nondemented elderly, those who developed DAT during a 3.6-year follow-up had significantly higher baseline levels [94]. Similarly, patients with DAT at baseline or individuals who developed DAT within 5 years after plasma collection were found to have higher levels of plasma A$\tilde{\beta}$42 levels than individuals who did not became demented, whereas a significant decline over 3 years of follow-up was observed in DAT patients [95]. In sum, findings on plasma A$\tilde{\beta}$42 levels indicate that this biomarker is not sensitive and specific for the diagnosis of early AD and DAT, though it might be used in selected cases for predicting AD risk and tracking progression.

6.2. APP Isoforms in Platelets

Among the different peripheral cells expressing APP, platelets are particularly interesting because they show concentrations of its isoforms equivalent to those found in the brain [96]. Some differences between these two cellular populations are nevertheless present both at mRNA and at protein levels: the isoform 695, lacking the Kuntiz Protease Inhibitor (KPI) domain, is by far the most abundant in neuronal tissue, whereas its expression is nearly undetectable in platelets in whom the major isoform is APP 770 [97]. After the platelets are activated, soluble forms of cleaved APP are released, analogous to processing in neurons.

Several hypothesis have been put forward on the possible physiological role of APP in blood: It is now known that APPs containing the KPI domain (sAPP-KPI+) are highly homologous to protease Nexin II, and the domain inhibits the activity of the blood coagulation factors IXa, Xa, and XIa [96, 98]. It also has been reported that APPs inhibit the activation of human Hageman factor (factor XII) and prolong the thromboplastin time. The inhibition of factor XII is independent from the presence of the KPI domain, indicating that other regions of the APP structure are involved.

The appropriateness of using platelets, as a cell mirroring some neurochemical processes, finds its rationale in the numerous similar features of platelets and neurones: Platelets store and release neurotransmitters and express appropriate neurotransmitter transporters and some neurone-related proteins such as NMDA receptors. Along these lines, different authors reported abnormalities in platelets' physiology and function in AD. At first, Zubenko and colleagues described an alteration in platelet membrane fluidity in AD patients [99]. Other studies, performed on platelets obtained from patients with moderate to severe AD, reported cytoskeletal abnormalities, cytochrome oxidase deficiency, abnormal cytoplasmic calcium fluxes, abnormal glutamate transporter activity, decreased phospholipase A, decreased phospholipase C activity, and increased cytosolic protein kinase C levels [100–104]. Moreover, a number of laboratories independently described alterations in APP metabolism/concentration in platelets of DAT patients when compared with control subjects matched for demographic characteristics [105–107]. In particular, two research groups have reported that AD is associated with a decrease in the amount of the higher (130-kDa) band compared with the lower (110-kDa) band [108, 109]. Reduction in the APP isoform ratio was found to be correlated with disease severity and progression [110, 111]. Sensitivity and specificity for AD diagnosis was in the 80%–95% range, based on post hoc cutoff scores [109]. A significant reduction was also observed in a sample fulfilling criteria for MCI, particularly in those MCI subjects who progressed to DAT 2 years later. In fact, sensitivity and specificity for prediction of conversion to DAT were 83% and 71%, respectively [112, 113]. In addition, there is evidence that alteration of the APP isoform ratio in platelets is accompanied by a significant modification of the activity of alpha (decreased) and beta-secretase (increased), since the early stages [114, 115], and is positively influenced by such drugs as statins and cholinesterase inhibitors [116, 117], thus supporting the view that platelets, although unlikely to be the source or to contribute to cerebral amyloid deposition, provide an easily accessible source of human material with which to study APP biochemistry and metabolism, both in physiological and pathological conditions.

7. Future Approaches in Preclinical Diagnosis of AD

Clinical diagnosis of AD is not difficult once the disease is established, but early diagnosis before dementia becomes overt is far from perfect, although it is likely that many therapies will be most effective in the early stages of AD. Thus, there is a great need for techniques to detect AD in the preclinical stages.

In the biological markers field, there may be particular merit to the use of approaches, such as proteomics, that simultaneously assay multiple biologic markers and their interactions [118, 119]. Some recent findings seem to indicate that this approach holds the potential to be a promising strategy [120]. A study using a combination multidimensional liquid chromatography and gel electrophoresis coupled to mass spectroscopy showed that this approach works well in identifying potential biomarkers in the realm of AD. In particular, different levels of several serum proteins were identified including alpha 1 acid glycoprotein, apolipoprotein E, complement C 3, transthyretin, factor H, haptoglobin alpha 2 chain, hemoglobin alpha chain, histidin-rich glycoprotein, vitronectin precursor, alpha-2 macroglobulin, and complement C4. Further, very recent data obtained by using surface-enhanced laser desorption ionization–time-of flight (TOF)–mass spectroscopy showed that, although no single marker was able to completely separate DAT cases from normal individuals, a five-peak pattern of markers did correctly classify 100% of DAT cases and 95% of controls.

These preliminary findings strongly support the view of using multiple markers to distinguish AD from normal and speak to the potential use of proteomics in developing tests for AD.

8. Conclusive Remarks

The last few years have witnessed impressive advances in many areas of basic and clinical neurosciences, but few fields experienced greater progress than those dealing with AD. Mutations in the APP, presenilin 1, and presenilin 2 genes that result in autosomal dominant familial AD have been identified. These discoveries made possible the development of transgenic animals, which provide useful available experimental models to study AD neuropathology and therapeutics. Furthermore, much has been learned about oxidative stress, inflammatory reactions, and cholesterol metabolism in brain tissue during the course of AD, and both epidemiologic and animal model–based experimental observations indicate that antioxidant, lipid-lowering, antiinflammatory or immunological strategies may be effective in preventing or slowing the onset and progression of this devastating disease. Although it remains to be understood whether current trials will result in useful treatments, there is an urgent need for better tools than those currently available to use these therapies with optimal effectiveness.

A wide range of biological markers with possible relevance to AD has been developed in the last decades, though in most cases results have been disappointing and still inconclusive. However, there is a vast literature on selected biomarkers more closely related to the main pathogenetic mechanisms of AD that indicate that either CSF or peripheral markers such as CSF Aβ42, CSF protein tau, CSF phospho-tau, and platelet APP isoform ratio might be of great aid in clinical practice, though it has to be acknowledged that no single laboratory diagnostic test yet identified permits accurate and reliable ante mortem diagnosis comparable to autopsy studies, which remain the gold standard for confirming the diagnosis. Given the multiplicity of pathophysiological processes implicated in AD and their complex overlapping with other neurodegenerative diseases, it might be well true that a single neurochemical marker unique to AD would never be identified, and that acceptable diagnostic accuracy would be reached only through the combination of different biomarkers tapping different pathogenetic steps. Future studies must confirm whether the challenge of early diagnosis of AD is not a chimera, but the promising perspective of the development of predictive markers of AD, in addition to emerging therapeutic strategies, seems more concrete than ever.

References

[1] von Strauss E, Viitanen M, De Ronchi D, Winblad B, Fratiglioni L. Aging and the occurrence of dementia: Findings from a population-based cohort with a large sample of nonagenarians. Arch Neurol 1999; 56(5):587–592.

[2] Di Carlo A, Baldereschi M, Amaducci L, et al. Incidence of dementia, Alzheimer's disease, and vascular dementia in Italy. The ILSA Study. J Am Geriatr Soc 2002; 50(1):41–48.

[3] Ghiso J, Frangione B. Amyloidosis and Alzheimer's disease. Adv Drug Deliv Rev 2002; 54(12):1539–1551.

[4] Petersen RC, Smith GE, Ivnik RJ, et al. Apolipoprotein E status as a predictor of the development of Alzheimer's disease in memory-impaired individuals. JAMA 1995; 273(16):1274–1278.

[5] Ritchie K, Lovestone S. The dementias. Lancet 2002; 360(9347):1759–1766.

[6] Cummings JL, Cole G. Alzheimer disease. JAMA 2002; 287(18):2335–2338.

[7] DeKosky ST, Marek K. Looking backward to move forward: Early detection of neurodegenerative disorders. Science 2003; 302(5646):830–834.

[8] Frisoni GB, Padovani A, Wahlund LO. The predementia diagnosis of Alzheimer disease. Alzheimer Dis Assoc Disord 2004; 18(2):51–53.

[9] Nestor PJ, Scheltens P, Hodges JR. Advances in the early detection of Alzheimer's disease. Nat Med 2004; 10(Suppl.):S34–S41.

[10] Petersen RC, Smith GE, Waring SC, Ivnik RJ, Tangalos EG, Kokmen E. Mild cognitive impairment: Clinical characterization and outcome. Arch Neurol 1999; 56(3):303–308.

[11] Schonheit B, Zarski R, Ohm TG. Spatial and temporal relationships between plaques and tangles in Alzheimer-pathology. Neurobiol Aging 2004; 25(6):697–711.

[12] Braak H, Braak E. Demonstration of amyloid deposits and neurofibrillary changes in whole brain sections. Brain Pathol 1991; 1(3):213–216.

[13] Galasko D. Biological markers and the treatment of Alzheimer's disease. J Mol Neurosci 2001; 17(2):119–125.

[14] Cummings BJ, Pike CJ, Shankle R, Cotman CW. Beta-amyloid deposition and other measures of neuropathology predict cognitive status in Alzheimer's disease. Neurobiol Aging 1996; 17(6):921–933.

[15] Alonso AD, Zaidi T, Novak M, Barra HS, Grundke-Iqbal I, Iqbal K. Interaction of tau isoforms with Alzheimer's disease abnormally hyperphosphorylated tau and *in vitro* phosphorylation into the disease-like protein. J Biol Chem 2001; 276(41):37967–37973.

[16] Gray EG, Paula-Barbosa M, Roher A. Alzheimer's disease: Paired helical filaments and cytomembranes. Neuropathol Appl Neurobiol 1987; 13(2):91–110.

[17] Masters CL, Simms G, Weinman NA, Multhaup G, McDonald BL, Beyreuther K. Amyloid plaque core protein in Alzheimer disease and Down syndrome. Proc Natl Acad Sci USA 1985; 82(12):4245–4249.

[18] Sommer B. Alzheimer's disease and the amyloid cascade hypothesis: Ten years on. Curr Opin Pharmacol 2002; 2(1):87–92.

[19] Selkoe DJ. The cell biology of beta-amyloid precursor protein and presenilin in Alzheimer's disease. Trends Cell Biol 1998; 8(11):447–453.
[20] Hardy J, Selkoe DJ. The amyloid hypothesis of Alzheimer's disease: Progress and problems on the road to therapeutics. Science 2002; 297(5580):353–356.
[21] Delacourte A, Sergeant N, Wattez A, et al. Tau aggregation in the hippocampal formation: An ageing or a pathological process? Exp Gerontol 2002; 37(10–11):1291–1296.
[22] Goate A, Chartier-Harlin MC, Mullan M, et al. Segregation of a missense mutation in the amyloid precursor protein gene with familial Alzheimer's disease. Nature 1991; 349(6311):704–706.
[23] Schellenberg GD, Payami H, Wijsman EM, et al. Chromosome 14 and late-onset familial Alzheimer disease (FAD). Am J Hum Genet 1993; 53(3):619–628.
[24] Sherrington R, Rogaev EI, Liang Y, et al. Cloning of a gene bearing missense mutations in early-onset familial Alzheimer's disease. Nature 1995; 375(6534):754–760.
[25] Levy-Lahad E, Wasco W, Poorkaj P, et al. Candidate gene for the chromosome 1 familial Alzheimer's disease locus. Science 1995; 269(5226):973–977.
[26] Renbaum P, Levy-Lahad E. Monogenic determinants of familial Alzheimer's disease: Presenilin-2 mutations. Cell Mol Life Sci 1998; 54(9):910–919.
[27] Kehoe P, Wavrant-De Vrieze F, Crook R, et al. A full genome scan for late onset Alzheimer's disease. Hum Mol Genet 1999; 8(2):237–245.
[28] Saunders AM, Schmader K, Breitner JC, et al. Apolipoprotein E epsilon 4 allele distributions in late-onset Alzheimer's disease and in other amyloid-forming diseases. Lancet 1993; 342(8873):710–711.
[29] Nalbantoglu J, Gilfix BM, Bertrand P, et al. Predictive value of apolipoprotein E genotyping in Alzheimer's disease: Results of an autopsy series and an analysis of several combined studies. Ann Neurol 1994; 36(6):889–895.
[30] Corder EH, Lannfelt L, Bogdanovic N, Fratiglioni L, Mori H. The role of APOE polymorphisms in late-onset dementias. Cell Mol Life Sci 1998; 54(9):928–934.
[31] Holtzman DM, Fagan AM, Mackey B, et al. Apolipoprotein E facilitates neuritic and cerebrovascular plaque formation in an Alzheimer's disease model. Ann Neurol 2000; 47(6):739–747.
[32] Strittmatter WJ, Weisgraber KH, Huang DY, et al. Binding of human apolipoprotein E to synthetic amyloid beta peptide: Isoform-specific effects and implications for late-onset Alzheimer disease. Proc Natl Acad Sci USA 1993; 90 (17):8098–8102.
[33] Locke PA, Conneally PM, Tanzi RE, Gusella JF, Haines JL. Apolipoprotein E4 allele and Alzheimer disease: Examination of allelic association and effect on age at onset in both early- and late-onset cases. Genet Epidemiol 1995; 12(1):83–92.
[34] Sorbi S, Nacmias B, Forleo P, et al. ApoE allele frequencies in Italian sporadic and familial Alzheimer's disease. Neurosci Lett 1994; 177(1–2):100–102.
[35] Borroni B, Archetti S, Agosti C, et al. Intronic CYP46 polymorphism along with ApoE genotype in sporadic Alzheimer Disease: From risk factors to disease modulators. Neurobiol Aging 2004; 25(6):747–751.
[36] Sparks DL. Coronary artery disease, hypertension, ApoE, and cholesterol: A link to Alzheimer's disease? Ann N Acad Sci 1997; 826:128–146.
[37] Breteler MM. Vascular risk factors for Alzheimer's disease: An epidemiologic perspective. Neurobiol Aging 2000; 21(2):153–160.
[38] Simons M, Keller P, De Strooper B, Beyreuther K, Dotti CG, Simons K. Cholesterol depletion inhibits the generation of beta-amyloid in hippocampal neurons. Proc Natl Acad Sci USA 1998; 95(11):6460–6464.

[39] Refolo LM, Malester B, LaFrancois J, et al. Hypercholesterolemia accelerates the Alzheimer's amyloid pathology in a transgenic mouse model. Neurobiol Dis 2000; 7(4):321–331.

[40] Wolozin B, Kellman W, Ruosseau P, Celesia GG, Siegel G. Decreased prevalence of Alzheimer disease associated with 3-hydroxy-3-methyglutaryl coenzyme A reductase inhibitors. Arch Neurol 2000; 57(10):1439–1443.

[41] Kawas C, Gray S, Brookmeyer R, Fozard J, Zonderman A. Age-specific incidence rates of Alzheimer's disease: The Baltimore Longitudinal Study of Aging. Neurology 2000; 54(11):2072–2077.

[42] Perry RJ, Hodges JR. Differentiating frontal and temporal variant frontotemporal dementia from Alzheimer's disease. Neurology 2000; 54(12):2277–2784.

[43] Loo DT, Copani A, Pike CJ, Whittemore ER, Walencewicz AJ, Cotman CW. Apoptosis is induced by beta-amyloid in cultured central nervous system neurons. Proc Natl Acad Sci USA 1993; 90(17):7951–7955.

[44] Marx JL, Kumar SR, Thach AB, Kiat-Winarko T, Frambach DA. Detecting Alzheimer's disease. Science 1995; 267(5204):1577.

[45] Marx J. Neurobiology. New clue to the cause of Alzheimer's. Science 2001; 292(5521): 1468.

[46] Mattson MP, Pedersen WA, Duan W, Culmsee C, Camandola S. Cellular and molecular mechanisms underlying perturbed energy metabolism and neuronal degeneration in Alzheimer's and Parkinson's diseases. Ann N Acad Sci 1999; 893:154–175.

[47] Seshadri S, Beiser A, Selhub J, Jacques PF, Rosenberg IH, D'Agostino RB, Wilson PW, Wolf PA. Plasma homocysteine as a risk factor for dementia and Alzheimer's disease. N Engl J Med 2002; 346(7):476–483.

[48] Clarke R, Smith AD, Jobst KA, Refsum H, Sutton L, Ueland PM. Folate, vitamin B12, and serum total homocysteine levels in confirmed Alzheimer disease. Arch Neurol 1998; 55(11):1449–1455.

[49] Lipton SA, Kim WK, Choi YB, et al. Neurotoxicity associated with dual actions of homocysteine at the N-methyl-D-aspartate receptor. Proc Natl Acad Sci USA 1997; 94(11):5923–5928.

[50] Kruman II, Culmsee C, Chan SL, et al. Homocysteine elicits a DNA damage response in neurons that promotes apoptosis and hypersensitivity to excitotoxicity. J Neurosci 2000; 20(18):6920–6926.

[51] Callahan CM, Hall KS, Hui SL, Musick BS, Unverzagt FW, Hendrie HC. Relationship of age, education, and occupation with dementia among a community-based sample of African Americans. Arch Neurol 1996; 53(2):134–140.

[52] Frisoni GB, Padovani A, Wahlund LO. The diagnosis of Alzheimer disease before it is Alzheimer dementia. Arch Neurol 2003; 60(7):1023.

[53] Petersen RC, Doody R, Kurz A, et al. Current concepts in mild cognitive impairment. Arch Neurol 2001; 58(12):1985–1992.

[54] Luis CA, Loewenstein DA, Acevedo A, Barker WW, Duara R. Mild cognitive impairment: Directions for future research. Neurology 2003; 61(4):438–444.

[55] Morris JC, Storandt M, Miller JP, et al. Mild cognitive impairment represents early-stage Alzheimer disease. Arch Neurol 2001; 58(3):397–405.

[56] Mufson EJ, Chen EY, Cochran EJ, Beckett LA, Bennett DA, Kordower JH. Entorhinal cortex beta-amyloid load in individuals with mild cognitive impairment. Exp Neurol 1999; 158(2):469–490.

[57] Chetelat G, Baron JC. Early diagnosis of Alzheimer's disease: Contribution of structural neuroimaging. Neuroimage 2003; 18(2):525–541.

[58] Fisk JD, Merry HR, Rockwood K. Variations in case definition affect prevalence but not outcomes of mild cognitive impairment. Neurology 2003; 61(9):1179–1184.
[59] Larrieu S, Letenneur L, Orgogozo JM, et al. Incidence and outcome of mild cognitive impairment in a population-based prospective cohort. Neurology 2002; 59(10):1594–1599.
[60] Petersen RC. Mild cognitive impairment: Transition between aging and Alzheimer's disease. Neurologia 2000; 15(3):93–101.
[61] Growdon JH, Selkoe DJ, Roses A, et al. Consensus report of the Working Group on Biological Markers of Alzheimer's Disease. Neurobiol Aging 1998; 19:109–116.
[62] Bosman GJ, Bartholomeus IG, de Man AJ, van Kalmthout PJ, de Grip WJ. Erythrocyte membrane characteristics indicate abnormal cellular aging in patients with Alzheimer's disease. Neurobiol Aging 1991; 12(1):13–18.
[63] Pirttila T, Mattinen S, Frey H. The decrease of CD8-positive lymphocytes in Alzheimer's disease. J Neurol Sci 1992; 107(2):160–165.
[64] Linder J, Nolgard P, Nasman B, Back O, Uddhammar A, Olsson T. Decreased peripheral glucocorticoid sensitivity in Alzheimer's disease. Gerontology 1993; 39(4):200–206.
[65] Ghanbari K, Ghanbari HA. A sandwich enzyme immunoassay for measuring AD7C-NTP as an Alzheimer's disease marker: AD7C test. J Clin Lab Anal 1998; 12(4):223–226.
[66] de Berker D, Jones R, Mann J, Harwood G. Hair abnormalities in Alzheimer's disease. Lancet 1997; 350(9070):34.
[67] Bonafe JL, Cambon L, Ousset PJ, Pech JH, Bellet H, Vallat C. Abnormal hair samples from patients with Alzheimer disease. Arch Dermatol 1998; 134(10):1300.
[68] Soininen H, Syrjanen S, Heinonen O, et al. Amyloid beta-protein deposition in skin of patients with dementia. Lancet 1992; 339(8787):245.
[69] Heinonen O, Soininen H, Syrjanen S, et al. beta-Amyloid protein immunoreactivity in skin is not a reliable marker of Alzheimer's disease. An autopsy-controlled study. Arch Neurol 1994; 51(8):799–804.
[70] Olsson A, Hoglund K, Sjogren M, et al. Measurement of alpha- and beta-secretase cleaved amyloid precursor protein in cerebrospinal fluid from Alzheimer patients. Exp Neurol 2003; 183(1):74–80.
[71] Sjogren M, Andreasen N, Blennow K. Advances in the detection of Alzheimer's disease—use of cerebrospinal fluid biomarkers. Clin Chim Acta 2003; 332(1–2):1–10.
[72] Palmert MR, Usiak M, Mayeux R, Raskind M, Tourtellotte WW, Younkin SG. Soluble derivatives of the beta amyloid protein precursor in cerebrospinal fluid: Alterations in normal aging and in Alzheimer's disease. Neurology 1990; 40(7):1028–1034.
[73] Van Nostrand WE, Wagner SL, et al. Decreased levels of soluble amyloid beta protein precursor in cerebrospinal fluid of live Alzheimer disease patients. Proc Natl Acad Sci USA 1992; 89(7):2551–2555.
[74] Andreasen N, Minthon L, Davidsson P, et al. Evaluation of CSF-tau and CSF Abeta42 as diagnostic markers for Alzheimer disease in clinical practice. Arch Neurol 2001; 58(3):373–379.
[75] Blennow K, Hampel H. CSF markers for incipient Alzheimer's disease. Lance Neurol 2003; 2(10):605–613.
[76] Riemenschneider M, Schmolke M, Lautenschlager N, et al. Cerebrospinal beta-amyloid (1–42) in early Alzheimer's disease: Association with apolipoprotein E genotype and cognitive decline. Neurosci Lett 2000; 284(1–2):85–88.
[77] Andreasen N, Hesse C, Davidsson P, et al. Cerebrospinal fluid beta-amyloid(1–42) in Alzheimer disease: Differences between early- and late-onset Alzheimer disease and stability during the course of disease. Arch Neurol 1999; 56(6):673–680.

[78] Maruyama M, Arai H, Sugita M, et al. Cerebrospinal fluid amyloid beta(1–42) levels in the mild cognitive impairment stage of Alzheimer's disease. Exp Neurol 2001; 172(2): 433–436.
[79] Hampel H, Teipel SJ, Fuchsberger T, et al. Value of CSF beta-amyloid1-42 and tau as predictors of Alzheimer's disease in patients with mild cognitive impairment. Mol Psychiatry 2004; 9(7):705–710.
[80] Skoog I, Davidsson P, Aevarsson O, Vanderstichele H, Vanmechelen E, Blennow K. Cerebrospinal fluid beta-amyloid 42 is reduced before the onset of sporadic dementia: A population-based study in 85-year-olds. Dement Geriatr Cogn Disord 2003; 15(3): 169–176.
[81] Sjogren M, Davidsson P, Tullberg M, et al. Both total and phosphorylated tau are increased in Alzheimer's disease. J Neurol Neurosurg Psychiatry 2001; 70(5):624–630.
[82] Sunderland T, Linker G, Mirza N, et al. Decreased beta-amyloid1-42 and increased tau levels in cerebrospinal fluid of patients with Alzheimer disease. JAMA 2003; 289(16): 2094–2103.
[83] Hulstaert F, Blennow K, Ivanoiu A, et al. Improved discrimination of AD patients using beta-amyloid(1–42) and tau levels in CSF. Neurology 1999; 52(8):1555–1562.
[84] Itoh N, Arai H, Urakami K, et al. Large-scale, multicenter study of cerebrospinal fluid tau protein phosphorylated at serine 199 for the antemortem diagnosis of Alzheimer's disease. Ann Neurol 2001; 50(2):150–156.
[85] Buerger K, Teipel SJ, Zinkowski R, et al. CSF tau protein phosphorylated at threonine 231 correlates with cognitive decline in MCI subjects. Neurology 2002; 59(4): 627–629.
[86] Hampel H, Buerger K, Zinkowski R, et al. Measurement of phosphorylated tau epitopes in the differential diagnosis of Alzheimer disease: A comparative cerebrospinal fluid study. Arch Gen Psychiatry 2004; 61(1):95–102.
[87] Galasko D, Clark C, Chang L, et al. Assessment of CSF levels of tau protein in mildly demented patients with Alzheimer's disease. Neurology 1997; 48(3):632–635.
[88] Schoonenboom NS, Pijnenburg YA, Mulder C, et al. Amyloid beta(1-42) and phosphorylated tau in CSF as markers for early-onset Alzheimer disease. Neurology 2004; 62(9):1580–1584.
[89] Maddalena A, Papassotiropoulos A, Muller-Tillmanns B, Jung HH, Hegi T, Nitsch RM, Hock C. Biochemical diagnosis of Alzheimer disease by measuring the cerebrospinal fluid ratio of phosphorylated tau protein to beta-amyloid peptide42. Arch Neurol 2003; 60(9):1202–1206.
[90] Mehta PD, Pirttila T, Mehta SP, Sersen EA, Aisen PS, Wisniewski HM. Plasma and cerebrospinal fluid levels of amyloid beta proteins 1–40 and 1–42 in Alzheimer disease. Arch Neurol 2000; 57(1):100–105.
[91] Tamaoka A, Fukushima T, Sawamura N, et al. Amyloid beta protein in plasma from patients with sporadic Alzheimer's disease. J Neurol Sci 1996; 141(1–2):65–68.
[92] Vanderstichele H, Van Kerschaver E, Hesse C, et al. Standardization of measurement of beta-amyloid(1–42) in cerebrospinal fluid and plasma. Amyloid 2000; 7(4):245–258.
[93] Fukumoto H, Tennis M, Locascio JJ, Hyman BT, Growdon JH, Irizarry MC. Age but not diagnosis is the main predictor of plasma amyloid beta-protein levels. Arch Neurol 2003; 60(7):958–964.
[94] Mayeux R, Tang MX, Jacobs DM, et al. Plasma amyloid beta-peptide 1–42 and incipient Alzheimer's disease. Ann Neurol 1999; 46(3):412–416.
[95] Mayeux R, Honig LS, Tang MX, et al. Plasma A[beta]40 and A[beta]42 and Alzheimer's disease: Relation to age, mortality, and risk. Neurology 2003; 61(9):1185–1190.

[96] Evin G, Zhu A, Holsinger RM, Masters CL, Li QX. Proteolytic processing of the Alzheimer's disease amyloid precursor protein in brain and platelets. J Neurosci Res 2003; 74(3):386–392.

[97] Tanzi RE, Gusella JF, Watkins PC, et al. Amyloid beta protein gene: cDNA, mRNA distribution, and genetic linkage near the Alzheimer locus. Science 1987; 235(4791):880–884.

[98] Smith RP, Higuchi DA, Broze GJ, Jr. Platelet coagulation factor XIa-inhibitor, a form of Alzheimer amyloid precursor protein. Science 1990; 248(4959):1126–1128.

[99] Zubenko GS, Wusylko M, Cohen BM, Boller F, Teply I. Family study of platelet membrane fluidity in Alzheimer's disease. Science 1987; 238(4826):539–542.

[100] Davies TA, Drotts D, Weil GJ, Simons ER. Flow cytometric measurements of cytoplasmic calcium changes in human platelets. Cytometry 1988; 9(2):138–142.

[101] Blass JP, Gibson GE. Nonneural markers in Alzheimer disease. Alzheime Dis Assoc Disord 1992; 6(4):205–224.

[102] Matsushima H, Shimohama S, Fujimoto S, Takenawa T, Kimura J. Changes in platelet phospholipase C protein level and activity in Alzheimer's disease. Neurobiol Aging 1995; 16(6):895–900.

[103] Ferrarese C, Begni B, Canevari C, et al. Glutamate uptake is decreased in platelets from Alzheimer's disease patients. Ann Neurol 2000; 47(5):641–643.

[104] Zoia C, Cogliati T, Tagliabue E, et al. Glutamate transporters in platelets: EAAT1 decrease in aging and in Alzheimer's disease. Neurobiol Aging 2004; 25(2):149–157.

[105] Di Luca M, Pastorino L, Cattabeni F, et al. Abnormal pattern of platelet APP isoforms in Alzheimer disease and Down syndrome. Arch Neurol 1996; 53(11):1162–1166.

[106] Davies TA, Long HJ, Sgro K, et al. Activated Alzheimer disease platelets retain more beta amyloid precursor protein. Neurobiol Aging 1997; 18(2):147–153.

[107] Rosenberg RN, Baskin F, Fosmire JA, et al. Altered amyloid protein processing in platelets of patients with Alzheimer disease. Arch Neurol 1997; 54(2):139–144.

[108] Baskin F, Rosenberg RN, Iyer L, Hynan L, Cullum CM. Platelet APP isoform ratios correlate with declining cognition in AD. Neurology 2000; 54(10):1907–1909.

[109] Padovani A, Pastorino L, Borroni B, et al. Amyloid precursor protein in platelets: A peripheral marker for the diagnosis of sporadic AD. Neurology 2001; 57 (12):2243–2248.

[110] Borroni B, Colciaghi F, Corsini P, et al. Early stages of probable Alzheimer disease are associated with changes in platelet amyloid precursor protein forms. Neurol Sci 2002; 23(5):207–210.

[111] Padovani A, Borroni B, Colciaghi F, et al. Platelet amyloid precursor protein forms in AD: A peripheral diagnostic tool and a pharmacological target. Mech Ageing Dev 2001; 122(16):1997–2004.

[112] Padovani A, Borroni B, Colciaghi F, et al. Abnormalities in the pattern of platelet amyloid precursor protein forms in patients with mild cognitive impairment and Alzheimer disease. Arch Neurol 2002; 59(1):71–75.

[113] Borroni B, Colciaghi F, Caltagirone C, et al. Platelet amyloid precursor protein abnormalities in mild cognitive impairment predict conversion to dementia of Alzheimer type: A 2-year follow-up study. Arch Neurol 2003; 60(12):1740–1744.

[114] Colciaghi F, Borroni B, Pastorino L, et al. [alpha]-Secretase ADAM10 as well as [alpha]APPs is reduced in platelets and CSF of Alzheimer disease patients. Mol Med 2002; 8(2):67–74.

[115] Colciaghi F, Marcello E, Borroni B, et al. Platelet APP, ADAM 10 and BACE alterations in the early stages of Alzheimer disease. Neurology 2004; 62(3):498–501.

[116] Baskin F, Rosenberg RN, Fang X, et al. Correlation of statin-increased platelet APP ratios and reduced blood lipids in AD patients. Neurology 2003; 60(12):2006–2007.

[117] Borroni B, Colciaghi F, Pastorino L, et al. Amyloid precursor protein in platelets of patients with Alzheimer disease: Effect of acetylcholinesterase inhibitor treatment. Arch Neurol 2001; 58(3):442–446.

[118] Butterfield DA. Proteomics: A new approach to investigate oxidative stress in Alzheimer's disease brain. Brain Res 2004; 1000(1–2):1–7.

[119] Puchades M, Hansson SF, Nilsson CL, Andreasen N, Blennow K, Davidsson P. Proteomic studies of potential cerebrospinal fluid protein markers for Alzheimer's disease. Brain Res Mol Brain Res 2003; 118(1–2):140–146.

[120] Choe LH, Dutt MJ, Relkin N, Lee KH. Studies of potential cerebrospinal fluid molecular markers for Alzheimer's disease. Electrophoresis 2002; 23(14):2247–2251.

ENDOTHELIAL MICROPARTICLES (EMP) AS VASCULAR DISEASE MARKERS

Joaquin J. Jimenez, Wenche Jy, Lucia M. Mauro, Laurence L. Horstman, Carlos J. Bidot, and Yeon S. Ahn

Department of Medicine, Division of Hematology Oncology, Wallace H. Coulter Platelet Laboratory, University of Miami School of Medicine, Miami, Florida 33136

1. Introduction

Endothelial cells (ECs) have long been implicated in the pathogenesis and pathophysiology of a myriad of vascular, thrombotic, and inflammatory disorders [1–3]. The prevalence of endothelial perturbation in vascular disease has underscored the need for noninvasive sensitive and specific markers

0065-2423/05 $35.00
DOI: 10.1016/S0065-2423(04)39005-0

to monitor endothelial status [4]. The popularity of endothelial microparticles (EMPs) as markers of perturbed has increased because of its promising clinical applications in the area of noninvasive EC monitoring [5]. This trend can be attributed to a significant increase in the information available about EMPs' role in health and disease and their presence in different disorders [5]. Concomitant with this wealth of information, new challenges have risen to the fore. Perhaps the most crucial, and most decidedly relevant to the field of clinical laboratory testing, has been the issue of methodology [5, 6]. Indeed, to this day, there is no consensus about which is the best protocol to determine EMP, and a number of different flow cytometric and immunologic methods with several variations have been adopted [6].

This review introduces the methodologies currently employed in the area of clinical EMP research. Moreover, issues that may have an effect on the clinical application of EMP assays are also discussed.

2. Characterization of EMPs

2.1. Background

An overview of the main aspects of endothelial dysfunction as they relate to thrombogenesis and inflammation is important to provide a background of the factors affecting EMP release and their phenotype. A large number of reviews are available on this subject; however, this section describes only the main features of EC perturbation and dysfunction as may be relevant to EMP.

2.1.1. *Endothelial Dysfunction*

Under resting conditions, ECs provide a nonthrombogenic and nonadhesive interface and also regulate vascular permeability and tone by the release of nitric oxide [7–10]. ECs are widely heterogeneous and exhibit distinctive morphological and behavioral differences, varying according to their different vascular bed of origin [11, 12]; however, their response to injury or perturbation has general features that are common to all ECs. The hallmark of perturbed ECs is a phenotypic shift toward a procoagulant and proadhesive state [3, 7, 8, 13, 14]. EC perturbation is a broad term that may encompass many pathologies; however, two large categories pertain to this review: endothelial activation, characterized by active EC processes, and EC apoptosis.

2.1.1.1. *EC Activation.* EC activation encompasses the active response of ECs to proinflammatory cytokines such as tumor necrosis factor-α (TNF-α), interleukin 1β (IL-1β), or interferon gamma (IFN-γ) [15, 16]. These responses are characterized as profound morphological and phenotypic changes [15–17]. The transcription of the nuclear factor Nf-kB and the ensuing activation of prosurvival genes of the iap and xiap family ensures the survival of ECs [17–20]. Simultaneously, it also signals that the expression of proadhesive molecules such as vascular cell adhesion molecule (VCAM-1), intracellular adhesion molecule (ICAM-1), and E-selectin [14–21]. Leukocyte adhesion is facilitated by the aforementioned molecules, wheras IL-8, IL-6, and monocyte chemoattractant protein 1 further stimulate the inflammatory response [22, 23]. TNF-α also elicits the release of large multimers of von Willebrand factor (vWF) and P-selectin from weibel-palade bodies. These multimers have been shown to be more proadhesive to platelets when compared to their monomeric counterparts, and they facilitate platelet aggregation [24, 25]. In addition, activated ECs can also negatively regulate fibrinolysis by releasing plasminogen activator inhibitor 1 (PAI-1) and by a decrease in tissue plasminogen activator expression [3, 8, 15]. Activated ECs can also initiate coagulation mediated by the expression of tissue factor (TF) [26].

2.1.1.2. *EC Apoptosis.* Apoptotic ECs have also been demonstrated to exhibit similar procoagulant properties, as the loss of membrane symmetry and the release of procoagulant membrane blebs generates a large amount of anionic phospholipid (PL) substrates, which in turn support thrombin generation [27, 28]. In addition, EC detachment resulting from apoptosis exposes the extracellular matrix, which potently activates platelet aggregation [29]. Despite the shared features between activated and apoptotic ECs, the intracellular mechanisms underlying each condition are distinct [29, 30]. Examples of known inducers of EC apoptosis include chemotherapeutic agents, high glucose and advanced glycation end-products, hypoxia, hyperoxia, and growth factor deprivation [31, 32].

2.1.2. *General Properties of Microparticles*

2.1.2.1. *Mechanisms of MP Generation.* Microparticles (MPs) have common attributes, regardless of their cellular origin [33, 34]. MPs generated during cellular activation may arise from common intracellular mechanisms involving calcium-mediated membrane-cytoskeleton interactions [33]. In this regard, it has been confirmed that calpain, calcium ionophore A23187, C5b-9, or oxidants induce MP shedding in a Ca^{2+}-dependent manner [33, 35–37]. In other cases, such as in erythrocytes, changes in intracellular pH have also been demonstrated as resulting in MP production [38, 39]. However, intracellular mechanisms underlying the process of membrane shedding in apoptosis appear to differ from activation. In apoptosis, contraction of an actin-myosin network on the cellular membrane is known to be a universal aspect of MP shedding [40]. Activation of Rho-associated kinase is believed to mediate this process [41]. Caspase 3, a known effector caspase, has been demonstrated to cleave Rho-associated kinase I, resulting in membrane skeletal changes and subsequent release of MPs [42].

2.1.2.2. *General Functions of MPs.* A major function of MPs is their procoagulant activity, which can be attributed in part to their role in providing a suitable PL environment for thrombin generation [27, 43]. Similarly, MPs can also actively initiate thrombosis resulting from TF found on their surface, as previously reviewed [33, 34]. However, some MPs have been shown to serve anticoagulant functions by being carriers of protein S and thrombomodulin (TM) and thereby activating the protein-C-S-TM anticoagulant pathway [44, 45]. In this regard, MPs are also known to be carriers of biologically active factors and have been demonstrated to interact with other cells [34]. For example, platelet MPs (PMPs) and leukocyte MPs (LMPs) are known to be capable of activating ECs, and both PMPs and EMPs can interact with leukocytes to induce their activation [46–48].

2.2. EMP

2.2.1. *Antigenic Characterization of EMP*

In 1990, Hamilton *et al.* demonstrated that C5b9 induced vesiculation of opsonized human umbilical vein endothelial cells (HUVEC) [49]. This was interpreted as a defensive function because the complement was shed from the cell on the EMP, allowing the cell to recover without lasting injury. It has since become clear that the concept of membrane blebbing, which had been observed with platelets and leukocytes, could be extrapolated to endothelial cells. Casciola-Rosen *et al.* observed the vesiculation of HUVEC during apoptosis and demonstrated that those vesicles were procoagulant [50]. The concept of EC membrane vesiculation was recapitulated in 1999 by Combes *et al.*, with their seminal work on activated HUVEC [51]. In their study, the researchers observed that on stimulation with TNF-α, HUVEC released MPs, termed EMPs, into the culture supernatants. EMPs were found to be $<1\ \mu$m in size. On the basis of extensive work on the cellular expression of different markers on the surface of ECs, the researchers measured EMPs and HUVEC after exposure to TNF-α or IL-1β [52]. Their findings indicated that EMPs express markers present on ECs such as CD31, CD51, CD54, and CD62E. In addition, these results correlated with the cellular expression of those molecules on the parent ECs. Following the characterization of activation-derived EMPs by Combes *et al.*, the release of EMPs was also characterized on apoptotic ECs [53–55]. It was observed that CD54, CD62E, and CD106 are expressed in TNF-α-treated but not apoptotic EC. In contrast, markers such as CD31, CD105, and annexin V (AV) were more significantly elevated in apoptosis (see Fig. 1).

2.2.1.1. *Other Inducers of EMP Release.* Since the study by Combes *et al.*, a number of studies have observed that EMPs are shed on stimulation with different agonists. Brodsky *et al.* have shown that PAI-1 induced the release of EMPs [56]. Plasma from patients with thrombotic thrombocytopenic purpura (TTP), multiple sclerosis (MS), and autoimmune disorders has also been demonstrated to induce the release of EMPs [53, 57–59]. Other inducers of apoptosis such as stimulation with mitomycin C, camptothecin, or growth factor withdrawal have also resulted in EMP generation [53–55].

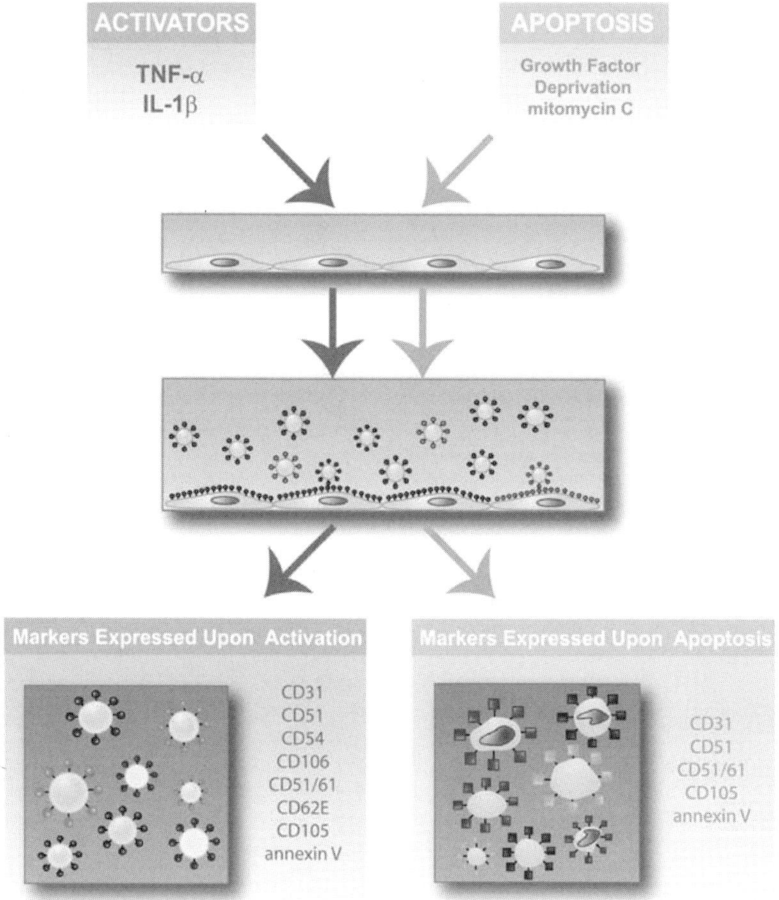

FIG. 1. Schematic representation of EMPs released from cultured endothelial cells (ECs) as measured by different fluorescent monoclonal antibodies. ECs cultured in the presence of tumor necrosis factor-α (TNF-α) or interleukin-1β (IL-1β) release EMPs expressing constitutive (CD31, CD51, and CD105) and inducible (CD54, CD62E, and CD106) markers (left). In contrast, EC deprived of growth factors or cultured in the presence of mitomycin C-release EMPs only express constitutive markers such as CD31 and CD105 (right). The number of EMP-expressing annexin V and constitutive markers is significantly larger in apoptosis than in activation. This pattern of EMP antigenic expression observed *in vitro*, if applicable to the clinical setting, may shed light on the pathophysiology of vascular and thrombotic disorders.

2.2.2. *Mechanism of EMP Release*

The exact mechanisms of EMP release remain to be completely elucidated. In this regard, Tramontano *et al.* have reported that EMPs arising from TNF-α-induced activation appear to be mediated through the Rho-kinase pathway [60]. The researchers also have found that the addition of fluvastatin resulted in a decrease in EMP counts. However, van Gorp *et al.* have demonstrated that peroxide-induced EMP release was mediated by glutathione oxidation [61, 62]. Another cellular mechanism observed in the setting of EMP release has been the clustering of different receptors before EMP shedding [63]. Whether the intracellular process of EMP release is distinct in activation or apoptosis is an important issue that remains to be clarified and could have important therapeutic implications.

2.2.3. *Functional Aspects of EMP*

2.2.3.1. *Procoagulant Activity.* EMPs, similar to other MPs, have been demonstrated to readily induce thrombogenesis [5, 33]. In this regard, EMPs have been shown to be procoagulant in a TF:FVII-dependent manner [51]. In addition, they have been shown to elicit thrombin generation and to provide platelet factor 3 activity, as reviewed [5]. EMPs have also been shown to carry other important proteins such as PA-1 and vWF on their surface [56, 57].

2.2.3.2. *Inhibition of Nitric Oxide–Derived Vasorelaxation.* In addition to their procoagulant activity, EMPs have been reported to have an inhibitory effect in vasorelaxation. Boulanger *et al.* reported that EMPs obtained from patients with ischemia decreased nitric oxide (NO) vasorelaxation in rat aortic endothelium [64]. Subsequently, Brodsky *et al.* reported that rat renal microvascular endothelial cells (MVEC) were deprived of growth factors, and EMPs were isolated by centrifugation [65]. Functional studies were performed in rat aortic rings using different concentrations of EMP. Endothelium-dependent vasorelaxation in response to acetylcholine and NO synthesis decreased in a dose-dependent manner on the addition of EMPs [65]. Moreover, EMP-mediated NO decrease was a result of increased oxidative stress, as observed in rat renal MVEC cultures. In this regard, EMPs have also been demonstrated to carry superoxide, additionally increasing the local production of superoxide in rat aortic rings. This effect could be inhibited by superoxide scavenging, also resulting in an increase in NO [65].

2.2.3.3. *Interaction of EMPs with Monocytes.* As observed with other MPs, EMPs have the ability to interact with other cell types. Sabatier *et al.* have demonstrated that EMPs bind to both THP-1 and monocytic cells and elicit their procoagulant activity [66]. EMP binding occurred in a time- and dose-dependent manner, with a maximal effect observed at a ratio of 50:1 EMPs. This effect appeared to be dependent on ICAM-1 and beta2 integrins, as blocking these receptors resulted in a decrease of the procoagulant effect.

We have observed similar findings using U937, a monocytic cell line, and have detected EMP–monocyte complexes in patients with MS [67].

2.2.3.4. *EMPs as Carriers of Molecules.* In addition to measurable antigens, EMPs as other MPs, are also carriers of molecules. Conformationally changed prion proteins, believed to aid in the propagation of spongiform encephalopathies, have been detected on EMPs released by infected ECs [68]. This finding indicates that EMPs may be active players in the infectious process of spongiform encephalopathies.

3. Methodology to Detect EMP

In the following section, the current clinical methodologies to detect EMPs and MPs as they pertain to EMPs are discussed. In addition, a summary of the most widely used clinical methods can be found in Table 1. The protocols described are circumscribed to clinical protocols, as they are representative of *in vitro* assays as well.

3.1. Other Methods for Detecting EC Injury Used in Clinical Studies

Clinical endothelial research has been hampered by the difficulty of accessing and sampling ECs in a noninvasive manner. The first approach to this problem consisted of the measurement of a variety of soluble molecules known to reflect perturbations of ECs by enzyme-linked immunoassay (ELISA) [69, 70]. These have prominently included VCAM-1, ICAM-1, E-selectin, TM, and vWF, as previously reviewed (71–79). Arguably the greatest disadvantage of these methods is lack of specificity [69–71]. ICAM-1, for example, is expressed by many lineages of activated cells [70]. Similarly, vWF is not specific for endothelial cells, as it is also released by megakaryocytes and platelets [71]. In contrast, although ELISA provides quantitative measurements of certain proteins, this methodology cannot differentiate soluble from MP-bound proteins [5].

A different approach to measuring EC damage has been circulating ECs (CECs). CECs have been described in patients with cardiovascular disease; namely, in myocardial infarction [72], in unstable angina [73, 74], and postangioplasty [75]. In addition, CECs have also been found to be elevated in Rickettsia Conorii infection [76] and hematological disorders such as sickle cell disease [77, 78], TTP [79], and systemic lupus erythematosus (SLE) [80].

The major disadvantage of this procedure is that CECs are present in quite small numbers in normal circulation. In addition, the very few CECs released to circulation may not be representative of the overall endothelial condition.

TABLE 1
Representative Protocols of Analysis of Endothelial Microparticles Used in Clinical Studies

	Markers	Plasma preparation	Frozen	MP concentration	Reference
Flow cytometry	CD31+/CD42b−CD51+	160 × g, 10 min; 1,500 × g, 6 min	No		[53]
					[81]
		160 × g, 10 min; 500 × g, 6 min			[58]
	CD31+/CD42b−	160 × g, 10 min; 1,500 × g, 6 min			[82]
	CD31+/CD42b−CD62E+ CD31+/CD42b−CD62E+	1,500 × g, 10 min			[83]
		160 × g, 10 min; 1,500 × g, 6 min			[84]
	CD31+/CD42b−CD62E+ /(vWf+)	160 × g, 10 min; 1,500 × g, 6 min			[57]
	CD62E+CD105+	5,000 × g, 5 min (×2)	−70 °C	17,000 g, 60 minutes	[85]

	CD31+/CD51+	1,500 × g, 15 min (×2); 13,000 × g, 1 min	No		[51]
	CD51+	1,500 × g, 5 min	−20 °C; −80 °C		[86]
	CD51+	1,500 × g, 5 min; 13,000 × g, 2 min.	No		[59]
					[87]
					[88]
	CD105+/CD41−/CD45−; CD105+/AV+	2,700 × g, 15 min; 2,700 × g, 5 min	LN_2	19,800 g, 10 minutes	[89]
	CD144; CD144+/TF+	15,000 × g, 10 min; 13,000 × g, 10 min	−80 °C	100,000 g, 60 minutes (×2); 31,000 g, 60 minutes (wash after ab. incubation)	[90]
	CD62E+ CD144+	1,500 × g, 20 min; 15,570 × g, 30 min	LN_2; −80 °C		[91]
Annexin V capture /prothrombinase assay	CD31+	1,500 × g, 15 min; 13,000 × g, 6 min	−80 °C	Immobilization onto annexin V	[96]
	CD31+ CD106+				[95]

3.2. EMP Assays Currently Employed

As of yet, there is no consensus as to which marker is more sensitive or specific to detect EMP. In addition, there are two main quantitative methodologies for detecting EMPs currently in use: flow cytometry (FC) and a combination of AV capture (AVC) and prothrombinase assay (PTA) with ELISA [5, 6]. Table 1 summarizes the main features of the clinical assays most commonly used so far.

3.2.1. *Flow Cytometry*

FC is the most traditional and widely used method to detect MPs. Although FC protocols vary with each flow cytometer and laboratory, all FC protocols have a set of parameters that enable the selection of EMP. These criteria generally rely on a combination of factors, such as antigenic expression as determined by a fluorescent threshold and size, relative to beads. Specifically, as EMPs have been demonstrated to be <1–1.5 μm in size, standard beads are used to establish a size limit, thereby eliminating larger, non-EMP material. Some laboratories mix MP samples with standard bead preparations, supplying an additional control parameter for each run in the event of a technical problem. Assay validation must be performed using EMPs generated *in vitro* to determine the fluorescent expression of the markers employed [51, 53, 57–59, 81–92]. When assaying clinical samples, monoclonal antibodies against platelets, leukocytes, or erythrocytes must be used to select nonendothelial MPs, which may also coexpress the marker tests. To this effect, two-color flow cytometry can be employed to differentiate EMPs from other EC antigens expressing MPs.

3.2.1.1. *Disadvantages of FC.* In this regard, a major limitation is the amount of dye that can be used in a single sample. To the best of our knowledge, only one laboratory has employed a protocol using three markers simultaneously [91]. In addition, the lack of commercially available fluorescent markers may also restrict the selection of monoclonal antibodies.

Another major problem with FC is the rigorous establishment controls. In addition to fluorescent isotype controls, whose values are subtracted upon data analysis, FC requires a large number of intra- and interassay controls. Even small variations such as contamination with particles found in the reagents can be a problem that requires special care [6]. Overall, the large number of parameters that must be in place render FC procedurally demanding, requiring a large amount of manpower and decreasing the possibilities of automation and mass-scale production.

3.2.1.2. *Advantages of FC.* However, depending on the protocol established for sample processing, FC results are faster than ELISA or assays requiring multiple steps and long incubation times. Sample preparation differences notwithstanding, EMP results are generally available promptly, deeming the implementation of EMP assay desirable in FC-enabled laboratories. In addition, results obtained with FC are more quantitative than with AVC/PTA and ELISA, as they provide a specific numerical count of MP.

A benefit of FC over AVC/PTA or ELISA methods is the ability to measure more than one marker simultaneously. This can be exploited not only in the area of MP population selection but also in protocols concerned with the interaction of EMPs with other cells. In this regard, new studies exploring the interaction of EMPs with leukocytes and platelets have exploited one of FC's major advantages over other methods: the ability to quantify MP–cellular complexes [67, 93, 94]. Similar methodologies have been employed with platelet microparticles (PMPs) or LMPs with great success and have contributed much of what we know about MPs' functional aspects [34]. The approach of these assays is reasonably straightforward and may one day be used for clinical testing.

3.2.2. *AVC/PTA*

AVC MPs are selected based on PL content by AV binding to detect MPs on the basis of their PL content [6]. After the initial process of capture onto the plates, MPs are assayed for their ability to generate thrombin. The former method consists of incubating immobilized MPs with factors Xa and Va as well as prothrombin and starting the reaction with the addition of calcium. Thrombin generation is then monitored by adding a chromogenic substrate. To deduce the quantity of PL expressed on MPs, samples are compared to a standard curve obtained standard liposomes with known amounts of PL, to convert absorbance values to equivalent PL units. The importance of this last procedure underscores the need for calibration of the PL standards and the blood factors used in the thrombogenic reaction to ensure that any changes in thrombin generation may be solely MPs dependent. Indeed, to avoid confounding activity that may affect MPs readings, factor Xa and thrombin in plasma samples is inactivated in plasma before the capture of MPs. Subsequently, EMPs and other MPs are identified according to the expression of different antigens, using ELISA.

3.2.2.1. *Disadvantages of AVC/PTA.* The most important disadvantage of this method is that it narrowly limits the detection of EMPs to those with a certain lipid content. Considering that AV preferentially binds normally in-facing anonic PL [43], AV may predominantly measure EMPs elicited by certain cellular pathways that involve changes in the lipid content of the EC membrane. In this regard, the expression of AV on EMPs has been demonstrated to vary according to the type of conditions eliciting EMP release. More specifically, it was observed that AV + EMP counts were significantly lower than EMPs expressing other antigenic markers such as CD31 or CD62E, especially in activated EC. Thus, in certain pathologies in which activation or another nonapoptotic process is predominant, using solely

AV as a parameter of EMP discrimination may result in deceptively low counts.

3.2.2.2. *Advantages of AVC/PTA.* However, when compared to FC, AVC/PTA presents some benefits. First, in addition to providing quantitative data, AVC/PTA also provides valuable information regarding the procoagulant activity of EMPs using a single method. Furthermore, it relies on widely used techniques without the need for specialized equipment other than a plate reader. Clearly, this provides a viable option to detect EMPs without incurring the cost usually associated with FC. In addition, despite being time consuming and requiring multiple steps, it provides the possibility of processing large quantities of samples per assay.

3.2.3. *Other Methodological Problems*

A problem common to both assay methods is obtaining and preparing clinical samples. Most studies employ trisodium citrate or acid citrate dextrose as anticoagulants, and with careful venipuncture, contamination arising from artifactual platelet activation can be avoided [5, 6]. An important aspect that can also have a significant effect on EMP counts is how samples are prepared to be assayed. As seen in Table 1, several protocols employ different centrifugation steps, and some even pellet MPs. Methodologies that do not involve step-centrifugation and pelleting of MPs tend to be faster and may yield higher levels of MPs, which may be lost while preparing the samples to be assayed. However, background may be higher because of nonspecific binding to other non-MP blood constituents, and results may be less specific than methods requiring multiple centrifugation steps.

A possible example of this problem can be observed when comparing studies published in preeclampsia (PE). vanWijk *et al.* [91] found no elevation of EMPs in PE patients, whereas Bretelle *et al.* showed a decrease in EMP levels when compared to normal pregnant controls [87]. In contrast, Gonzalez-Quintero *et al.* documented the presence of elevated EMPs in PE patients [82, 84]. These conflicting results are very likely a result of the differences in methodology, ranging from three diverging centrifugation protocols to the selection of markers.

3.2.4. *Antigenic Markers Employed in Clinical EMP Studies*

Several antigenic markers commonly used for clinical EMP assay have been listed in Table 1. The antigens tested can be widely classified into two groups based on the pattern of EC expression: inducible, expressed only on EC activation, and constitutive, also expressed under resting conditions. In the first group, CD54 (ICAM-1), CD62E (E-selectin), CD106 (VCAM-1), anti-TF, and anti-MCP-1 are included. Similarly, constitutive markers CD51 and CD51/CD61 (vitronectin receptor, $\alpha v\beta 3$ integrin complex), CD31

(PECAM-1), CD105 (endoglin), CD144 (VE-cadhedrin), or CD146 (S-endo 1) have also been employed. In several studies, inducible antigens have been reported to be more elevated during the acute phase of many diseases when compared to constitutive antigens. However, some of these markers, most notably CD54 and TF, are widely expressed by other activated cells. In contrast, constitutive markers, although widely expressed, tend to be more specific. A marked exception is CD31, which is widely expressed on EC and EMP, yet has elevated coexpression by platelets and leukocytes. Nevertheless, all markers are, to a certain degree, present in other hematological cells. Thus, when using FC, the use of antibodies to detect MPs of nonendothelial origin is preferable and often a necessity. In some select cases, such as with CD62E or CD51, the number of non-EC MPs expressing such antigen is low, and the use of antibodies to measure other MPs, although preferable, does not necessarily have a significant effect on measurements [59, 83–85, 87, 88]. Conversely, when using markers such as CD31, a desirable antigen because of its impressive expression of EMPs, the use of antibodies against leukocytes and, most important, platelets is vital. MPs of leukocyte origin can be effectively detected using CD45, a panleukocyte marker with a high affinity to all leukocytes. In this regard, CD41 remains the preferred marker for the measurement of PMP because of its abundance in platelets; however, CD42b is also used.

3.2.5. *A Step Toward Standardization*

In this regard, a forum was recently hosted by the International Society of Thrombosis and Haemostasis [6]. In a definitive move toward standardizing the assay of EMP, six laboratories submitted detailed protocols, which were then published in the *Journal of Thrombosis and Haemostasis*. This effort is a milestone in the process of reaching a consensus between the different laboratories measuring EMPs and may signal a new era in the area of clinical EMP research.

4. EMPs in Clinical Disease

4.1. Introduction

The body of published clinical EMP reports remains relatively small and varies widely in scope, mainly because the clinical application of the assay of EMPs is still in its early stages. In all disorders studied, the presence of EC perturbation or dysfunction has been implicated by the presence of soluble markers of EC injury. The reports reviewed herein strongly indicate that EMPs are good indicators of EC status *in vivo*. To simplify the presentation of results, reports have been organized into broader categories based on the clinical presentation of the disease studied. Clinical reports as they pertain to EMPs published to this date have been summarized in Table 2.

TABLE 2
REPRESENTATIVE FINDINGS OF LEVELS OF ENDOTHELIAL MICROPARTICLES IN DIFFERENT DISORDERS

Category	Reference	Disease	Findings
Cardiovascular Disease	[95]	SA; UA; MI	↑ EMP in MI and UA. = SA. EMP ⊬ MI vs UA or Cont vs. SA
	[81]	SA;UA;MI	↑ EMP in MI and UA.= SA. CD31 not CD51+ EMP / MI vs UA, Cont vs. SA and 1st MI.
	[96]	MI	VitC induced ↓ EMP in DL and HRF, = DM
	[94]	SA;MI	↑ EMP–PLT in SA and UA. ↓MI, increasing 48hr later to SA levels. MCP-1 ↑↑MI and ↑SA.
Diabetes Mellitus	[88]	DM1 DM2	↑ EMP and MP with PCA in DM1. EMP / Hem_{a1}.↑ MP no PCA, = EMP in DM2
	[97]	DM	↑ EMP and MP DM. EMP / Neph and Ox-LDL titers. Sap induced ↓ EMP and PMP
Hypertension and Preeclampsia	[83]	MHTN ModHTN	↑↑MHTN ModHTN. EMP smoking, BP, DM sVCAM-1 and sCD62E.
	[82]	PE	↑ EMP in NPC and ↑↑ PE. EMP / BP.
	[84]	PE;GH	↑ EMP in NPC, ↑↑ GH and ↑↑↑ PE. EMP / BP and Uprot.
	[91]	PE	= EMP in PE.↑ PMP and LMP. EMP / LMP.
	[87]	PE; IUGR	↑ MP with PCA in NPC, PE and IUGR. = EMP in PE and IUGR.
Inflammatory and Infectious Disorders	[85]	PSV	↑ CD105+ and ↑↑ CD62E+ EMP in PSV vs. FC. EMP / APRP.
	[86]	CM; UM; A	↑ EMP CM vs. UM or A. EMP / coma and clinical course.
	[93]	SIRS	↑ EMP and EMP–NC in SIRS. EMP ⊬ sCD62E, TM or PAI-1.
Autoimmune Disorders	[51]	LA	↑ EMP in LA
	[59]	LA; SLE; APS	↑ EMP in LA, APS and APL+ SLE. =APL–SLE and APL–T.
	[92]	SLE	↑ EMP and PMP in SLE
Hematological Disorders	[89]	PNH;AA;SCD	↑ EMP in PNH, ↑↑ SCD and = AA.
	[90]	SCD	↑ MP with PCA in steady SCD and ↑ ↑ in crisis. ↑ ↑ ↑ RBCMP , ↑ ↑ EMP and ↑ LMP. ↑↑ TF+ LMP ↑ EMP with PCA.

TABLE 2 (*Continued*)

Category	Reference	Disease	Findings
	[53]	TTP; ITP	↑ CD31+/42b− and CD51+ EMP in TTP, = in R or ITP. Both EMP / PltCt and LDH.
	[57]	TTP	↑ CD31+/42b−↑↑ CD62E+ EMP in TTP, = in TTP-R or ITP. Both EMP / PltCt and LDH.↑ CD62E+/vWF+ EMP in TTP, = in TTP−R.
Neurological Disorders	[58]	MS	↑ CD31+/42b-EMP in MS, = in R. ↑ CD51+ EMP in MS and R. Both subsets EMP / Gad+ MRI.
	[67]	MS	↑ EMP in MS, = in R. ↑ Mono−EMP in MS = in R. EMP and Mono−EMP / Gad+ MRI.

SA, stable angina; UA, unstable angina; MI, myocardial infarction; DM(1;2), diabetes mellitus (type I; type II); (M;Mod)HTN, (malignant; moderate) hypertension; NPC; normal pregnant controls; PE, preeclampsia; GH, gestational hypertension; IUGR, intrauterine growth restriction; PSV, pediatric systemic vasculitis; CM, cerebral malaria; UM, uncomplicated malaria; A, anemia; SIRS, systemic inflammatory response syndrome; LA, lupus anticoagulant; SLE, systemic lupus erythematosus; APL, antiphospholipid antibody; APS, antiphospholipid antibody syndrome; PNH, paroxysmal nocturnal hemoglobinuria; AA, aplastic anemia; SCD, sickle cell disease; TTP, thrombotic thrombocytopenic purpura; ITP, immune thrombocytopenic purpura; MS, multiple sclerosis; EMP, endothelial microparticles; MP, microparticles; PMP, platelet microparticles; LMP, leukocyte microparticles; AV, annexin V; TF, tissue factor; VitC, vitamin C; BP, blood pressure; Uprot, urine protein levels; T; thrombosis; APRP, acute protein reaction products; PltCt, platelet count; LDH, lactate dehydrogenase; Gad+, gadolinium enhancement; MRI, magnetic resonance imaging; Neph, nephropathy; Sap, sapogrelate HCL; OX-LDL, oxidized LDL; ↑, elevated; ↑↑, more elevated; ↑↑↑, most elevated; ↓, decreased; =, no change/equal; /, correlated with; ⁄̸, did not correlate with.

4.2. Cardiovascular Disease

The presence of EMPs was first reported in acute coronary syndromes (ACS) by Mallat *et al.* [95]. The patient population studied consisted of patients with stable angina and patients with unstable angina and myocardial infarction (MI). In this study, procoagulant phospholipid-rich MPs of leukocyte, platelet, and endothelial origin were measured; EMPs were identified by AV+/CD31+ or AV+CD146+ expression. The researchers' main finding was a marked increase in MPs and EMPs bearing both markers in ACS over stable angina or controls. A positive correlation was observed between CD31+ and CD146+ EMP, but not between PMP and EMP. EMP levels did not discriminate SA from controls or, in ACS, unstable angina from MI.

In patients with MI, EMPs were significantly increased over other MP subsets. In addition, the elevation in EMP levels persisted for days after the acute episode, in accordance with the propensity for thrombosis generally observed in these within this time period.

Lending further support to this work, Bernal *et al.* also documented the presence of elevated EMPs in patients with ACS [81]. Two different subsets of EMPs, CD31+/CD42b− and CD51+, and PMPs were quantitated in MI and unstable angina. CD31+/CD42b− EMPs as well as PMPs were elevated in patients with ACS over SA and controls. However, a significant difference between CD51+ EMP levels in ACS versus SA was not recorded. Similarly, in patients with MI, CD31+/CD42b− but not CD51+ EMP was elevated in patients with a first MI versus patients with previous coronary events. When comparing SA to controls, EMPs but not PMPs were found to be discriminative. On the basis of these data, the authors hypothesized that CD31+/CD42b− EMPs are indicative of acute endothelial injury, whereas CD51+ EMPs could account for chronic EC injury. Taken together, these results indicate that procoagulant EMPs are present in ACS and may play a role in the thrombotic events in MI.

In a prospective, randomized study, Morel *et al.* [96] evaluated the effect of vitamin C (vit C) on annexin V+/CD31+ EMP generation in MI. Within this group, patients were classified according to the following criteria: diabetes mellitus (DM), dyslipidemia (DL), and patients with high risk factors. Vit C administration had no significant effect on EMP release in patients with DM. However, EMP counts in patients with DL and in high-risk patients were significantly decreased in treated groups versus placebo. Thus, oxidation may underlie the endotheliopathy in patients with DL and at high risk, whereas EC injury in DM may arise from different mechanisms.

More recently, Heloire *et al.* have demonstrated the presence of increased circulating EMP–platelet complexes (EMP-PLTC) in SA and MI patients [94]. EMP-PLTCs were defined as aggregates coexpressing platelet markers such as CD41a or CD42b as well as endothelial markers such CD105 and CD62E and MCP-1. EMP-PLTCs were elevated approximately twofold in SA over controls. In MI patients, EMP-PLTCs were decreased during onset and following reperfusion, thereafter increasing 48 hours later to values comparable to SA. An increase in expression of MCP-1 expression in EMP-PLTC, in SA, and most significantly, in MI patients indicated that EMPs present in EMP-PLTC were derived of activated EC. The observation indicates that persistent EC injury, the release of EMP, and the formation of EMP-PLTC are present in SA. Moreover, the decrease in EMP-PLTC during an MI indicates that these complexes may play a role in the acute thrombotic events of CAD.

4.3. Diabetes Mellitus

In a study conducted by Sabatier *et al.*, annexin V+ MP, CD41+ PMP, and CD51+ EMP were studied in patients with DM type 1 (DM1) and DM type 2 (DM2) [88]. In addition, the procoagulant activity of MPs was also evaluated. In DM1, annexin V MP, EMPs, and PMP were elevated. In addition, the procoagulant activity of MPs was elevated. In this subset of patients, procoagulability correlated with the glycosylated hemoglobin1c. However, although DM2 exhibited an elevation in AV+MPs V+ MPs, EMPs, or PMPs were not elevated. Interestingly, the MPs quantitated in these patients did not have increased procoagulant activity. These results indicate that in DM1, unlike DM2, EC perturbation and platelet activation are present. Moreover, the procoagulability observed may be associated to glucose imbalance. More recently, Nomura *et al.* have documented an elevation of EMPs, PMPs, soluble P-selectin, and CD40L, as well as anti-oxLDL antibodies [97]. The concentration of all the above markers was elevated in DM patients. EMP levels correlated with nephropathy and anti-oxLDL titers. On treatment with sarprogrelate, HCL, EMP, PMP, and soluble markers decreased significantly. Their results strongly implicate anti-oxLDL antibodies as an injurious agent and a potent inducer of EMPs and PMPs.

4.4. Hypertension and Preeclampsia

In a cross-sectional study, Preston *et al.* have shown an elevation in EMPs in hypertension (HTN) [83]. CD31+/CD42b-EMPs and CD41+ PMPs were studied in three patient populations: severe malignant HTN (MHTN; diastolic blood pressure levels of $\geq$120), moderate HTN (ModHTN; diastolic blood pressure levels of >95–100), and normotensive controls. CD31+/CD42b− EMPs were most elevated in MHTN and ModHTN versus controls. However, PMPs were increased in MHTN but not ModHTN when compared with controls. Neither PMPs nor EMPs could discriminate between MHTN versus ModHTN or ModHTN versus control. EMP counts also correlated with systemic blood pressure, diastolic blood pressure, DM, and smoking, whereas PMP counts only correlated with systemic blood pressure and diastolic blood pressure. When comparing soluble VCAM-1, CD62E, ICAM-1, or vWF with EMP, only VCAM-1 and CD62E were found to significantly correlate with EMP. These findings indicate that EC injury is present in both MHTN and ModHTN, being more significant in MHTN.

In related studies, Gonzalez-Quintero *et al.* have studied EMPs in PE [82]. In their prospective study, CD31+/CD42b- and CD51+ EMPs were studied in normal pregnant women and PE. EMP counts were found to be significantly elevated in PE patients over healthy controls, which in turn were higher than in normal, nonpregnant individuals. In addition, as observed with patients with hypertension, EMP counts correlated with mean arterial pressure. More recently, this same group detected the presence of CD31+/CD42b− and CD62E+ EMP in a larger number of patients with PE and gestational hypertension [84]. EMP levels were significantly more elevated in PE over gestational hypertension, which levels in turn were elevated over normal pregnant controls. As documented in their previous study, the authors observed a correlation between mean arterial pressure and both subsets of EMP. In addition, EMP counts were found to parallel urine protein levels. In contrast, VanWijk *et al.* demonstrated that CD62E+ and CD144+ EMPs were not elevated in PE [91]. Instead, an elevation in PMPs and LMPs was observed; EMP counts correlated with LMP levels. In a study comparing normal pregnant controls, PE, and intrauterine growth restriction, Bretelle *et al.* demonstrated the presence of CD51+ EMP, CD41+ PMP, and procoagulant annexin V+ MP [87]. Interestingly, both EMPs and PMPs were decreased in PE and intrauterine growth restriction, whereas procoagulant AV+ MPs were not. The authors concluded that the presence of EMPs and PMPs may be accountable for the inflammatory and procoagulable response in normal pregnancy. Similarly, in PE and intrauterine growth restriction, procoagulant MPs may be important in the thrombotic events observed in these disorders. The discrepancies in results among these four studies can be explained on close examination of the methodology to detect EMPs and has been discussed previously [5].

4.5. Inflammatory and Infectious Disease

Brogan *et al.* have studied CD62E+ and CD105+ EMPs and PMPs in pediatric systemic vasculitis (PSV) [85]. An elevation in CD62E+ EMPs and, to a lesser extent, CD105+ EMPs in patients with PSV over patients who underwent remission, control, or febrile controls was observed. EMPs correlated with vasculitis activity and the presence of acute phase reaction products in PSV; however, there was no correlation between those parameters in febrile controls. The elevation in EMPs provides further evidence of the role of EC injury in acute PSV. Moreover, the presence of CD62E+ EMP indicates that EC activation may underlie the endotheliopathy in PSV.

In a different study, Combes *et al.* has demonstrated an increase in CD51+ EMP in pediatric patients with uncomplicated malaria, cerebral malaria, and severe malaria with and without coma [86]. EMPs were elevated on admission in patients with cerebral malaria over controls or uncomplicated malaria, decreasing on follow-up. Severe malaria patients presenting with coma had the highest EMP count on admission. Thus, the finding of elevated EMPs in cerebral malaria and severe malaria with coma underscores the presence of EC injury in these disorders, supporting previous evidence of EC involvement in *Plasmodium falciparum* infection.

In a recent study conducted by Ogura *et al.*, CD31+/CD54+ EMP and CD31+/CD62E+ EMP-polymorphonuclear leukocyte (PMNL) complexes (EMP-PMNLC) were evaluated in patients with severe systemic inflammatory response syndrome (SIRS) [93]. Patients exhibited elevated platelet and leukocyte counts, high C-reactive protein levels, and increased soluble markers of PCA (thrombin-antithrombin II and plasmin/α plasmin inhibitor complexes, D-dimer). Soluble CD62E, TM, and PAI-1 as well as leukocyte oxidative activity were also elevated in SIRS. Circulating EMPs were significantly increased in SIRS patients; however, no significant correlation between soluble CD62E, TM, or PAI-1 and EMPs was recorded. EMPs from SIRS were found to bind preferentially PMNL, forming complexes that were elevated in SIRS. On the basis of these findings, it was hypothesized that EMP, arising after EC injury, may play a role in SIRS, not only because of their known PCA but also by binding PMNL, possibly enhancing the inflammatory response.

4.6. Autoimmune Disorders

Combes *et al.* first quantitated in the plasma of 30 normal controls and 30 patients with lupus anticoagulant (LA) [51]. CD31+/CD51+ EMP counts in LA were significantly elevated when compared to normal controls. In addition, it was observed that those patients with thrombotic events had EMP counts approximately elevated by twofold when compared to patients who did not develop thrombosis. No correlation was observed with age, sex, or treatment with anticoagulant agents; however, patients undergoing treatment with coagulants exhibited elevated EMP levels.

More recently, Dignat-George *et al.* have quantitated EMPs in patients with antiphospholipid syndrome, SLE with and without titers of antiphospholipid antibodies (APL), in addition to non APL-associated thrombosis and normal controls [59]. Their study has shown that EMPs were elevated in the plasma patients with antiphospholipid syndrome and SLE with APL titers, but not in patients with APL-SLE, non APL-associated thrombosis. In accordance with these findings, Abid Hussein *et al.* have also shown an increase in C42b+ CD61+ PMP as well as CD51+ CD62E+ EMP [92]. It was also observed that *in vitro*, only stimulation with SLE plasma induced the release of EMPs in EC cultures. Thus, only in antiphospholipid syndrome do APL antibodies reacting directly with EC appear to be the causative agents of EC damage. However, in APL+ SLE, EMPs resulting indirectly from antibodies found in these diseases may aggravate the procoagulant response.

4.7. Hematological Disorders

Simak *et al.* have measured EMPs in patients with paroxysmal nocturnal hemoglobinuria (PNH), sickle cell disease (SCD), aplastic anemia (AA), and normal controls [89]. EMPs were measured using a combination of CD105, CD144, CD54, and annexin V. The researchers' findings indicate that a subset of EMPs coexpressing CD105 and annexin V were significantly elevated in patients with SCD over PNH, AA, or healthy donors. EMPs were elevated in PNH, however, when compared to AA or normal controls. In contrast, EMPs expressing CD105 and CD54 were more significantly elevated. As observed with other markers, CD144+ EMPs were increased in SCD and PNH over controls and AA. Shet *et al.* further characterized the role of EMPs in SCD [90]. In their study, CD144+ EMP counts were compared to AV+ MP and PMPs, LMPs, and red blood cell MPs. In addition, procoagulant activity was assessed and TF expression measured. In patients with SCD during crisis and steady state, all MP subsets were significantly increased over normal controls. MPs were more elevated during crisis than steady state. When considering the relative groups of MPs, it was observed that red blood cell MPs were the most abundant, followed by EMPs, PMPs, and to a lesser degree, LMPs. It was also observed that EMPs and LMPs coexpressing TF were also significantly increased; however, TF+ LMPs were more abundant than TF+ EMPs. In functional studies, MPs from SCD but not controls exhibited increased procoagulant activity, which was partially abrogated by blocking. Both studies indicate that procoagulant MPs play a role in the thrombotic events in SCD crisis, also contributing to the prothrombotic state observed during the steady state of this disease.

The presence of procoagulant EMPs in patients with TTP was first reported by Jimenez *et al.* [53]. When compared with counts in normal controls or ITP patients, CD31+/CD42b− and CD51+ EMP were significantly elevated in TTP patients. EMP counts positive for both markers were found to correlate with platelet counts and lactate dehydrogenase. More recently, CD62E+ and CD31+/CD42b− EMP were compared in TTP patients during the acute phase versus remission [64]. As previously observed, EMPs were significantly elevated over controls and returned to values comparable to normal, correlating with the clinical course of the disease. CD62E+ EMPs were markedly more elevated than CD31+CD42b-EMPs in all patients.

More recently, a subset of CD62E+ EMPs coexpressing vWF was found to be elevated during the acute phase of TTP but not upon remission.

4.8. Neurological Disease

Minagar *et al.* first documented elevated levels of EMPs in patients with exacerbation, patients in clinical remission, and controls [58]. Two sets of phenotypic were assayed, CD31+/CD42b− EMP and CD51+ EMP. Gadolinium enhancement in MRI (Gad+ MRI) was correlated with EMP counts. CD31+/CD42b− EMPs correlated well with gadolinium enhancement and were significantly elevated when compared to controls during exacerbation, but not on remission. In contrast, CD51+ EMPs remained elevated over controls in both patient groups. In a later study, Jy *et al.* demonstrated the presence of EMP-monocyte (EMP-mono) complexes in MS patients [67]. EMP-mono complexes were measured by flow cytometry using a combination of CD45 and CD54 or CD62E. *In vitro*, it was observed that monocytes preferentially bound EMPs over neutrophils or lymphocytes. EMPs also activated monocytes and increased transendothelial migration in a tissue culture model. In MS patients, these complexes were elevated during exacerbation over patients in remission or controls. As previously documented, EMPs were elevated over both remission and control groups. Interestingly, only CD62E+ but not CD54+ EMPs were present. Finally, a positive correlation was observed between EMP-mono and EMP and gadolium enhancement in MRI lesions. This study provided a possible mechanism of action for EMPs in MS, further implicating them in the process of blood–brain barrier breaching, an important event in the pathogenesis of this disease.

5. Perspective

In sum, EMPs have emerged as a preferred direct method for assessing EC injury in different disorders. EMP analysis could provide insight into the actual status of the endothelium *in vivo* by a simple blood analysis. However, there is a need for refinement and standardization of the assay method. Overall, most groups have relied on flow cytometry for the measurement of EMPs; nevertheless, other methods such as ELISA are available and may be an option in the future. The main challenge remains in the selection of specific and sensitive monoclonal antibodies that may yield consistent results between different laboratories. In addition, the protocols for sample handling and storage need to be clearly delineated. The assay is still a relatively new, and we are confident that with refinement it will become a valuable tool for the practicing physician.

Acknowledgments

This work was supported by the Wallace H. Coulter Foundation. We are also grateful for the Charles and Jane Bosco Research Fund and the Mary Beth Weiss Research Fund.

References

[1] Cines DB, Pollak ES, Buck CA, et al. Endothelial cells in physiology and in the pathophysiology of vascular disorders. Blood 1998; 9:3527–3561.

[2] Aird WC. Endothelial cell dynamics and complexity theory. Crit Car Med 2002; 30: S180–S185.

[3] Becker BF, Heindl B, Kupatt C, Zahler S. Endothelial function and hemostasis. Z Cardiol 2000; 89:160–167.

[4] Verma S, Buchanan MR, Anderson TJ. Endothelial function testing as a biomarker of vascular disease. Circulation 2003; 108:2054–2059.

[5] Horstman LL, Jy W, Jimenez JJ, Ahn YS. Endothelial microparticles as markers of endothelial dysfunction. Front Biosci 2004; 9:1118–1135.

[6] Jy W, Horstman LL, Jimenez JJ, et al. Measuring circulating cell-derived microparticles. J Thromb Haemost 2004; 2:1842–1843.

[7] Rodgers G. Hemostatic properties of normal and perturbed vascular cells. FASE J 1988; 2:16–123.

[8] Bombeli T, Mueller M, Haeberli A. Anticoagulant properties of the vascular endothelium. Thromb Haemost 1997; 77:408–423.

[9] Thomas SR, Chen K, Keaney JF, Jr. Oxidative stress and endothelial nitric oxide bioactivity. Antioxid Redox Signa 2003; 5:181–194.

[10] Bates DO, Harper SJ. Regulation of vascular permeability by vascular endothelial growth factors. Vascul Pharmacol 2002; 39:225–237.

[11] Garlanda C, Dejana E. Heterogeneity of endothelial cells. Specific markers. Arterioscler Thromb Vasc Biol 1997; 17:1193–1202.

[12] Gerritsen ME. Functional heterogeneity of vascular endothelial cells. Biochem Pharmacol 1987; 36:2701–2711.

[13] Nawroth PP, Stern DM. Endothelial cell procoagulant properties and the host response. Semin Thromb Hemost 1987; 13:391–397.

[14] Carlos TM, Harlan JM. Leukocyte-endothelial adhesion molecules. Blood 1994; 84:2068–2101.

[15] Pober JS, Gimbrone MA, Jr., Lapierre LA, Mendrick DL, Fiers W, Rothlein R, Springer TA. Overlapping patterns of activation of human endothelial cells by interleukin-1, tumor necrosis factor and immune interferon. J Immunol 1986; 137:1893–1896.

[16] Pober JS. Effects of tumor necrosis factor and related cytokines on vascular endothelial cells. Cib Foun Sympos 1987; 131:170–184.

[17] Madge LA, Pober JS. TNF signalling in vascular endothelial cells. Exp Mol Pathol 2001; 70:317–325.

[18] Yilmaz A, Bieler G, Spertini O, Lejeune FL, Ruegg C. Pulse treatment of human vascular endothelial cells with high doses of tumor necrosis factor and interferon-gamma results in simultaneous synergistic and reversible effects on proliferation and morphology. Int J Cancer 1998; 77:592–599.

[19] Stehlik C, deMartin R, Kumabashiri I, Schmid JA, Binder BR, Lipp J. Nuclear factor (NF)-kappaB-regulated X-chromosome-linked iap gene expression protects endothelial cells from tumor necrosis factor alpha-induced apoptosis. J Exp Med 1998; 188:211–216.

[20] Polunovsky VA, Wendt CH, Ingbar DH, Peterson MS, Bitterman PB. Induction of endothelial cell apoptosis by TNF alpha: Modulation by inhibitors of protein synthesis. Exp Cell Res 1994; 214:584–594.

[21] Carlos T, Kovach N, Schwartz B, et al. Human monocytes bind to two cytokine-induced adhesive ligands on cultured human endothelial cells: Endothelial-leukocyte adhesion molecule-1 and vascular cell adhesion molecule-1. Blood 1991; 77:2266–2271.

[22] Yeh M, Leitinger N, de Martin R, Onai N, Matsushima K, Vora DK, Berliner JA, Reddy ST. Increased transcription of IL-8 in endothelial cells is differentially regulated by TNF-alpha and oxidized phospholipids. Arterioscler Thromb Vasc Biol 2001; 21:1585–1591.

[23] Watson C, Whittaker S, Smith N, Vora AJ, Dumonde DC, Brown KA. IL132-6 acts on endothelial cells to preferentially increase their adherence for lymphocytes. Clin Exp Immunol 1996; 105:112–119.

[24] Paleolog EM, Carew MA, Pearson JD. Effects of tumour necrosis factor and interleukin-1 on von Willebrand factor secretion from human vascular endothelial cells. Int J Radiat Biol 1991; 60:279–285.

[25] Moake JL, Rudy CK, Troll JH, Weinstein MJ, Colannino NM, Azocar J, Seder RH, Hong SL, Deykin D. Unusually large plasma factor VIII: von Willebrand factor multimers in chronic relapsing thrombotic thrombocytopenic purpura. N Engl J Med 1982; 307:1432–1435.

[26] Mulder AB, Hegge-Paping KS, Magielse CP, Blom NR, Smit JW, van der Meer J, Hallie MR, Bom VJ. Tumor necrosis factor alpha-induced endothelial tissue factor is located on the cell surface rather than in the subendothelial matrix. Blood 1994; 84:1559–1566.

[27] Freyssinet JM, Toti F, Hugel B, Gidon-Jeangirard C, Kunzelmann C, Martinez MC, Meyer D. Apoptosis in vascular disease. Thromb Haemost 1999; 82:727–735.

[28] Zwaal RFA, Schroit AJ. Pathophysiologic implications of membrane phospholipid asymmetry in blood cells. Blood 1997; 89:1121–1132.

[29] Ruf A, Morgenstern E. Ultrastructural aspects of platelet adhesion on subendothelial structures. Semin Thromb Hemost 1995; 21:119–122.

[30] Stefanec T. Endothelial apoptosis: Could it have a role in the pathogenesis and treatment of disease? Chest 2000; 117:841–854.

[31] Hogg N, Browning J, Howard T, Winterford C, Fitzpatrick D, Gobe G. Apoptosis in vascular endothelial cells caused by serum deprivation, oxidative stress and transforming growth factor-beta. Endothelium 1999; 7:35–39.

[32] Min C, Kang E, Yu SH, Shinn SH, Kim YS. Advanced glycation end products induce apoptosis and procoagulant activity in cultured human umbilical vein endothelial cells. Diabete Res Clin Pract 1999; 46:197–202.

[33] Freyssinet JM. Cellular microparticles: What are they bad or good for? J Thromb Haemost 2003; 1:1655–1662.

[34] Horstman LL, Ahn YS. Platelet microparticles: A wide-angle perspective. Crit Rev Oncol Hematol 1999; 30:111–142.

[35] Basse F, Gaffet P, Bienvenue A. Correlation between inhibition of cytoskeleton proteolysis and anti-vesiculation effect of calpeptin during A23187-induced activation of human platelets: Are vesicles shed by filopod fragmentation? Biochim Biophys Act 1994; 1190:217–224.

[36] Bellomo G, Mirabelli F, Salis A, Vairetti M, Richelmi P, Finardi G, Thor H. Orrenius S. Oxidative stress-induced plasma membrane blebbing and cytoskeletal alterations in normal and cancer cells. Ann NY Acad Sci 1988; 551:128–130.

[37] Houle F, Rousseau S, Morrice N, Luc M, Mongrain S, Turner CE, Tanaka S, Moreau P, Huot J. Extracellular signal-regulated kinase mediates phosphorylation of tropomyosin-1 to promote cytoskeleton remodeling in response to oxidative stress: Impact on membrane blebbing. Mol Biol Cell 2003; 14:1418–1432.

[38] Iglic A, Hagerstrand H, Kralj-Iglic V, Bobrowska-Hagerstrand M. A possible physical mechanism of red blood cell vesiculation obtained by incubation at high pH. J Biomech 1998; 31:151–156.

[39] Bobrowska-Hagerstrand M, Hagerstrand H, Iglic A. Membrane skeleton and red blood cell vesiculation at low pH. Biochim Biophys Acta 1998; 1371(1):123–128.

[40] Mills JC, Stone NL, Erhardt J, Pittman RN. Apoptotic membrane blebbing is regulated by myosin light chain phosphorylation. J Cell Biol 1998; 140:627–636.

[41] Coleman ML, Sahai EA, Yeo M, Bosch M, Dewar A, Olson MF. Membrane blebbing during apoptosis results from caspase-mediated activation of ROCK I. Nat Cell Biol 2001; 3:339–345.

[42] Sebbagh M, Renvoize C, Hamelin J, Riche N, Bertoglio J, Breard J. Caspase-3-mediated cleavage of ROCK I induces MLC phosphorylation and apoptotic membrane blebbing. Nat Cell Biol 2001; 3:346–352.

[43] Dachary-Prigent J, Toti F, Satta N, Pasquet JM, Uzan A, Freyssinet JM. Physiopathological significance of catalytic phospholipids in the generation of thrombin. Semin Thromb Hemost 1996; 22:157–164.

[44] Satta N, Freyssinet JM, Toti F. The significance of human monocyte thrombomodulin during membrane vesiculation and after stimulation by lipopolysaccharide. Br J Haematol 1997; 96:534–542.

[45] Dahlback B, Wiedmer T, Sims PJ. Binding of anticoagulant vitamin K dependent protein S to platelet-derived microparticles. Biochemistry 1992; 31:12769–12777.

[46] Nomura S, Tandon NN, Nakamura T, Cone J, Fukuhara S, Kambayashi J. High-shear-stress-induced activation of platelets and microparticles enhances expression of cell adhesion molecules in THP-1 and endothelial cells. Atherosclerosis 2001; 158:277–287.

[47] Barry OP, Pratico D, Savani RC, FitzGerald GA. Modulation of monocyte-endothelial cell interactions by platelet microparticles. J Clin Invest 1998; 102:136–144.

[48] Mesri M, Altieri DC. Endothelial cell activation by leukocyte microparticles. J Immunol 1998; 161:4382–4387.

[49] Hamilton KK, Hattori R, Esmon CT, Sims PJ. Complement proteins C5b-9 induce vesiculation of the endothelial plasma membrane and expose catalytic surface for assembly of the prothrombinase enzyme complex. J Biol Chem 1990; 265:3809–3814.

[50] Casciola-Rosen L, Rosen A, Petri M, Schlissel M. Surface blebs on apoptotic cells are sites of enhanced procoagulant activity: Implications for coagulation events and antigenic spread in systemic lupus erythematosus. Proc Natl Acad Sci USA 1996; 93:624–1629.

[51] Combes V, Simon AC, Grau GE, Arnoux D, Camoin L, Sabatier F, Mutin M, Sanmarco M, Sampol J, Dignat-George F. *In vitro* generation of endothelial microparticles and possible prothrombotic activity in patients with lupus anticoagulant. J Clin Invest 1999; 104:93–102.

[52] Mutin M, Dignat-George F, Sampol J. Immunologic phenotype of cultured endothelial cells: Quantitative analysis of cell surface molecules. Tissue Antigens 1997; 50:449–458.

[53] Jimenez JJ, Jy W, Mauro LM, Horstman LL, Ahn YS. Elevated endothelial microparticles in thrombotic thrombocytopenic purpura findings from brain and renal microvascular cell culture and patients with active disease. Brit J Haematol 2001; 112:81–90.

[54] Simak J, Holada K, Vostal JG. Release of annexin V-binding membrane microparticles from cultured human umbilical vein endothelial cells after treatment with camptothecin. BM Cell Biol 2002; 3:11.

[55] Jimenez JJ, Jy W, Mauro LM, Soderland C, Horstman LL, Ahn YS. Endothelial cells release phenotypically and quantitatively distinct microparticles in activation and apoptosis. Thromb Res 2003; 109:175–180.

[56] Brodsky SV, Malinowski K, Golightly M, Jesty J, Goligorsky MS. Plasminogen activator inhibitor-1 promotes formation of endothelial microparticles with procoagulant potential. Circulation 2002; 106:2372–2378.

[57] Jimenez JJ, Jy W, Mauro LM, Horstman LL, Soderland C, Ahn YS. Endothelial microparticles released in thrombotic thrombocytopenic purpura express von Willebrand factor and markers of endothelial activation. Br J Haematol 2003; 123:896–902.

[58] Minagar A, Jy W, Jimenez JJ, Sheremata WA, Mauro LM, Horstman LL, Ahn YS. Elevated plasma endothelial microparticles in multiple sclerosis. Neurology 2001; 56:1319–1324.

[59] Dignat-George F, Camoin-Jau L, Sabatier F, et al. Endothelial microparticles: A potential contribution to the thrombotic complications of the antiphospholipid syndrome. Thromb Haemost 2004; 91:667–673.

[60] Tramontano AF, O'Leary J, Black AD, Muniyappa R, Cutaia MV, El-Sherif N. Statin decreases endothelial microparticle release from human coronary artery endothelial cells: Implication for the Rho-kinase pathway. Biochem Biophys Res Commun 2004; 320:34–38.

[61] van Gorp RM, Broers JL, Reutelingsperger CP, Bronnenberg NM, Hornstra G, van Dam-Mieras MC, Heemskerk JW. Peroxide-induced membrane blebbing in endothelial cells associated with glutathione oxidation but not apoptosis. Am J Physiol 1999; 277:C20–C28.

[62] van Gorp RM, Heeneman S, Broers JL, Bronnenberg NM, van Dam-Mieras MC, Heemskerk JW. Glutathione oxidation in calcium- and p38 MAPK-dependent membrane blebbing of endothelial cells. Biochim Biophys Acta 2002; 1591:129–138.

[63] Jy W, Jimenez JJ, Mauro LM, Ahn YS, Newton KR, Mendez AJ, Arnold PI, Schultz DR. Agonist-induced capping of adhesion proteins and microparticle shedding in cultures of human renal microvascular endothelial cells. Endothelium 2002; 9:179–189.

[64] Boulanger CM, Scoazec A, Ebrahimian T, Henry P, Mathieu E, Tedgui A, Mallat Z. Circulating microparticles from patients with myocardial infarctions cause endothelial dysfunction. Circulation 2001; 104:2649–2652.

[65] Brodsky SV, Zhang F, Nasjletti A, Goligorsky MS. Endothelium-derived microparticles impair endothelial function *in vitro*. Am J Physiol Heart Circ Physiol 2004; 286: H1910–H1915.

[66] Sabatier F, Roux V, Anfosso F, Camoin L, Sampol J, Dignat-George F. Interaction of endothelial microparticles with monocytic cells *in vitro* induces tissue factor-dependent procoagulant activity. Blood 2002; 99:3962–3970.

[67] Jy W, Minagar A, Jimenez JJ, Sheremata WA, Mauro LM, Horstman LL, Bidot C, Ahn YS. Endothelial microparticles (EMP) bind and activate monocytes: Elevated EMP-monocyte conjugates in multiple sclerosis. Front Biosci 2004; 9:3137–3144.

[68] Simak J, Holada K, D'Agnillo F, Janota J, Vostal JG. Cellular prion protein is expressed on endothelial cells and is released during apoptosis on membrane microparticles found in human plasma. Transfusion 2002; 42:334–342.

[69] Blann A, Seigneur M. Soluble markers of endothelial cell function. Clin Hemorheol Microcirc 1997; **17**:3–11.

[70] Andre AJH, Newman W. Circulating adhesion molecules in disease. Immunol Toda 1993; 14:506–512.

[71] Mannucci PM. Von Willebrand factor: A marker of endothelial damage? Arterioscler Thromb Vasc Biol 1998; 18:1359–1362.

[72] Hladovec J, Prerovsky V, Stanek, Fabian J. Circulating endothelial cells in acute myocardial infarction and angina pectoris. Klin Wochenschr 1978; 56:1033–1036.

[73] Mutin M, Canavy I, Blann A, Bory J, Sampol J, Dignat-George F. Direct evidence of endothelial injury in acute myocardial infarction and unstable angina by demonstration of circulating endothelial cells. Blood 1999; 93:2951–2958.

[74] Dignat-George F, Sampol J. Circulating endothelial cells in vascular disorders: New insights into an old concept. Eur J Haematol 2000; 65:215–220.

[75] George F, Brisson C, Poncelet P, et al. Rapid Isolation of human endothelial cells from whole blood using S-Endo1 monoclonal antibody coupled to immuno-magnetic beads: Demonstration of endothelial injury after angioplasty. Thromb Haemost 1992; 67:147–163.

[76] George F, Brouqui M, Boffa M, et al. Demonstration of Rickettsia-conorii-induced endothelial injury in vivo by measuring circulating endothelial cells, thrombomodulin and von Willebrand factor in patients with Mediterranean spotted fever. Blood 1993; 82:2109–2116.

[77] Sowemimo-Coker SO, Meiselman HJ, Fracis Jr, B. Increased circulating EC in sickle cell crisis. Am J Hematol 1989; 31:263–265.

[78] Solovey A, Gui L, Ramakrishnan S, Steinber MH, Hebbel RP. Sickle cell anemia as a possible state of enhanced anti-apoptotic tone: Survival effect of vascular endothelial growth factor on circulating endothelial cells and unanchored endothelial cells. Blood 1999; 93:3824–3830.

[79] Lefevre P, George F, Durand JM, Sampol J. Detection of circulating endothelial cells in thrombotic thrombocytopenic purpura. Thromb Haemost 1993; 69:522.

[80] Clancy RM. Circulating endothelial cells and vascular injury in systemic lupus erythematosus. Curr Rheumatol Rep 2000; 2:39–43.

[81] Bernal-Mizrachi L, Jy W, Jimenez JJ, Pastor J, Mauro LM, Horstman LL, de Marchena E, Ahn YS. High levels of circulating endothelial microparticles in patients with acute coronary syndromes. Am Heart J 2003; 145:962–970.

[82] Gonzalez-Quintero VH, Jimenez JJ, Jy W, Mauro LM, Hortman L, O'Sullivan MJ, Ahn Y. Elevated plasma endothelial microparticles in preeclampsia. Am J Obstet Gynecol 2003; 189:589–593.

[83] Preston RA, Jy W, Jimenez JJ, Mauro LM, Horstman LL, Valle M, Aime G, Ahn YS. Effects of severe hypertension on endothelial and platelet microparticles. Hypertension 2003; 41:211–217.

[84] Gonzalez-Quintero VH, Smarkusky LP, Jimenez JJ, Mauro LM, Jy W, Hortsman LL, O'Sullivan MJ, Ahn YS. Elevated plasma endothelial microparticles: Preeclampsia versus gestational hypertension. Am J Obstet Gynecol 2004; 191:1418–1424.

[85] Brogan A, Shah V, Brachet C, Harnden A, Mant D, Klein N, Dillon MJ. Endothelial and platelet microparticles in vasculitis of the young. Arthritis Rheum 2004; 50:927–936.

[86] Combes V, Taylor TE, Juhan-Vague I, Mege JL, Mwenechanya J, Tembo M, Grau GE, Molyneux ME. Circulating endothelial microparticles in Malawian children with severe falciparum malaria complicated with coma. JAMA 2004; 291:2542–2544.

[87] Bretelle F, Sabatier F, Desprez D, et al. Circulating microparticles: A marker of procoagulant state in normal pregnancy and pregnancy complicated by preeclampsia or intrauterine restriction. Thromb Haemost 2003; 89:486–492.

[88] Sabatier F, Darmon P, Hugel B, Combes V, Sanmarco M, Velut JG, Arnoux D, Charpiot P, Freyssinet JM, Oliver C, Sampol J, Dignat-George F. Type 1 and type 2 diabetic patients display different patterns of cellular microparticles. Diabetes 2002; 51:2840–2845.

[89] Simak J, Holada K, Risitano AM, Zivny JH, Young NS, Vostal JG. Elevated circulating endothelial membrane microparticles in paroxysmal nocturnal haemoglobinuria. Br J Haematol 2004; 125:804–813.

[90] Shet AS, Aras O, Gupta K, Hass MJ, Rausch DJ, Saba N, Koopmeiners L, Key NS, Hebbel RP. Sickle blood contains tissue factor-positive microparticles derived from endothelial cells and monocytes. Blood 2003; 102:2678–2683.

[91] VanWijk MJ, Nieuwland R, Boer K, van der Post JA, VanBavel E, Sturk A. Microparticle subpopulations are increased in preeclampsia: Possible involvement in vascular dysfunction? Am J Obstet Gynecol 2002; 187:450–456.

[92] Abid Hussein MN, Meesters EW, Osmanovic N, Romijn FP, Nieuwland R, Sturk A. Antigenic characterization of endothelial cell-derived microparticles and their detection *ex vivo*. J Thromb Haemost 2003; 1:2434–2443.

[93] Ogura H, Tanaka H, Koh T, Fujita K, Fujimi S, Nakamori Y, Hosotsubo H, Kuwagata Y, Shimazu T, Sugimoto H. Enhanced production of endothelial microparticles with increased binding to leukocytes in patients with severe systemic inflammatory response syndrome. J Traum 2004; 56:823–830.

[94] Heloire F, Weill B, Weber S, Batteux F. Aggregates of endothelial microparticles and platelets circulate in peripheral blood. Variations during stable coronary disease and acute myocardial infarction. Thromb Res 2003; 110:173–180.

[95] Mallat Z, Benamer H, Hugel B, Benessiano J, Steg PG, Freyssinet JM, Tedgui A. Elevated levels of shed membrane microparticles with procoagulant potential in the peripheral circulating blood of patients with acute coronary syndromes. Circulation 2000; 101:841–843.

[96] Morel O, Jesel L, Hugel B, Douchet MP, Zupan M, Chauvin M, Freyssinet JM, Toti F. Protective effects of vitamin C on endothelium damage and platelet activation during myocardial infarction in patients with sustained generation of circulating microparticles. J Thromb Haemost 2003; 1:171–177.

[97] Nomura S, Shouzu A, Omoto S, Nishikawa M, Iwasaka T, Fukuhara S. Activated platelet and oxidized LDL induce endothelial membrane vesiculation: Clinical significance of endothelial cell-derived microparticles in patients with type 2 diabetes. Clin Appl Thromb Hemost 2004; 10(3):205–215.

ADVANCES IN CLINICAL CHEMISTRY, VOL. 39

PROTEOMICS IN CLINICAL LABORATORY DIAGNOSIS

Stacy H. Shoshan and Arie Admon

Department of Biology, Technion-Israel Institute of Technology, Haifa 32000, Israel

Two-dimensional polyacrylamide gel electrophoresis (2D-PAGE), high-performance liquid chromatography (HPLC), mass spectrometry (MS), tandem mass spectrometry (MS/MS), post-translational modifications (PTMs), time of flight (TOF), multi-dimensional protein identification technology (MudPIT), enzyme-linked immunosorbent assay (ELISA), cancer antigen 125 (CA125), laser capture microdissection (LCM), cholamidoprophy dimethalammonio-1-propanesufonate (CHAPS), polymerase chain reaction (PCR), Human Proteome Organization (HUPO), Proteomics Standards Initiative (PSI), hematoxylin and eosin (H&E), Terry's Polychrome, (TP), Toluidine Blue (TB), Nuclear Fast Red (NFR), Cresyl Violet (CV), Methylene Blue (MB), tumor-specific antigens (TSAs), tumor-associated antigen (TAA), hepatocellular carcinoma (HCC), prostate specific antigen (PSA), graft-versus-host disease (GVHD), CD8 anti-viral factor (CAF), human immunodeficiency virus (HIV).

0065-2423/05 $35.00
DOI: 10.1016/S0065-2423(04)39006-2

1. Introduction to Proteomics

Proteomics is a technological and scientific methodology commonly used for the analysis of complex protein samples in basic research that is rapidly becoming one of the most innovative approaches for clinical diagnosis. Replacing the long-established techniques that examined one protein at a time, proteomics imparts the simultaneous characterization of large numbers of proteins. Much of its strength stems from the recent developments in two-dimensional polyacrylamide gel electrophoresis (2D-PAGE), high-performance liquid chromatography (HPLC), mass spectrometry (MS), protein arrays, and bioinformatics. Considering the multifactorial causes of diseases, particularly cancer, a single biomarker that can characterize the entire pathologic process may not always exist. Therefore, multiparameter biomarkers, such as the protein patterns generated by 2D-PAGE and MS, are more likely to accurately reflect complex diseases. Searching for biomarkers in tissue, serum, and other body fluids is becoming more commonplace since it has become possible to directly interface biological samples with proteomic instruments. The complexity of many samples has made preparing and resolving them technically challenging, but recent advances in the technology have facilitated more accurate and high-throughput analyses. Modern proteomic technologies have the capacity to synergistically aid histopathology techniques in improving diagnostic sensitivity and specificity. Although proteomics is a powerful tool that has the potential to generate knowledge that is complementary to several other emerging fields of study, many of its technologies are currently limited to use in a research laboratory setting. However, the near future holds promise for bringing proteomics from the bench to the bedside as revolutionary diagnostic tools that can be used for early detection, diagnosis, and prognosis [reviewed recently in 1, 2].

2. Modern Proteomic Technology

Biological samples that are useful for clinical diagnosis often contain very complex mixtures of proteins, as up to a few thousand may be present in tissue extract or body fluids, and each of them can have a variety of posttranslational modifications (PTMs). Before the analysis of individual proteins, an essential first step in proteomics is carrying out multidimensional separations to fractionate the sample into its individual constituents. Fractionations are needed to obtain high analytic sensitivity, which is done by increasing the relative concentration and purity of the individual components (as discussed in Section 5). To facilitate the use of small sample sizes and to prevent loss of individual proteins during processing, these initial fractionations should involve a minimal number of steps and as little handling as possible.

The main technologies of proteomics, namely 2D-PAGE, HPLC, MS, protein arrays, and bioinformatics, facilitate both the detection and analyses of individual component proteins and peptides in complex biological samples. The recent advancements in these technologies have enabled researchers to resolve the protein content of body fluids and tissue samples in one or two simple steps. As a result, not only can the individual peptides be detected and analyzed but, at the same time, specific protein patterns unique to any given sample are also generated. The most significant effect of proteomics is not the implementation of each of these technologies individually into the laboratory but the combination of them together to form streamline processes that can be highly automated and that can carry out different steps in a short time with high sensitivity [3].

2.1. 2D-PAGE

Since the introduction of 2D-PAGE in 1975 [4, 5], long before the term "proteomics" was even invented [reviewed recently in 1, 6], it has evolved to become one of the most powerful techniques for separating proteins and generating protein patterns. It has become widely applied to a variety of biological samples and therefore promises to have a major effect on medical diagnostics.

To run a 2D-PAGE, proteins are separated in the first dimension, usually using (commercially available) immobilized pH gradient gel strips for isoelectric focusing [7], and then in the second dimension using SDS-PAGE (for which both linear and gradient precast gels are also commercially available). Because of standardization and the covalent immobilization of the ampholytes into the acrylamide, the pH range of the isoelectric focusing, the length of the strips, and the shape of the pH gradient can be precisely defined by the commercial manufacturers. Thus, for both dimensions the resolution has been vastly improved and the reproducibility has been dramatically increased.

Because of standardization and commercially available gel strips, 2D-PAGE has become more user friendly and no longer requires highly skilled and experienced operators. The procedures can be run by technicians after a relatively short training period. Covalent attachment of ampholytes to the acrylamide has allowed for loading of larger amounts of proteins onto each strip, facilitating the detection and analysis of the less abundant proteins. Moreover, as the gel strips do not need to be prerun to establish the pH gradient, as was needed in the past, the proteins can be loaded onto them just by rehydrating them in a protein solution.

The net result of a 2D-PAGE is a 2D "fingerprint." The resulting spots, even those containing nanogram amounts of individual proteins, are then color-stained so they can easily be detected. Available stains include Coomassie blue, silver, and zinc-imidazole, as well as fluorescent ones such as Sypro-ruby [8–10]. The protein fingerprints can be digitally recorded and then analyzed with image processing software to detect unique patterns and specific proteins for further comparisons between different samples. Of particular interest are the PTMs, which affect both the isoelectric points and the molecular masses of proteins, thus shifting the location in the gel of the altered proteins relative to the unmodified forms. Shifts resulting from the PTMs of proteins can add another layer of information about the disease because abnormal PTMs are often either a cause or an effect of the pathological state, and therefore may be important for diagnosis or prognosis. Once a fingerprint protein pattern for a given disease is established, or a deviation from a normal pattern can be recognized, the data can be incorporated into a diagnostic kit.

Although 2D-PAGE has the potential to be used daily in a clinical setting, it is currently not amenable to complete automation, and this limitation hinders its use as a separation technology for routine laboratory diagnosis. However, it is only a matter of time before 2D-PAGE becomes fully automated, as all of the necessary steps can employ ready-made reagents. Already, some groups have maximized the use of robotics in running gels. Because two different proteins can be hidden in one single protein spot, an automated approach to analyzing each spot can be very useful. First, "spot-picking robots" transfer the gel pieces to microtiter plates. Then, the plates are processed in automated washing stations, followed by automated trypsin digestion and peptide spotting on MS target plates. As many as 15,000 spots per day are processed because of this fully automated gel spot analysis system [3].

2.2. HPLC

One-, two-, and multidimensional capillary HPLC is becoming an attractive alternative to 2D-PAGE as a preparatory step for large-scale identifications and comparisons of protein repertoires in complex biological samples. Most commonly, the first dimension of HPLC separation is strong cation exchange, with reversed-phase usually serving as the second (or last) dimension. Chromatography is amenable to automation and can be connected directly to MS through an electrospray interface. The main limitation in the use of chromatography as a separation method relates to the low recovery rate for some of the proteins from the columns. Most proteins are too large and too hydrophobic for standard reversed-phase chromatography; consequently, they irreversibly "stick" to the columns. Small peptides, however, are more amenable to chromatography, including reversed-phase chromatography. Therefore, the most common approach is based on first digesting the entire protein content of the biological sample, or a set of individual proteins, into small peptides. The proteins are digested most often with an enzyme such as trypsin, which cuts at specific amino acids (lysine and arginine). The resulting peptides are then separated by one- or two-dimensional HPLC, resulting in peptide maps that are very reproducible. Their patterns can then be compared between samples with the aim of identification of disease markers [11]. As described below (see section 2.3), the individual peptides in the mixtures can be identified using MS, their relative amounts can be quantified, the PTM patterns can be defined, and the patterns of the peptides can be compared between different samples [12].

When a limited amount of a sample is available, proteomic analyses require the scaling-down of the chromatography columns and flow rates to accommodate the subpicomole and even subfemtomole amounts of proteins. Capillary columns made from fused silica are the solution, as they offer a very smooth internal surface and high-pressure resistance and can easily be cut to the desired size. The columns require packing with chromatography adsorbents; thus, many commercial sources have made available capillaries prepacked with a variety of beads and, more recently, with monolithic materials polymerized into the columns [13]. The main problem with the capillaries is their tendency to clog and break. HPLC pumps designed for capillary flow rates are also available but are not always required for all applications, as the flow of regular HPLC can be split with only a fraction of it directed into the capillary columns.

2.3. Mass Spectrometry

Mass spectrometers are used not only to detect the masses of proteins and peptides, but also to identify the proteins, to compare patterns of proteins and peptides, and to scan tissue sections for specific masses. MS is able to do this by giving the mass-to-charge ratio of an ionized species as well as its relative abundance. For biological sample analysis, mass spectrometers are connected to an ionizing source, which is usually matrix-assisted laser desorption ionization (MALDI) [14], surface-enhanced laser desorption/ionization (SELDI, a modified form of MALDI) [15], or electrospray ionization [16]. These interfaces enable the transfer of the peptides or proteins from the solid or liquid phase, respectively, to the gas (vacuum) phase inside the mass spectrometer. Both MALDI and electrospray ionization can be connected to different types of mass analyzers, such as quadrupole, quadruple-ion-traps, time of flight (TOF), or hybrid instruments such as quadrupole-TOF or Fourier transform-ion cyclotron resonance. Each of these instruments can measure masses, ranging from small pieces of peptides to very large proteins, relatively accurately.

Exact mass measurements are particularly useful when applied to proteolytic peptides as a way of identifying the proteins from which they were derived. After treating a protein with a specific protease, such as trypsin, the masses of the resulting peptides are determined. This list of measured masses is then compared, using computer databanks, to a theoretical list of masses calculated for each protein based on the specific protease that was used (discussed in Section 2.5). The main caveat is that the use of peptide mass fingerprinting leads to correct protein identification only if the mass spectrometer is very accurate and of high resolution. Otherwise, erroneous identifications are declared. When mixtures of proteins are proteolyzed, the danger of incorrect identifications is more acute, as the list of masses might have a better match with an incorrect protein.

Most of the modern mass spectrometers are also capable of fragmenting peptides into smaller pieces by collision-induced dissociations with gas atoms in the vacuum phase and then measuring the masses of the resultant fragments. This process is called tandem mass spectrometry (MS/MS). A "fingerprint" of mass fragments for each peptide will be generated according to their amino acid sequence and PTMs. Therefore, MS/MS, which is the combination of proteolysis of the proteins, separation of the resulting peptides, and gas-phase fragmentation, has become the most common approach for protein identification and PTM characterization [12, 17].

The MS/MS generated data consist of the intact masses of each peptide and its fragments, which is usually sufficient information to identify the peptide and therefore its source protein. The identifications are usually performed with dedicated software tools (see Section 2.5) that compare the measured masses from the MS/MS data to those of all the possible peptides in protein databanks to identify the peptides (without having to actually sequence the peptides). The sensitivity of modern mass spectrometers is very high, reaching the femtomole, attomole, and even smaller ranges of proteins (less than nanogram amounts), thereby making it possible to analyze even minute amounts of tissues [18]. In addition, the throughput of the instruments is sufficiently high so that hundreds or even thousands of proteins can be analyzed daily.

When 2D-PAGE and HPLC are interfaced to MS, unique peptide maps for each biological sample of interest can be generated. The pattern of each map can be compared between samples, and by implementing pattern recognition tools, they can become indicators of a disease. Starting with 2D-PAGE, stained gel pieces are excised, the proteins within them are proteolyzed, and the resulting peptides are eluted and subsequently analyzed by HPLC and MS/MS. This process also allows for each of the stained spots to be identified and then analyzed for PTMs. It can be concluded that 2D-PAGE, HPLC, and MS can all be connected to generate remarkably accurate and sensitive data. Together, they can not only detect particular proteins or peptides in a biological sample but also quantify them, establish their PTMs, identify molecular patterns, and even compare samples [recently reviewed in 1, 12].

2.4. Protein Arrays

To evaluate massive amounts of complex mixtures of proteins and be able to detail changes in the content as indicators of disease, proteins arrays are the key. Protein arrays are made by physically attaching proteins (or antibodies directed against them) to glass or plastic slides at a high surface density. The modern tool of "printing" proteins on slides is sufficiently powerful such that thousands of different proteins can be separately attached at different locations on slides the size of regular microscope slides. There are commercially available glass slides with different premade chemical functionalities or preactivated groups for covalent attachment of the proteins. This supports the construction of protein or antibody arrays for large-scale analyses of proteins in body fluids and tissue extracts. The binding of the soluble proteins within the biological mixtures to the bound proteins (or antibodies) on the arrays can then be detected by optical means, such as fluorescent labeling. Other ways to check for binding are to add secondary antibodies directed against the initial soluble ones, and then use plasmon resonance [19, 20] or MS to scan the array [21–24, reviewed in 25–27].

2.5. Bioinformatics

All four of the proteomic technologies mentioned above (2D-PAGE, HPLC, MS, and protein arrays) are dependent on the use of bioinformatics as a tool for data mining and elucidation. Often, MS runs generate lists of thousands of potential peptide biomarkers, and only with the help of dedicated software tools can the data be analyzed. Computer searches involving databanks of peptides and proteins are used to compare the lists of masses of the proteolytic peptides to theoretical proteolytic products. Matches between the observed mass and the calculated mass can serve as a way of identifying proteins of interest [28–32].

Bioinformatics tools involving computer-based statistical analyses are essential for data management and analysis. When a complex biological sample containing thousands of different proteins is analyzed by multifaceted approaches, such as multidimensional protein identification technology, the identification of the proteins in the mixture is extremely complicated. Even multiple peptide identification methods, such as using both MS and

MS/MS, are often not sufficiently accurate that they can be faithfully relied on [33]. In these cases, a statistical analysis should be introduced that would include more parameters that could assist in protein and peptide identification. For example, the observed and calculated relative retention times from reversed-phase columns could be correlated to provide further validation [34, 35].

Pattern recognition tools under the umbrella of bioinformatics are just beginning to emerge to help deal with the challenge of distinguishing molecular patterns in different samples. Already there is a need for dedicated software tools to compare patterns generated from the analysis of normal tissue extracts to those of diseased specimens [36]. Other software tools are being developed with the aim of helping to organize MS data, such as clustering algorithms. They help deal with the massive amounts of MS/MS generated data combining similar spectra throughout an entire project. Clustering data both reduces the amount of data that needs to be processed and also increases the signal-to-noise ratio to a large extent ([33] and references therein). Software tools designed for clustering data, such as Pep-Miner (http://www.haifa.il.ibm.com/projects/verification/bioinformatics), reduce it to a manageable size for further analysis. For example, in one large-scale project using different lung cancer cell lines, Pep-Miner reduced 517,000 mass spectra to 20,900 clusters and identified about 830 peptides [35].

2.6. Technology Portability, Reproducibility, and Performance Characteristics

When considering the practical applications of the aforementioned proteomic technologies in a clinical setting, the issues of portability, reproducibility, and performance characteristics are also important (see Section 6). To address portability, one approach is to develop miniaturized, microfabricated devices, such as "lab-on-a-chip" [reviewed in 37]. These chips integrate multiple steps of different analytical procedures and only require submicroliter amounts of reagents and samples. These types of devices offer high efficiency and reproducibility and would allow for such technologies to be accessible not only to large centralized laboratories but also to small-scale research centers.

For proteomic technologies to be of widespread use, the reproducibility of each method must be determined. Several studies have been conducted to investigate the coefficient of variance associated with different technologies. One group examined a fully automated HPLC 9.4-tesla Fourier transform–ion cyclotron resonance MS designed for unattended proteomics research 24 hours per day to investigate the instruments overall performance.

Reproducibility was found to deviate from 1% to 5% in uncorrected elution times, and repeatability was found to have less than 20% deviation in the detected abundances for more abundant peptides from the same aliquot analyzed a few weeks apart [38]. Another study aimed to evaluate the intraday and interday reproducibility of selected plasma samples using magnetic bead separation and a MALDI-TOF MS (OmniFLEX, Bruker Daltonics). The average intraday coefficient of variance of each mass spectra peak area was determined to be 18%, whereas the average interday coefficient of variance (evaluated on three different days) was 26% [39]. When considering SELDI spectra using ProteinChip WCX2 (Ciphergen Biosystems), reproducibility can be analyzed by examining the mass location and intensity from array to array on a single chip (intraassay) and between chips (interassay). One group found that the intraassay and interassay coefficients of variance for normalized intensity (peak height or relative concentration) were 12% and 18%, respectively [40].

Technology performance characteristics include factors such as automation to maximize reproducibility, efficiency, speed, and throughput and make the investigative process minimally labor intensive [reviewed in 41]. The need for time- and cost-effective biological testing has encouraged the development of combinatorial chemistry products that overcome the obstacles posed by small amounts of each protein within complex samples. Ideally, these tools would purify, analyze, and identify the protein of interest in a rapid and robust manner. Although improvements have been made in the currently available technologies, there is no single technique for protein analysis that is optimal for all biological samples (see Table 1) [reviewed in 42].

3. Proteins, Proteomics, and Patterns

Proteins are the functional units of a cell. They have constantly changing expression levels, locations, and PTMs that may be associated with the onset, progression, and remission of disease. Proteomic technologies (as described in Section 2) are used to detect patterns of differentially expressed proteins that have numerous potential clinical applications. Techniques such as 2D-PAGE and HPLC combined with MS can be used to identify proteins for early detection, diagnosis, prognosis, and response to treatment.

Whereas past techniques for studying proteins, such as enzyme-linked immunosorbent assay, were a slow and laborious process of examining one protein at a time, the currently used proteomic analyses of thousands of proteins simultaneously rapidly displays results in terms of protein patterns. The mass spectra generated by analyzing clinical samples from healthy and

TABLE 1
TECHNOLOGY PERFORMANCE CHARACTERISTICS FOR DIFFERENT MASS SPECTROMETERS

	Accuracy	Resolution	Sensitivity	Relative time per sample
ESI				
Q	Low	Low	High	Slow
QIT	Low	Low	Higher	Slow
Q-TOF	High	High	High	Slow
FT-ICR	Very High	Very High	Higher	Slow
MALDI				
Reflectron (high-end)	High	High[a]	High	Fast
Linear (low-end)	Medium	Low	High	Fast
SELDI	Low	Low	High	Fast

[a] Resolution decreases with increasing sample complexity or contamination.
ESI–electrospray ionization.
Q–quadrupole.
QIT–quadrupole-ion-trap.
Q-TOF–quadrupole time of flight.
FT–ICR–Fourier transform-ion cyclotron resonance.
MALDI–matrix-assisted laser desorption ionization.
SELDI–surface-enhanced laser desorption/ionization.

diseased individuals can be compared to define an optimum discriminatory pattern. This concept was the foundation of the bioinformatics tool developed to distinguish neoplastic from nonneoplastic disease within the ovary [43]. Using a preliminary set of mass spectra from the analysis of serum from 50 unaffected women and 50 patients with ovarian cancer, a proteomic pattern that completely discriminated cancer from noncancer was identified. The discriminatory pattern, comprising many individual proteins, none of which could autonomously differentiate diseased from healthy individuals, was then used to classify an independent set of 116 masked serum samples. All 50 ovarian cancer cases were correctly identified, and 63 of the 66 cases of nonmalignant disease were recognized as not cancer. Overall, with a 95% confidence interval, this yielded a sensitivity of 100% (93–100), specificity of 95% (87–99), and positive predictive value of 94% (84–99). One of the newer discriminatory patterns was shown to be both 100% sensitive and specific in a blinded set of 52 healthy women and 92 patients with ovarian cancer [44]. The test is currently lacking the proper clinical validation and software licensing which it requires before use for screening in conjunction with cancer antigen 125 (CA125). At present, the most widely used biomarker for ovarian cancer is CA125, which has a positive predictive value of less than 10% as a single marker and improves to about 20% with the addition of ultrasound [45]. Therefore, mass spectra–generated diagnostic patterns may have higher accuracy than traditional biomarkers of cancer detection. This approach should be performed carefully to ascertain that each peptide is characterized and identified for its role in the carcinogenic process [45a].

Similar to the potential utility of patterns generated from mass spectra analyses, patterns of protein expression observed by 2D-PAGE may also be used for diagnostic purposes [6, 46, 47]. 2D-PAGE resolves complex protein mixtures into numerous unique spots, forming a distinctive pattern. The pattern formed by the protein spots from normal and diseased tissue can then be visually compared. Proteins of interest can be excised from the gel, digested using trypsin, and analyzed by MS to reveal their identity.

Just as if a novel single biomarker was being evaluated for its utility, for potential proteomic patterns to be used as biomarkers they must be validated for their specificity, sensitivity, and predictive value. When considering a single biomarker, a major concern is the issue of false positives causing unnecessary invasive and costly testing on healthy individuals, and false negatives resulting in morbidity and mortality associated with undiagnosed or misdiagnosed disease [48]. The use of a panel of biomarkers would enhance the positive predictive value of a test and minimize false positives and false negatives [49]. Logically, multiple biomarkers would have a higher level of discriminatory information compared to a single biomarker, especially for sizable, diverse patient populations.

Past biomarker discovery efforts have used a one-by-one approach to search for the single elusive protein in the blood. Because there are thousands of intact, modified, and cleaved protein isoforms in human serum, most of which have not been elucidated, finding the lone biomarker is like searching for a needle in a haystack, as each candidate peptide must be identified and checked individually. Given the complex nature of diseases that are most often multistep processes, it is also very likely that one solitary biomarker does not exist for many of them. Thus, proteomics, with its main thrust focusing on high-throughput protein characterization, is perfectly suited to tackle complex biological samples to identify discriminatory patterns [reviewed in 48, 50, 51].

4. Accessing Biological Samples

Ideally, biomarkers should be measurable in a versatile and easily accessible body fluid, such as serum or urine, to maximize clinical use in terms of patient compliance and readiness to interface with the diagnostic equipment. To advance the widespread use of novel diagnostic techniques at the clinical level, a patient's willingness to undergo an unfamiliar and innovative test is an important consideration. Several factors, including the associated level of discomfort and procedural risks, must be taken into account. Another important parameter is patient classification, such as high-risk status resulting from a family history of a disease or a genetic predisposition, compared with routine screening of the general population. Most people are willing to come in for a peripheral blood draw or to give a urine specimen without any signs, symptoms, or definitive indications other than the request of a physician. That makes these fluids prime targets for finding proteomic patterns of clinical usefulness. Biological samples whose collection involves enduring minor amounts of discomfort or minimally invasive procedures include saliva, sweat, and tears. Invasive procedures are required to collect cerebral spinal fluid, amniotic fluid, peritoneal fluid, pleural fluid, synovial fluid, and nipple aspirate fluid, and therefore demand a definitive purpose other than general population screening. With the advances in separation techniques and modern proteomic technologies, all of these body fluids can be prepared such that they can be directly interfaced with medical diagnostics equipment [52].

4.1. Invasion Interface

In any disease state, the deranged molecular processes do not remain confined to individual cells. Cancerous growths, for example, begin with foci of malignant cells that extend outward, creating a tumor–host interface. The tumor microenvironment involves a variety of cells and surrounding stromal and vascular compartments through which blood and local body fluids circulate [53]. Thus, factors that are secreted or shed by malignant cells circulate locally as well as throughout the body. Although the stimulating and promoting factors are still to be defined, body fluids—regional ones as well as serum—are an information reservoir that can be exploited.

As a result of the recent advances in MS, proteomic spectra from a small volume of blood can be generated in literally seconds. A routine peripheral blood draw provides more than an adequate sample, making proteomic studies involving serum amenable to widespread clinical use. However, tissue specimens containing only a few hundred cells as the starting point for analysis may be more problematic because a protein amplification system analogous to PCR is not available. In that sense, researchers working with nucleic acids have the advantage of using an amplification method to overcome the problem of small amounts of material—an option that is not available with proteins. Thus, novel microproteomic technologies that can employ minute amounts of cellular material need to be developed.

4.2. Laser Capture Microdissection

After acquiring a minimal amount of specimen, ranging from a few hundred to a few million cells, current proteomic analyses are able to generate MS patterns in a relatively short time span. In one study, less than 2 hours was needed to process human breast tissue that had been removed from the patient in the operating room until the MS data had been acquired. Mammary tissue containing normal breast epithelium and invasive carcinoma was compared, and over 40 peaks were identified that significantly differed in intensity [54]. This study employed LCM to acquire the tissue specimens.

LCM allows for the isolation of specific cell populations from a mixture of cell types under direct microscopic visualization [55]. In the past, tissue heterogeneity complicated biologically meaningful analysis of bulk solid tumor samples and therefore severely hindered the discovery of disease markers (because of the inability of researchers to isolate pure cell populations). Fortunately, this technique procures samples with a precision of 3–5 μm, where the exact morphologies of both the captured cells and the surrounding tissue are preserved. Within minutes, LCM can procure one to several thousand cells with the DNA, RNA, and proteins remaining intact and unperturbed. Using appropriate methodology, the proteins from all compartments of the cells can be readily obtained in their native conformations and assayed for activity. Frozen tissue samples, surgical- or autopsy-archived paraffin-embedded tissues, cytology cell preparations fixed with formalin or alcohol, and hematoxylin and eosin–stained or unstained tissues can all be extracted by LCM. Thus, LCM is perfectly suited for selectively studying proteins within a specific cell population given a limited quantity of tissue. Coupled to MS, protein analysis from pure cell populations numbering fewer than 100 are now a reality.

In general, LCM is used to capture a very small number of cells, which then undergo purification and subfractionation steps to isolate the proteins. Once the proteins have been run through 2D-PAGE and MS, a proteomic pattern is generated and the specific peptides of interest are identified. It is usually necessary to then go back to the tissue sample to harvest a bigger population of cells containing the desired proteins, so that a larger quantity of them will be available for more in-depth studying.

The difficulties involving sample preparation are exemplified in working with human brain material [56]. The brain is highly complex and heterogeneous, and profiling entire cellular proteomes without contaminants is technically challenging. Furthermore, postmortem tissues undergo protein degradation and artifact generation. Thus, integrating proteomic technologies into the study of neurological and psychiatric disorders is also limited by access to brain and nerve tissue from living subjects.

5. Preparing and Resolving Complex Specimens

Just as tissues need to be separated into their constituent cell parts and then the proteins within further purified, body fluids are complex biological samples with wide ranges of components that require subcellular fractionation to increase protein identification. Given that the dynamic range of proteins in biological systems can reach parts per million or lower, isolating low-abundance proteins poses a particular challenge. Two complementary approaches that aid in detecting proteins expressed at lower levels are implementing cellular fractionation methods that reduce complexity and employing affinity purification for selective enrichment. After beginning with prefractionation of whole-cell lysates, multiple modes of separation in succession (e.g., ion exchange followed by reversed-phase chromatography) are often used. Strategies that reduce proteome complexity facilitate analysis by increasing the dynamic range of protein detection [reviewed in 57].

One technique involves using a SELDI probe for extracting, purifying, and amplifying an analyte within a complex biological solution before MS analysis. The affinity surface of the SELDI probe is designed such that its chemical and physical properties maximize binding of the desired peptides, while all the other molecules can be removed via sequential gradient washing steps. Binding characteristics of the probe may be wide-ranging, similar to chromatographic media, such as ion-exchange, reversed-phase, and immobilized metal. Alternatively, the probe may be very specific for certain antibodies, enzymes, or ligands [58].

5.1. Optimizing Biological Samples

Every biological sample should be optimally prepared to interface with the target diagnostic equipment to obtain the highest quality data. Plasma and urine, the archetypal body fluids for proteomic analysis, each pose intrinsic difficulties in effectively purifying the protein constituents. Plasma has an abundance of lipids and salts; urine has a relative excess of salts and diluted concentrations of proteins. One group of researchers successfully devised a protein extraction method using a CHAPS-based (cholamidoprophy dimethalammonio-1-propanesufonate) buffer to remove the lipids, combined with a centrifugal filter device to exclude the salts. This sample preparation method was applied to plasma as well as other body fluids to reproducibly yield over 90% of the solubilized proteins [59].

Another completely different strategy for increasing the dynamic range of proteins is by removing the highly expressed proteins to improve detection of those expressed at lower levels. Within a given sample, individual protein concentrations often span many orders of magnitudes, with most of the proteins of diagnostic interest present in relatively small amounts. Analyses of complex samples, such as serum, that contain both abundant (albumin and antibodies) and very rare (e.g., proteins released because of disruption of cells) proteins are technically difficult because of the masking of the rare proteins by the abundant ones. Only 22 proteins constitute 99% of the serum protein concentration, which means that characterizing the other 1% of the proteins, which would include peptides shed or secreted by healthy, diseased and dying cells, is quite a technical and analytical challenge [60]. Given 1 mL of blood, finding a peptide at a concentration of 10 pg/mL, such as interleukin 6, among the albumin molecules present at 55 mg/mL, is like finding one individual human being by searching through the population of the entire world: 1 in 6.2 billion [61].

5.2. Subfractionation of Protein Mixtures

Clearly, extensive subfractionation efforts need to be carried out before analyzing a sample. For 2D-PAGE, loading sufficient sample to detect trace proteins invariably means excessive amounts of high-abundance proteins. The widely spread pattern of albumin and immunoglobulins on the gel obscures proteins with similar isoelectric points and molecular weights. Thus, a variety of techniques can be employed to remove a few of the very high concentration proteins to improve detection of the trace ones. Several reagent kits are available to specifically remove albumin or to bind and remove albumin and immunoglobulins [62]. One technique, immunoaffinity-based protein subtraction chromatography, was applied before running a gel and resulted in the detection of more than 350 additional lower-abundance proteins [63]. This method removed several overbearing molecules, including albumin and immunoglobulins G and A, effectively and reproducibly. However, caution must be used in applying fractionation methods for this purpose. In one study, after removing many of the interfering proteins using centrifugal ultrafiltration, it was found that a large number of other proteins of potential clinical interest were unintentionally eliminated as well [64].

Rather than avoiding the problems associated with eliminating carrier molecules, such as albumin, some groups take advantage of the fact that biomarkers may exist in association with the high-abundance proteins: They study both the carrier molecules and their cargo. One group combined immunoprecipitation with microcapillary reversed-phase liquid chromatography and MS/MS to investigate the low–molecular weight proteins that associate with six of the most abundant serum proteins. Using this targeted isolation technique, 210 proteins were identified, of which 73% and 67% were not found in previous studies of the low–molecular weight or whole-serum proteomes, respectively [65]. Many of the identified proteins were recognized as clinically useful biomarkers with a variety of associations from meningiomas to pregnancy to coagulation factor precursors.

A different study purely focused on isolating carrier molecules and their bound proteins to search for biomarkers of clinical interest. It was not only found that circulating carrier proteins were reservoirs for the accumulation and amplification of putative disease markers but also that the low–molecular mass proteins that bound to albumin were distinct from those bound to nonalbumin carriers. Using SELDI-TOF, it was further verified that albumin bound peptides associated with ovarian cancer. This demonstrated that albumin capture was an effective method for harvesting disease-relevant biomarkers [66].

6. Standardization of Proteomic Approaches

Proteomic technologies are currently and increasingly being applied to a broad range of medical fields, and thus turning out enormous data sets at progressively faster rates. For this tremendous amount of potentially useful information to be incorporated into routine clinical practice, several standardization issues must be resolved. The international Human Proteome Organization is already trying to deal with many of them (http://www.hupo.org). Established by the Human Proteome Organization in April 2002, the Proteomics Standards Initiative aims to define community standards for data representation in proteomics and to facilitate data comparison, exchange, and verification [67]. In line with these principles, protocols must be developed regarding the proper use of the technology and equipment to achieve any given objective. Technical variables in specimen collecting, handling, processing, and storing need to be meticulously evaluated and normalized. Variances in the quality of the data also need to be systematically accounted for. Sets of stringent rules should be globally implemented regarding criteria for data reporting and validation to ensure accuracy and avoid publications of erroneous information. No single set of data is definitive. Multiple sets of data using alternative methods of analyses should be incorporated before drawing any firm conclusions. Ultimately, a worldwide information management system needs to be initiated, such that reference values can be integrated and made readily accessible to all laboratories, whether they are centralized facilities or in remote locations.

This daunting process should begin by establishing a comprehensive map of the healthy human proteome [68]. This formidable challenge involves organizing the data into a user-friendly format so that the information can easily be retrieved according to a variety of parameters: biological system, tissue type, cell type, and subcellular compartment. To define "normal," numerous variables must be accounted for including gender, age, time of day, health status, and other yet-unrecognized factors. This baseline information is crucial for detecting the subtle variations that would distinguish a normal from a diseased state. Therefore, current proteomic research efforts extend from identifying the primary sequence of peptides and proteins, including their isoforms and PTMs, to three-dimensional structure characterization and cellular location, function, and interactions with other proteins and molecules.

7. Combining Modern Proteomic Technologies with Traditional Histopathology Techniques

Just as pathologists must know normal histology to recognize abnormal pathology, for a clinician to routinely use mass spectra, the data must be interfaced with computerized pattern recognition algorithms to unambiguously establish control versus diseased states. Until such comprehensive and properly validated tools are available for use as novel "gold standard" diagnostic procedures, proteomic patterns are more likely to be used in the near future as supplements to current medical work-up schemes. Therefore, many research groups have been working on devising protocols that would allow for both a visual evaluation using a light microscope and protein analyses by scanning MALDI-MS on the same tissue section.

Examining biological samples to detect proteins and peptides using MALDI is like transforming the mass spectrometer into a scanning device, much like a microscope. It is already being used as a modern tool for pathology diagnostics [69–71], as it forms 2D mass images of tissue specimens. The labeling of fixed tissues sections with antibodies, as is currently practiced in pathology, is limited to looking for molecules against which the antibodies are already available. The use of scanning MS, as a novel type of microscope, is not restricted to previously identified proteins. Thus, it permits obtaining extensive amounts of information on the molecules distributed throughout the tissue sections and serves as a diagnostic tool of immense power.

One of the most commonly used staining procedures involves hematoxylin and eosin, but it was found that the dyes interfere with either the proteins or the MALDI process. As a consequence, the quality of the mass spectra is significantly compromised [54]. When five other staining protocols (Terry's Polychrome, Toluidine Blue, Nuclear Fast Red, Cresyl Violet, and Methylene Blue) were compared to a control section (unstained tissue rinsed in 70% and 100% ethanol), Cresyl Violet and Methylene Blue had the highest degrees of profile similarities in terms of both staining quality and MS results. In particular, Cresyl Violet offered high nuclear/cytoplasm contrast, which is essential for adequate histological analysis of cancerous tissue [69]. Using these staining protocols, histopathological and MS analyses can be used synergistically to provide sensitive and reliable results.

An alternative approach that is compatible with LCM microscopy involves 2D-PAGE and MALDI-TOF-MS, which use fixed but unstained tissue. This technique, called navigated LCM, uses an adjacent stained section to direct the dissection of an unstained area of interest. The target region must be fixed but remains unstained, thus avoiding the associated generation of artifacts. This technology is ideal for microdissection of the brain because many of the nuclei are too small or irregularly shaped to precisely isolate manually. In one study, a section of rat brain was fixed using 70% ethanol for 30 seconds, followed by navigated LCM, and compared to fresh, unfixed tissue dissected by hand. The 2D-PAGE protein patterns showed 14 matched spots that were subsequently excised, digested with trypsin, and then analyzed by MALDI-TOF-MS. The same 15 proteins were identified in both gels, thereby showing no statistically significant difference in protein coverage between the two different tissue processing methods [72]. This demonstrates that navigated LCM can be used to obtain samples without affecting protein integrity and interfering with downstream MS analysis.

The key to making an accurate diagnosis is in identifying the pathognomonic features of a specimen that unmistakably distinguish it. Using a microscope, this can be done by visually inspecting a tissue section using appropriate stains to highlight and contrast specific features. For MS, it is not the individual proteins but, rather, the pattern that they together form that gives the diagnosis. The discriminatory pattern alone can be used as the diagnostic tool, without the prerequisite of establishing the identification, function, or even clinical relevance of each protein independently. Although identifying each and every peptide is the ultimate goal, it is not obligatory before the technology can be put to clinical use.

8. Practical Clinical Applications

In addition to the potential diagnostic power of a given MS pattern, the advantages of delineating the exact relationship between the various proteins and the disease are seemingly infinite. The identified peptides and proteins can be used for screening, diagnosis, staging, prognosis, monitoring treatment response, or detection of tumor recurrence [73]. Proteomic technology has relevance to every field of medicine and science, making a complete review of the current applications beyond the scope of this chapter. Instead, a few remarkable examples demonstrating the breadth of possibilities are presented, including cancer and infectious diseases [reviewed in 49, 74, 75]. Other topics of interest that are not covered include evaluating drug toxicities [76, 77]; profiling of neuropsychiatric disorders such as schizophrenia, Alzheimer disease, and Parkinson disease [reviewed in 78]; immunophenotyping of leukemias with a long-term goal of monitoring minimal residual disease [79], and characterizing human spermatozoa surface antigens in relation to immunological infertility [25].

8.1. Tumor Markers

Because 2D-PAGE and MS patterns can be used to differentiate between healthy and cancer patients, the proteins in these spectra contain tumor markers. If they could be isolated and identified, they would be useful for innumerable purposes, most notably for early detection. If a tumor marker can warn of the presence of disease at an early stage, before the cancer can be visualized using imaging studies, this would increase the chances of treating at a curative stage.

Tumor antigens can be classified as tumor-specific antigens if they are exclusive to a particular type of cancer cell, or tumor-associated antigens if they are not unique to malignant cells, but are expressed to a significantly greater extent in tumor relative to normal cells. Tumor-associated antigens and tumor-specific antigens include mutated oncogene proteins (p53, c-myc) [80, 81], embryonic proteins (alpha-fetal protein, carcinoembryonic antigen) [82] peptides from immunoprivileged sites (cancer-testis antigens) [83], and overexpressed proteins (HER2/neu) [84–86].

Many tumor markers have been identified in recent years since the improvements in proteomic technologies. For example, one study analyzed human serum with 2D-PAGE and MALDI-TOF-MS to identify a distinct repertoire of autoantibodies associated with hepatocellular carcinoma. These proteins differentiated the cancer patients from those chronically infected with hepatitis B or C, which constitute a high-risk group for developing hepatocellular carcinoma. Thus, proteomic-based technology was used to identify a set of four proteins that may have utility in early diagnosis of hepatocellular carcinoma [87].

A different study used 2D-PAGE and MS to detect proteins from lung adenocarcinoma tissue associated with patient survival. Importantly, protein profiles were used to predict survival of stage I tumor patients. The current standard of care for these patients is surgical resection alone. Therefore, if high-risk stage I individuals could be singled out, they could receive more aggressive adjuvant therapy, which could increase survival rates [88].

Treatment regimens are often based on the predicted clinical outcome; thus, physicians differentially administer aggressive or conservative therapies. This is especially true for prostate cancer, where some men develop advanced disease at a young age and others are merely under observation for prolonged periods of time. To better characterize the immune response in prostate cancer patients, a study was conducted using HPLC and protein microarrays. 2D liquid chromatography was used to separate proteins from the prostate cancer cell line LNCaP into 1760 fractions. These fractions were spotted onto microarrays, which were then incubated with serum samples from men with prostate cancer and male controls. A bioinformatics-based decision tree with two levels of partitioning classified the samples with 98% accuracy as either prostate cancer or control. These results indicated that patterns of immune recognition generated from microarrays of fractionated proteins could be used for prostate cancer diagnosis [89].

Although efforts are underway to identify markers in serum and prostate tissue, the question arose as to whether metastatic prostate cancer cell lines accurately represent *in vivo* disease. It was found that *in vitro* cell cultures (LnCaP and PC3) shared less than 20% of proteins when compared to *in vivo* LCM procured malignant prostate cancer. 2D-PAGE protein profiles were used to compare normal and malignant cells to immortalized cells from the same patient. Protein expression patterns were dramatically altered when cells were grown in culture and immortalized; most notably, a loss of prostate specific antigen expression was observed [18]. Thus, caution must be used when working with immortalized cell lines to discover potential disease markers.

8.2. Disease-Related Applications

Current diagnostic methods rely on invasive procedures, such as biopsies, to accurately assess many conditions. For many disorders, visualization of the pathology via imaging modalities is only possible at an advanced stage. One of the main goals of proteomics is to use readily accessible body fluids to make an accurate diagnosis earlier in the course of the disease. One issue is whether all diseases have associated cells that secrete or shed markers that can be reliably detected through serum or urine profiling, thus obviating the need to perform more invasive procedures. It is still not known which pathologies can be diagnosed by analyzing body fluids, and which ones can be detected only by examining the tissues themselves.

Differential diagnosis of graft-versus-host disease currently depends on organ biopsy to distinguish it from other common complications associated with transplantation. As a means of avoiding repeated biopsies, one group analyzed the urine of patients after hematopoietic stem cell transplantation to generate a peptide pattern that could be used to diagnose graft-versus-host disease. Capillary electrophoresis and electrospray ionization–TOF-MS were used to evaluate the protein patterns in urine from 40 posttransplant patients and compared them to those of five patients with sepsis. Peptides present or absent with an absolute difference of more than 50% or with a more than 10-fold difference in the MS signal intensity between the two groups were accepted as significant disease markers. Sixteen graft-versus-host-disease-specific and 13 sepsis-specific peptides were identified, showing that this technology may be used as a powerful diagnostic tool. Early and accurate identification of patients developing complications after transplantation can lead to more timely and appropriate therapeutic interventions, ultimately translating into reduced morbidity [90].

Since 1986, scientists had been trying to identify the CD8 antiviral factor that was found in greater than normal amounts in certain HIV-1-infected individuals [91]. The elusive CD8 antiviral factor was of tremendous clinical interest because it conferred immunological stability, characterizing this group as long-term nonprogressors. The breakthrough discovery of CD8 antiviral factor being α–defensin 1, 2, and 3 was made possible with the use of SELDI-TOF-MS and searching through protein databases. Only with the help of modern proteomic technology was the extensive 16-year search effectively ended with a greater understanding of the soluble peptide factors suppressing HIV-1 replication [92].

9. Future Directions

A resource infrastructure is already being developed to accommodate the shift from current modalities to proteomic technologies used for clinical laboratory diagnosis. The future of cost-effective, fast, highly sensitive, and reliable diagnoses will be based on 2D-PAGE, HPLC, MS, protein arrays, and bioinformatics. Protein patterns generated from 2D-PAGE and MS will more accurately point to early cancer diagnoses than the currently used histological grading and staging systems. Panels of protein biomarkers in serum and urine will be used for population screening. Collecting these and other readily accessible body fluids will replace invasive biopsies and imaging studies as information sources related to treatment response and disease recurrence. Instead of first- and second-line drug regimens, patient-tailored therapies will become the standard because of individualized protein profiles that can predict the most appropriate medication schemes.

To move from the research phase to the product phase, new generations of proteomic equipment will continue to be developed. This includes novel microproteomic technologies that use tiny amounts of material with extraordinary detection levels for target proteins and peptides. Automation with robotics and standardized machinery will replace manual labor to bring down costs, increase the rate of data analysis, and ensure reproducibility. Biomarker candidates, specifically disease-associated protein patterns, will be verified for their sensitivity, specificity, and reliability of detection. Potential new drug therapies will likewise be tested for their safety and efficacy. Proteomic technologies have the power to bring about all of these objectives, and with time and imagination they will become our new reality.

Acknowledgment

Research in the laboratory of A. Admon is funded by the Greta Koppel Small Cell Lung Carcinoma Fund.

References

[1] Hanash S. Disease proteomics. Nature 2003; 422(6928):226–232.

[2] Patterson SD, Aebersold RH. Proteomics: The first decade and beyond. Nat Genet 2003; 33(Suppl.):311–323.

[3] Zolg JW, Langen H. How industry is approaching the search for new diagnostic markers and biomarkers. Mol Cell Proteomics 2004; 3(4):345–354.

[4] O'Farrell PH. High resolution two-dimensional electrophoresis of proteins. J Biol Chem 1975; 250(10):4007–4021.

[5] Klose J. Protein mapping by combined isoelectric focusing and electrophoresis of mouse tissues. A novel approach to testing for induced point mutations in mammals. Humangenetik 1975; 26(3):231–243.

[6] Celis JE, Gromov P. Proteomics in translational cancer research: Toward an integrated approach. Cancer Cell 2003; 3(1):9–15.

[7] Gorg A, Obermaier C, Boguth G, et al. The current state of two-dimensional electrophoresis with immobilized pH gradients. Electrophoresis 2000; 21(6):1037–1053.

[8] Lauber WM, Carroll JA, Dufield DR, Kiesel JR, Radabaugh MR, Malone JP. Mass spectrometry compatibility of two-dimensional gel protein stains. Electrophoresis 2001; 22(5):906–918.

[9] White IR, Pickford R, Wood J, Skehel JM, Gangadharan B, Cutler P. A statistical comparison of silver and SYPRO Ruby staining for proteomic analysis. Electrophoresis 2004; 25(17):3048–3054.

[10] Kang C, Kim HJ, Kang D, Jung DY, Suh M. Highly sensitive and simple fluorescence staining of proteins in sodium dodecyl sulfate-polyacrylamide-based gels by using hydrophobic tail-mediated enhancement of fluorescein luminescence. Electrophoresis 2003; 24(19–20):3297–3304.

[11] Wittke S, Kaiser T, Mischak H. Differential polypeptide display: The search for the elusive target. J Chromatogr B 2004; 803(1):17–26.

[12] Aebersold R, Mann M. Mass spectrometry-based proteomics. Nature 2003; 422 (6928):198–207.

[13] Le Gac S, Carlier J, Camart JC, Cren-Olive C, Rolando C. Monoliths for microfluidic devices in proteomics. J Chromatogr B Analyt Technol Biomed Life Sci 2004; 808(1):3–14.
[14] Karas M, Hillenkamp F. Laser desorption ionization of proteins with molecular masses exceeding 10,000 daltons. Anal Chem 1988; 60(20):2299–2301.
[15] Hutchens TW, Yip T-T. New desorption strategies for the mass spectrometric analysis of macromolecules. Rapid Commun Mass Spectrom 1993; 7(7):576–580.
[16] Fenn JB, Mann M, Meng CK, Wong SF, Whitehouse CM. Electrospray ionization for mass spectrometry of large biomolecules. Science 1989; 246(4926):64–71.
[17] Biemann K, Scoble HA. Characterization by tandem mass spectrometry of structural modifications in proteins. Science 1987; 237(4818):992–998.
[18] Ornstein DK, Gillespie JW, Paweletz CP, et al. Proteomic analysis of laser capture microdissected human prostate cancer and *in vitro* prostate cell lines. Electrophoresis 2000; 21(11):2235–2242.
[19] Homola J. Present and future of surface plasmon resonance biosensors. Anal Bioanal Chem 2003; 19:19.
[20] McDonnell JM. Surface plasmon resonance: Towards an understanding of the mechanisms of biological molecular recognition. Curr Opin Chem Biol 2001; 5(5):572–577.
[21] de Wildt RM, Mundy CR, Gorick BD, Tomlinson IM. Antibody arrays for high-throughput screening of antibody-antigen interactions. Nat Biotechnol 2000; 18(9):989–994.
[22] MacBeath G. Protein microarrays and proteomics. Nat Genet 2002; 32(Suppl.):526–532.
[23] Michaud GA, Salcius M, Zhou F, et al. Analyzing antibody specificity with whole proteome microarrays. Nat Biotechnol 2003; 21(12):1509–1512.
[24] Walter G, Bussow K, Lueking A, Glokler J. High-throughput protein arrays: Prospects for molecular diagnostics. Trends Mol Med 2002; 8(6):243–250.
[25] Robinson WH, DiGennaro C, Hueber W, et al. Autoantigen microarrays for multiplex characterization of autoantibody responses. Nat Med 2002; 8(3):295–301.
[26] Cutler P. Protein arrays: The current state-of-the-art. Proteomics 2003; 3(1):3–18.
[27] Zhu H, Bilgin M, Snyder M. Proteomics. Annu Rev Biochem 2003; 72:783–812.
[28] Henzel WJ, Billeci TM, Stults JT, Wong SC, Grimley C, Watanabe C. Identifying proteins from two-dimensional gels by molecular mass searching of peptide fragments in protein sequence databases. Proc Natl Acad Sci USA 1993; 90(11):5011–5015.
[29] Mann M, Hojrup P, Roepstorff P. Use of mass spectrometric molecular weight information to identify proteins in sequence databases. Biol Mas Spectrom 1993; 22(6):338–345.
[30] Pappin DJC, Hojrup P, Bleasby AJ. Rapid identification of proteins by peptide-mass fingerprinting. Curr Biol 1993; 3(6):327–332.
[31] James P, Quadroni M, Carafoli E, Gonnet G. Protein identification by mass profile fingerprinting. Biochem Biophys Res Comm 1993; 195(1):58–64.
[32] Yates JR, Speicher S, Griffin PR, Hunkapiller T. Peptide mass maps: A highly informative approach to protein identification. Anal Biochem 1993; 214(2):397–408.
[33] Nesvizhskii AI, Aebersold R. Analysis, statistical validation and dissemination of large-scale proteomics datasets generated by tandem MS. Drug Discov Today 2004; 9(4):173–181.
[34] Palmblad M, Ramstrom M, Markides KE, Hakansson P, Bergquist J. Prediction of chromatographic retention and protein identification in liquid chromatography/mass spectrometry. Anal Chem 2002; 74(22):5826–5830.
[35] Beer I, Barnea E, Ziv T, Admon A. Improving large-scale proteomics by clustering of mass spectrometry data. Proteomics 2004; 4(4):950–960.

[36] Radulovic D, Jelveh S, Ryu S, et al. Informatics platform for global proteomic profiling and biomarker discovery using liquid-chromatography-tandem mass spectrometry. Mol Cell Proteomics 2004; 21:21.

[37] Gambari R, Borgatti M, Altomare L, et al. Applications to cancer research of "lab-on-a-chip" devices based on dielectrophoresis (DEP). Technol Cancer Res Treat 2003; 2(1):31–40.

[38] Belov ME, Anderson GA, Wingerd MA, et al. An automated high performance capillary liquid chromatography-Fourier transform ion cyclotron resonance mass spectrometer for high-throughput proteomics. J Am Soc Mass Spectrom 2004; 15(2):212–232.

[39] Zhang X, Leung SM, Morris CR, Shigenaga MK. Evaluation of a novel, integrated approach using functionalized magnetic beads, bench-top MALDI-TOF-MS with prestructured sample supports, and pattern recognition software for profiling potential biomarkers in human plasma. J Biomol Tech 2004; 15(3):167–175.

[40] Wong YF, Cheung TH, Lo KW, et al. Protein profiling of cervical cancer by protein-biochips: Proteomic scoring to discriminate cervical cancer from normal cervix. Cance Lett 2004; 211(2):227–234.

[41] Quadroni M, James P. Proteomics and automation. Electrophoresis 1999; 20(4–5):664–677.

[42] Triolo A, Altamura M, Cardinali F, Sisto A, Maggi CA. Mass spectrometry and combinatorial chemistry: A short outline. J Mas Spectrom 2001; 36(12):1249–1259.

[43] Petricoin EF, Ardekani AM, Hitt BA, et al. Use of proteomic patterns in serum to identify ovarian cancer. Lancet 2002; 359(9306):572–577.

[44] Conrads TP, Fusaro VA, Ross S, et al. High-resolution serum proteomic features for ovarian cancer detection. Endocr Relat Cancer 2004; 11(2):163–178.

[45] Cohen LS, Escobar PF, Scharm C, Glimco B, Fishman DA. Three-dimensional power Doppler ultrasound improves the diagnostic accuracy for ovarian cancer prediction. Gynecol Oncol 2001; 82(1):40–48.

[45a] Diamandis EP. Point: Proteomic patterns in biological fluids: Do they represent the future of cancer diagnostics? Clin Chem 2003; 49(8):1272–1275.

[46] Hanash SM. Global profiling of gene expression in cancer using genomics and proteomics. Curr Opin Mol Ther 2001; 3(6):538–545.

[47] Hirsch J, Hansen KC, Burlingame AL, Matthay MA. Proteomics: Current techniques and potential applications to lung disease. Am J Physiol Lung Cell Mol Physiol 2004; 287(1):L1–L23.

[48] Rosenblatt KP, Bryant-Greenwood P, Killian JK, et al. Serum proteomics in cancer diagnosis and management. Annu Rev Med 2004; 55:97–112.

[49] Srinivas PR, Verma M, Zhao Y, Srivastava S. Proteomics for cancer biomarker discovery. Clin Chem 2002; 48(8):1160–1169.

[50] Wulfkuhle JD, Liotta LA, Petricoin EF. Proteomic applications for the early detection of cancer. Nat Rev Cancer 2003; 3(4):267–275.

[51] Petricoin E, Wulfkuhle J, Espina V, Liotta LA. Clinical proteomics: Revolutionizing disease detection and patient tailoring therapy. J Proteome Res 2004; 3(2):209–217.

[52] Turner BJ, Weiner M, Yang C, TenHave T. Predicting adherence to colonoscopy or flexible sigmoidoscopy on the basis of physician appointment-keeping behavior. Ann Intern Med 2004; 140(7):528–532.

[53] Liotta LA, Kohn EC. The microenvironment of the tumour-host interface. Nature 2001; 411(6835):375–379.

[54] Xu BJ, Caprioli RM, Sanders ME, Jensen RA. Direct analysis of laser capture microdissected cells by MALDI mass spectrometry. J Am Soc Mass Spectrom 2002; 13(11):1292–1297.

[55] Emmert-Buck MR, Bonner RF, Smith PD, et al. Laser capture microdissection. Science 1996; 274(5289):998–1001.

[56] Choudhary J, Grant SG. Proteomics in postgenomic neuroscience: The end of the beginning. Nat Neurosci 2004; 7(5):440–445.

[57] Dreger M. Proteome analysis at the level of subcellular structures. Eur J Biochem 2003; 270(4):589–599.

[58] Tang N, Tornatore P, Weinberger SR. Current developments in SELDI affinity technology. Mas Spectrom Rev 2004; 23(1):34–44.

[59] Joo WA, Lee DY, Kim CW. Development of an effective sample preparation method for the proteome analysis of body fluids using 2-D gel electrophoresis. Biosci Biotechnol Biochem 2003; 67(7):1574–1577.

[60] Yu LR, Zhou M, Conrads TP, Veenstra TD. Diagnostic proteomics: Serum proteomic patterns for the detection of early stage cancers. Dis Marker 2003; 19(4–5):209–218.

[61] Anderson NL, Anderson NG. The human plasma proteome: History, character, and diagnostic prospects. Mol Cell Proteomics 2002; 1(11):845–867.

[62] Ahmed N, Barker G, Oliva K, et al. An approach to remove albumin for the proteomic analysis of low abundance biomarkers in human serum. Proteomics 2003; 3(10):1980–1987.

[63] Pieper R, Su Q, Gatlin CL, Huang ST, Anderson NL, Steiner S. Multi-component immunoaffinity subtraction chromatography: An innovative step towards a comprehensive survey of the human plasma proteome. Proteomics 2003; 3(4):422–432.

[64] Georgiou HM, Rice GE, Baker MS. Proteomic analysis of human plasma: Failure of centrifugal ultrafiltration to remove albumin and other high molecular weight proteins. Proteomics 2001; 1(12):1503–1506.

[65] Zhou M, Lucas DA, Chan KC, et al. An investigation into the human serum "interactome." Electrophoresis 2004; 25(9):1289–1298.

[66] Mehta AI, Ross S, Lowenthal MS, et al. Biomarker amplification by serum carrier protein binding. Dis Marker 2003; 19(1):1–10.

[67] Orchard S, Hermjakob H, Apweiler R. The proteomics standards initiative. Proteomics 2003; 3(7):1374–1376.

[68] Wilkins MR, Sanchez JC, Gooley AA, et al. Progress with proteome projects: Why all proteins expressed by a genome should be identified and how to do it. Biotechnol Genet Eng Rev 1996; 13:19–50.

[69] Chaurand P, Schwartz SA, Caprioli RM. Assessing protein patterns in disease using imaging mass spectrometry. J Proteome Res 2004; 3(2):245–252.

[70] Chaurand P, Schwartz SA, Billheimer D, Xu BJ, Crecelius A, Caprioli RM. Integrating histology and imaging mass spectrometry. Anal Chem 2004; 76(4):1145–1155.

[71] Schwartz SA, Weil RJ, Johnson MD, Toms SA, Caprioli RM. Protein profiling in brain tumors using mass spectrometry: Feasibility of a new technique for the analysis of protein expression. Clin Cance Res 2004; 10(3):981–987.

[72] Mouledous L, Hunt S, Harcourt R, Harry J, Williams KL, Gutstein HB. Navigated laser capture microdissection as an alternative to direct histological staining for proteomic analysis of brain samples. Proteomics 2003; 3(5):610–615.

[73] Poon TC, Johnson PJ. Proteome analysis and its impact on the discovery of serological tumor markers. Clin Chim Acta 2001; 313(1–2):231–239.

[74] Jain KK. Role of proteomics in diagnosis of cancer. Technol Cancer Res Treat 2002; 1(4):281–286.

[75] Rai AJ, Chan DW. Cancer proteomics: Serum diagnostics for tumor marker discovery. Ann NY Acad Sci 2004; 1022:286–294.

[76] Kennedy S. Proteomic profiling from human samples: The body fluid alternative. Toxicol Lett 2001; 120(1–3):379–384.

[77] Mian S, Ball G, Hornbuckle J, et al. A prototype methodology combining surface-enhanced laser desorption/ionization protein chip technology and artificial neural network algorithms to predict the chemoresponsiveness of breast cancer cell lines exposed to Paclitaxel and Doxorubicin under *in vitro* conditions. Proteomics 2003; 3(9):1725–1737.

[78] Kim SI, Voshol H, van Oostrum J, Hastings TG, Cascio M, Glucksman MJ. Neuroproteomics: Expression profiling of the brain's proteomes in health and disease. Neurochem Res 2004; 29(6):1317–1331.

[79] Belov L, de la Vega O, dos Remedios CG, Mulligan SP, Christopherson RI. Immunophenotyping of leukemias using a cluster of differentiation antibody microarray. Cance Res 2001; 61(11):4483–4489.

[80] Iba T, Kigawa J, Kanamori Y, et al. Expression of the c-myc gene as a predictor of chemotherapy response and a prognostic factor in patients with ovarian cancer. Cancer Sci 2004; 95(5):418–423.

[81] Wang JY, Hsieh JS, Chen CC, et al. Alterations of APC, c-met, and p53 genes in tumor tissue and serum of patients with gastric cancers. J Surg Res 2004; 120(2):242–248.

[82] Perkins GL, Slater ED, Sanders GK, Prichard JG. Serum tumor markers. Am Fam Physician 2003; 68(6):1075–1082.

[83] Scanlan MJ, Gure AO, Jungbluth AA, Old LJ, Chen YT. Cancer/testis antigens: An expanding family of targets for cancer immunotherapy. Immunol Rev 2002; 188:22–32.

[84] Parmiani G, Castelli C, Dalerba P, et al. Cancer immunotherapy with peptide-based vaccines: What have we achieved? Where are we going? J Natl Cancer Inst 2002; 94(11):805–818.

[85] Mocellin S, Rossi CR, Nitti D, Lise M, Marincola FM. Dissecting tumor responsiveness to immunotherapy: The experience of peptide-based melanoma vaccines. Biochim Biophys Acta 2003; 1653(2):61–71.

[86] Disis ML, Gooley TA, Rinn K, et al. Generation of T-cell immunity to the HER-2/neu protein after active immunization with HER-2/neu peptide-based vaccines. J Clin Oncol 2002; 20(11):2624–2632.

[87] Le Naour F, Brichory F, Misek DE, Brechot C, Hanash SM, Beretta L. A distinct repertoire of autoantibodies in hepatocellular carcinoma identified by proteomic analysis. Mol Cel Proteomics 2002; 1(3):197–203.

[88] Chen G, Gharib TG, Wang H, et al. Protein profiles associated with survival in lung adenocarcinoma. PNAS 2003; 100(23):13537–13542.

[89] Bouwman K, Qiu J, Zhou H, et al. Microarrays of tumor cell derived proteins uncover a distinct pattern of prostate cancer serum immunoreactivity. Proteomics 2003; 3(11):2200–2207.

[90] Kaiser T, Kamal H, Rank A, et al. Proteomics applied to the clinical follow-up of patients after allogeneic hematopoietic stem cell transplantation. Blood 2004; 104(2):340–349.

[91] Walker CM, Moody DJ, Stites DP, Levy JA. CD8+ lymphocytes can control HIV infection *in vitro* by suppressing virus replication. Science 1986; 234(4783):1563–1566.

[92] Zhang L, Yu W, He T, et al. Contribution of Human alpha-Defensin 1, 2, and 3 to the Anti-HIV-1 Activity of CD8 Antiviral Factor. Science 2002; 298(5595):995–1000.

MOLECULAR DETERMINANTS OF HUMAN LONGEVITY

Francesco Panza,* Alessia D'Introno,* Anna M. Colacicco,* Cristiano Capurso,*,† Rosa Palasciano,* Sabrina Capurso,* Annamaria Gadaleta,* Antonio Capurso,* Patrick G. Kehoe,‡ and Vincenzo Solfrizzi*

***Department of Geriatrics, Center for the Aging Brain, Memory Unit, University of Bari, 11-70124, Bari, Italy**
†Department of Geriatrics, University of Foggia, 71100 Foggia, Italy
‡Department of Care of the Elderly, University of Bristol, Bristol BS16 1LE, England

1. Introduction

Aging can be considered the product of an interaction between genetic, environmental, and lifestyle factors, which in turn influence longevity that varies between and within species [1]. Given the high complexity of the phenomenon, several theories have been proposed providing an insight in the role of genetic and environmental factors in the process of aging [2, 3].

0065-2423/05 $35.00
DOI: 10.1016/S0065-2423(04)39007-4

The "disposable soma theory" [4, 5], the most attractive theory, as well as other evolutionary theories state that aging is not only under genetic control and can also be considered a result of the failure of homeostasis. Therefore, although most studies agree that genetics influence longevity in humans, the magnitude of this effect is debated [6]. A study on children of nonagenarians indicated a strong relationship between genetic influences and longevity [7], as did a study that compared the life span of adopted children with those of their adoptive and biological parents [8]. Nonetheless, studies on twins reared together and twins reared apart indicated a small genetic influence on longevity [9, 10]. In fact, in one study, genetic factors explained no more than 30% of the variance in longevity [10], and in another study, this variance was even less [9]. However, these studies did not analyze the oldest survivors, nor did they compare the longer-living with the shorter-living subjects. In fact, a strong relationship between genetics and longevity was demonstrated when centenarians were included [7, 8], suggesting that genetic control of longevity is greatest in the oldest adults. A recent study demonstrated that the siblings of centenarians are three to four times more likely to survive to the 10th decade of life, compared with siblings of noncentenarians [11]. Furthermore, immediate ancestors of Jeanne Calment from France, who died at the age of 123 years, were shown to be 10 times more likely to reach age 80 years than the ancestral cohort [12]. These studies support the concept that longevity is a familial trait likely to be inherited and points to extreme age as the phenotype for an initial approach in identifying chromosomal regions that harbor longevity-assurance genes. Therefore, although the debate of the importance on genetics in determining the reaching of extreme longevity is open, it is mandatory to study the role of genetic determinants of longevity in humans, and several studies have been focused on healthy centenarians [2]. In fact, these exceptionally long-living individuals represent a model—not discussed—of successful aging, having escaped the major age-associated diseases, and with most of them maintaining good cognitive and functional status.

When age is plotted against the log of mortality rate, it gives a straight line as the mortality rate increases exponentially with age [13]. However, the log of mortality rate falls below the expected at the ages of 95–100 years, indicating that mortality rate is no longer increasing exponentially in this age group [14, 15]. Although the mortality rate from cancers increases by approximately 10% per decade, it actually decreases after the age of 90 years. Thus, those individuals who have achieved an age of 90 years or older seem to be biologically unique; they have escaped disease-related mortality and get the biological make-up for successful aging. Interestingly, because the ratio of women to men at age 100 is 5 to 1, the female phenotype contributes to longevity independent of other genetic characteristics. Thus, based on biological distinctions, differences in mortality pattern, and the marked decrease in cancer, centenarians are likely to possess the strongest genetic determinants of longevity.

The identification of gene variants involved in aging and longevity presents an interesting challenge [6]. The discovery of genetic variations that explain even 5%–10% of the variation in survival to extreme old age could yield important clues about the cellular and biochemical mechanisms that affect the aging process and susceptibility to age-associated diseases [16]. Candidate gene approaches, in which a gene is chosen based on function and the presumption that alteration in its function may affect the phenotype, have met with some success. Nonetheless, the definition of longevity and its associated intermediate phenotypes is still being debated. Furthermore, in the absence of detailed genealogies and prospective data, it is not possible to know definitively which individuals are or will be long-lived, and which are or will not be long-lived. Finally, individuals who died at early ages in accidents or war, or even from diseases resulting from environmental or other genetic factors, may still have harbored longevity-assurance genes [6].

Recently, Richard Miller proposed a classification of longevity genes in different categories [17] (Table 1); the first one includes genes that cause or accelerate aging, even though it is a point of debate whether or not genetic mutations exist in nature that actually either cause or accelerate aging (e.g., P53 gene, telomerase gene). The second category concerns genes that increase the risk of a specific illness early in life but do not appear to be related to aging (e.g., CF gene and cystic fibrosis), or alternatively, genes that increase the risk of specific illness that resembles some of the consequences of aging.

TABLE 1

HYPOTHETICAL CLASSIFICATION OF LONGEVITY GENES OR THEIR ALTERNATIVES[a]

1. Genes that cause aging (P53?)
2a. Genes that increase the risk of a specific disease early in life but do not appear to be related to aging (e.g., CF gene and cystic fibrosis)
2b. Genes that increase the risk of specific disease that resembles some of the consequences of aging (e.g., Werner's syndrome)
3. Genes that influence or cause age-related diseases (e.g., Alzheimer's disease and apolipoprotein E ϵ4 allele)
4. Low-fitness genes that extend maximum life span, probably by slowing down aging (as observed in lower organism mutations; e.g., *daf* genes)
5. Polymorphic genetic loci that influence the rate of aging (many quantitative trait loci with varying influences on aging and age-associated diseases)
6. Genes that influence differences in life-span among species (e.g., longevity-enabling genes)[a]

[a]Adapted from Miller RA. A position paper on longevity genes. Document URL: http://sageke.sciencemag.org/cgi/content/full/sageke;2001/9/vp6 and from: Miller RA. A position paper on longevity genes. Sci Aging Knowledge Environ 2001, vp6 (2001) (17).

The third category consists of genes that influence or cause age-related diseases (e.g., Alzheimer disease [AD] and apolipoprotein E [APOE] ϵ4 allele). Because variations in these genes are also associated with increased mortality risk, it is likely that centenarians do not have many of these predisposing variations. However, because the frequency of disease alleles is reduced in centenarians versus younger controls selected from the population, the statistical power of an association study between centenarians and subjects with a specific disease should be increased. This should be particularly true when searching for alleles that have a relatively high frequency in the general population. The fourth class includes low-fitness genes that extend maximum life span, probably by slowing down aging, as observed in lower organism mutations. One approach to determining the significance of such genes in humans is to screen for polymorphisms of their human homologs and to determine the allele frequencies among specific human phenotypes such as centenarians and to compare them to ethnically matched younger controls or controls predisposed to premature mortality [18]. The fifth and sixth categories concern, respectively, polymorphic genetic loci that influence the rate of aging, and genes that influence differences in life span among species (e.g., longevity-enabling genes). A useful approach to finding these life span genes may be association studies using centenarian sibships. Families highly clustered for longevity, with five or more centenarian siblings and multiple centenarian cousins, provide the potential opportunity to perform linkage studies, linking the extreme longevity phenotype to a specific gene or genes [19].

The aim of this chapter is to examine in depth the current knowledge of the role of different genetic determinants in modulating aging and in reaching of extreme longevity in humans, with particular interest in centenarian studies. We reviewed studies belonging to the third Miller's class of longevity genes, concerning those genes that influence or cause age-related diseases [17]. First, we focused our attention on genes involved in vascular risk and vascular-related diseases and discussed the evidence that genetic factors, likely to be linked to both vascular disease and AD, may have an additional role in determining human longevity. Second, we reviewed principal findings on genetic factors linked to inflammation (interleukin 6 [IL-6] gene and other cytokine genes) and regulating immune response.

2. Vascular Genetic Factors and Human Longevity

Complex interrelationships between age-associated illnesses such as vascular disease and AD indicate that biological and genetic pathways may be worthy of examination in centenarian populations to provide insights into human longevity. This is also borne out by the involvement of lipoprotein metabolism and a number of vascular genetic risk factors [20, 21]. The search for factors involved in aging and longevity has progressed extensively in recent years because of increased human life expectancy and elevation of the number of elderly people, which in turn results in increased prevalence of age-related illnesses. Different genetic and nongenetic factors have been examined in the quest to understand the biological basis of human longevity. For example, centenarians are characterized by marked delay or escape from age-related diseases, such as coronary artery disease (CAD), cerebrovascular disease (CVD), and AD, which respectively are the first, the third, and the fourth largest causes of mortality in Western populations. Thus one can suggest that genes and biochemical factors likely to be implicated in these disorders may have a role in human longevity.

Among the genetic markers, the APOE gene has been the most widely examined in centenarian populations [22–24], because of its well-documented role in AD [25] as well as vascular diseases [26]. In fact, the APOE ϵ4 allele is associated with high serum total cholesterol, low-density lipoprotein cholesterol (LDL-C), and apolipoprotein B gene (*APOB*) levels in many populations [27] and has been found to increase risk for CAD, myocardial infarction [26], and AD [25]. Another gene, the angiotensin 1 converting enzyme (ACE1), a suggested risk factor for CAD [28] and late-onset AD [29], has also been associated with longevity [24]. However, the latter longevity findings have not been replicated in independent populations [30–33]. A small number of other genes linked to lipoprotein metabolism and vascular disease have been investigated as putative markers for longevity, although generally with negative or inconsistent association results. However, it remains too early to make any clear conclusions about the involvement of these genetic and nongenetic factors in successful aging and longevity, as more studies in different centenarian populations are required.

2.1. APOE Gene Polymorphism and Longevity

APOE has three common isoforms, namely, E2, E3, and E4, which are coded by the APOE alleles ϵ2, ϵ3, and ϵ4 at a single locus on chromosome 19. The APOE gene has been extensively examined in populations worldwide for its association with an increased risk for CAD and AD [25, 34] and has also been evaluated in centenarian populations in different studies.

2.1.1. *APOE Studies on Centenarians and Longevity*

In a recent study we showed that the frequency of the ϵ4 allele was significantly higher in middle-aged subjects than in centenarians, indicating selection against ϵ4 alleles. In the same study the ϵ2 allele was found to be increased in centenarians, although not significantly [23] (Table 2). Our study

TABLE 2
PRINCIPAL STUDIES ON GENE POLYMORPHISMS LIKELY INVOLVED IN HUMAN LONGEVITY

Apolipoprotein E (APOE) polymorphism studies

Reference	Results and conclusions
Shachter *et al.*, 1994	The APOEϵ4 allele frequency was significantly decreased in centenarians versus middle-aged controls (0.05 vs. 0.11; $p < .001$), while the APOEϵ2 allele frequency was significantly increased (0.13 versus 0.07; $p < .01$).
Louhija *et al.*, 1994	The APOEϵ2 allele frequency was significantly higher (0.07 versus 0.04; $p < .05$) and the APOEϵ4 allele lower in centenarians than in control populations (0.08 versus 0.23; $p < .001$).
Panza *et al.*, 1999	The APOEϵ4 allele frequency was significantly higher in middle-aged subjects than in centenarians (0.08 versus 0.02; $p < .05$), while APOEϵ2 allele frequency did not reach significance.

Angiotensin I converting enzyme 1 (ACE1) polymorphism studies

Reference	Results and conclusions
Shachter *et al.*, 1994	Increased ACE1*DD genotype was observed in centenarians compared to middle-aged subjects (0.4 versus 0.26; $p < .01$).
Rahmutula *et al.*, 2001	Within the Uighur group the frequency of the D allele was significantly higher in the older age group (90–113 years) (0.45) than in the younger age group (59–70 years) (0.36) ($p < .04$).

Other relevant gene polymorphism

Methyltetrahydrofolatereductase (MTHFR) polymorphism

Matsushita *et al.*, 1997	The homozygous mutation occurred significantly less often in both the older (55–79 years) and oldest ($\geq$80 years) groups than in the younger group, especially among males (total: younger = 0.19, older = 0.14, oldest = 0.07; $p = .006$ for differences among the three age groups).

Paraoxonase 1 (PON1) polymorphism

Bonafè *et al.*, 2002	The percentage of carriers of the R allele at codon 192 was higher in centenarians than in controls (0.54 versus 0.45).
Rea *et al.*, 2004	The oldest subjects in Italy demonstrated an increased frequency of the PON1 192 R allele (0.32 versus 0.26) ($p = .02$).

3′Apolipoprotein B (APOB)-VNTR polymorphism

De Benedictis *et al.*, 1997	There was an association between the APOB locus and longevity. In particular, the frequency of 3′APOB-VNTR alleles with fewer than 35 repeats was significantly lower in cases than in controls.
De Benedictis *et al.*, 1998	The frequency of SS (small, 26–34 repeats) in the genotype pool increased from the group aged 10–19 years (3.06) to that aged 40–49 years (8.51). Then it declined, reaching the minimum value in centenarians (1.58).

TABLE 2 (*Continued*)

PAI-1 4G/5G polymorphism	
Mackness *et al.*, 1998	In centenarians there was a significantly higher frequency of the 4G allele and of the homozygous 4G/4G genotype associated with high PAI-1 levels.
Apolipoprotein CIII (APOCIII)-SstI polymorphism	
Louhija *et al.*, 1999	The APOC-III S2 allele (SstI restriction site present) occurred more often in the centenarians (0.13) than in the youngest reference population (0.09) ($p < .05$).
Apolipoprotein CIII (APOCIII) T-455C polymorphism	
Anisimov *et al.*, 2001	A greater frequency of the APOC-III-455C allele was demonstrated with aging ($p < .005$).

confirmed the findings from previous studies from France [24] and Finland [22] showing a decrease of the APOEϵ4 allele frequency and a concomitant increase in APOEϵ2 allele in centenarians compared with patients from younger age groups (Table 2). A similar, but not significant, trend was found in a smaller cohort from Japan [35–37]. A decrease in frequency of APOEϵ4 allele was also observed in studies on octogenarians and nonagenarians [38, 39]. Interesting and perhaps even surprising was the observation from a Northern Ireland population that, despite a high intrinsic incidence of cardiovascular disease in nonagenarian subjects, the ϵ4 allele frequency was still reduced and the ϵ3 unchanged and ϵ2 increased [40]. Furthermore, in a related study involving 1562 Han Chinese subjects (20–108 years of age), the APOEϵ4 allele frequency in the oldest age group ($\geq$85 years) was significantly lower from that in the young (20–39 years), middle (40–59 years), and old (60–84 years) age groups (2.5% versus 8.4%, 7.9%, and 7.6%, respectively) [41].

2.1.2. *APOE Allele Geographical Trends and APOE Serum Levels in Centenarians*

In our study, we reported for the first time a significant geographic trend for increasing ϵ2 allele and decreasing ϵ4 allele prevalence in centenarian populations from northern and southern regions of Europe [23]. We suggested that this geographical trend could influence the strength of association between the ϵ2 allele and longevity, as confirmed by a Finnish centenarian study showing a significant trend for increased ϵ2 allele frequency in persons aged 100–101 (9%), 102–103 (21%), and 104 years and older (25%) [42]. However, in the same population of centenarians who were identified from the Finnish National Population Register in 1991, a previous study had showed no significant survival differences after a 3-year follow-up among different APOE allele carriers, indicating that in this very select group the protection/

risk effect associated with the APOE polymorphism might be attenuated [43]. In another recent study conducted on 177 Danish centenarians and data from a study of 40-year-old Danish men, 21% of the centenarians were APOEϵ2-carriers and 15% were ϵ4-carriers compared to 13% and 29%, respectively, of the young men. The relative mortality risk-values were 0.95 for ϵ2-carriers and 1.13 for ϵ4-carriers, using ϵ_3/ϵ_3 and ϵ_4/ϵ_2 genotypes as reference, thus supporting their view that the APOE gene is a "frailty gene" and not a "longevity gene" [44]. Furthermore, we have recently evaluated serum *APOE* levels in centenarian population compared to young healthy adults and demonstrated a significant trend for decreasing serum *APOE* levels in healthy centenarians and young healthy adults from APOE ϵ2- to ϵ4-carriers; we also observed significant differences in serum *APOE* levels with respect to age in APOEϵ4-carriers either in young healthy adults compared to elderly age-matched controls for AD subjects or in young healthy adults compared to centenarians, but only after adjustment for serum high-density lipoprotein cholesterol levels [45]. The role of serum *APOE* concentration in extreme longevity may be explained by the relevance of this factor in CVD [46], which may be linked with observation that serum *APOE* concentration modulates lipid metabolism [47], or related to genetic association with inheritance of APOE variants, as has also been suggested with respect to CAD [48].

2.2. ACE1 Gene and Centenarians

ACE1, located on chromosome 17, has an insertion (I allele)/deletion (D allele) genetic polymorphism in its 16th intron [49]. The physiological importance of ACE1 I/D polymorphism relies on the fact that subjects bearing the D allele have higher circulating and tissue *ACE* levels than those with I allele.

The potential role of the ACE I/D polymorphism in longevity was initially suggested in a French study by Shachter *et al.* [24], who reported a higher prevalence of the ACE1*D allele in centenarians compared with younger subjects aged 20–70 years (Table 2). The greater representation of the D allele in the very old subjects was an apparent paradox because previous studies had shown that patients with CAD bear the D allele more frequently [28]. The apparent dichotomy might be explained by the possibility that there is an advantage from the inheritance of the D allele that might be linked to the involvement of the ACE1 polymorphism in a wider range of cellular functions than previously thought, such as repair to damaged tissues or resistance to neoplasia or infection, all of which would favor increased survival rates [50, 51]. However, although this remains a speculation, the unexpected association of ACE1*D/D genotype with longevity [24] might also be related to the recent suggested relationship between ACE1*D allele and reduced risk of AD development [29].

Since the original ACE1–longevity association was reported [24], several studies have attempted to replicate the findings, but with little success

[30–33]. A small study on British subjects aged 84 years and older reported a significant depletion of the ACE1*I/I genotype, but only in elderly males compared to elderly women and controls, indicating an influence of gender on the penetrance of ACE1 locus [52].

A retyping of the ACE1 locus in the original centenarian cohort of Schachter [24] revealed a 5% difference with the original frequencies of the ACE1 alleles. These differences were homogeneously distributed across genders in both populations of centenarians and controls [31]. It is notable that the French centenarians and controls of the original study [24] were not matched for geographical origin, and 20% of the centenarians were born outside of France or had unknown birthplaces, indicating heterogeneity in the geographical distribution of the ACE1 polymorphism. In a recent study by our group, we found no evidence to support the suggested role of the ACE1 I/D polymorphism in human longevity [33], as did previous studies from Denmark [30] and Korea [32]. However, we did provide the first report that the ACE1*I allele frequency decreases according to a geographical gradient from Northern to Southern Europe in centenarians, while conversely, the ACE1*D allele frequency appears to increase in the opposite direction (from south to north) [33]. These data reflect similar results from other separate studies on ACE1 involvement in CAD, where closer inspection of ACE1 data in the cohorts revealed regional differences in the ACE1*D allele frequency (3% within France and 7% between Northern Ireland and Southern France) [28].

In the Asian population, the only association supporting the role of ACE1*D allele in longevity was found in a certain ethnic group, where the frequency of the D allele was significantly higher in the older age group (0.448) than in the younger age group (0.355) [53] (Table 2).

While it appears that APOE may play a role in longevity in different populations [22–24, 31, 35–37], the majority of evidence on ACE1 [30–33, 54] does not, to date, support its involvement in longevity. It is clear that only large longitudinal studies specifically designed for addressing the effect of the ACE1 polymorphism on extreme longevity could provide more robust evidence. However, although it is noteworthy that significant geographical differences exist in ACE1*I allele frequency in centenarians [33], which could effect, with other environmental factors (e.g., differing diet, lifestyle), on variation in human longevity, one might also have to question the value for further studies of ACE1 in longevity with the overwhelming evidence indicating otherwise.

2.3. Current Search for Other Vascular Susceptibility Genes Related to Longevity

The development of vascular disease, a major cause of mortality in Western countries, involves the interaction of multiple genetic factors and environmental influences. Comparison of genotype frequencies among age-stratified healthy populations may be a useful strategy to identify polymorphisms associated with common human diseases, such as vascular diseases (CAD or CVD) or AD. Indeed some of the vascular susceptibility genes that have been independently associated with both vascular disease and AD have been investigated with respect to their putative roles in longevity.

2.3.1. *Nonapolipoprotein Genes*

An example of one such gene is the methyltetrahydrofolatereductase gene (*MTHFR*), which is involved in the metabolism of homocysteine. A common mutation in this gene (C677T) has been identified that results in thermolability and mildly reduced activity of the encoded enzyme, which in turn can cause hyperhomocystinemia [55], that has been suggested to be an independent risk factor for CAD and stroke [56]. The hypothesis of MTHFR involvement in longevity is based on the premise that, if the MTHFR TT genotype is associated with higher levels of homocysteine and premature death, its prevalence should decrease with advancing age. This hypothesis was supported by two different studies, in which a decrease in MTHFR TT genotype or T allele frequency was observed in healthy elderly group compared with the younger controls [57, 58] (Table 2). A trend toward a lower frequency of the mutant allele in centenarians compared to a control group was also reported by a French study [59]. In contrast, no statistically significant differences in MTHFR allele or genotype frequency were observed between elderly aged and younger controls in other independent populations [30, 54, 60–63]. These conflicting results indicate that although MTHFR genetic variation cannot be wholly excluded in longevity mediation, the effect of this genetic polymorphism on longevity is not as important as previously thought.

The links between some polymorphic genetic factors involved in hemostasis and blood pressure regulation and both vascular diseases and AD have prompted investigations into their role in longevity. In a Danish study, the analysis of several genetic polymorphisms linked to vascular risk, in addition to both MTHFR C677T and ACE1 (blood coagulation factor VII [FVII] [R/Q353 and intron 7(37bp)n], β-fibrinogen [BcIII and –455G/A], plasminogen activator inhibitor type I [PAI-1] [-675(4G/5G)], tissue plasminogen activator [intron 8 ins311], platelet receptor glycoprotein IIb/IIIa [L/P33], prothrombin [20210G/A], angiotensinogen [M/T 235]) in 187 unselected centenarians and 201 healthy blood donors, revealed no differences in the frequencies of the high–vascular risk alleles of all these polymorphisms in centenarians and blood donors [6, 30]. Another study conducted in Danish centenarians versus young individuals investigated three genetic polymorphisms (angiotensinogen M/T235, FVII R/Q353, and FVII-323ins10) and showed significant influences on survival in males, with reduced hazards of death in subjects bearing the angiotensinogen M235 allele, the FVII Q353 allele, and the FVII-323P10 allele [64]. The R/Q353 polymorphism of the FVII gene was also analyzed in a Scottish group of subjects with age of 90 years compared with younger controls. A trend, although nonsignificant, was revealed toward overrepresentation of the Q allele in the nonagenarians group [65]. Thus, it is possible that the FVII gene might exert just a weak influence on longevity. Three gene polymorphisms (factor VII [R/Q353], β-fibrinogen [−455G/A] and PAI-1), all associated with the plasma levels of coagulation and fibrinolysis proteins, were evaluated in a group of Italian centenarians, and differences were only found between centenarians and young controls in the allele and genotype frequency distribution of the PAI-1 polymorphism [66]. However, the association of this PAI-1 polymorphism with longevity has not since been replicated [30, 67, 68].

In addition to the MTHFR gene, five other vascular-related polymorphisms (factor V G1691A Leiden, factor II G20210A, glycoprotein Ia C807T, GPIIIa T1563C, and ACE1) were examined. Only the PL^{A2} allele frequency of the GPIIIa polymorphism decreased from younger to older age groups, especially in ACE1*DD carriers [54].

Two polymorphisms in the gene encoding paraoxonase 1 (*PON1*) (a leucine [L allele] to methionine [M allele] substitution at codon 55, and a glutamine [Q allele] to arginine [R allele] substitution at codon 192), previously identified and linked with CAD [69, 70], were assessed in 308 centenarians and 579 people aged 20–65 years [71]. In this study the percentage of carriers of the R allele at codon 192 was higher in centenarians than in controls, while no significant difference was observed for the PON1 L55M polymorphism. The authors proposed that the B allele decreases mortality in carriers, but that the effect of PON1 variability on the overall population mortality is rather slight, and they suggest that PON1 is one of the genes affecting the individual adaptive capacity and, therefore, the rate and quality of aging [71] (Table 2).

A recent study including a large combined group of Italian centenarians and octo/nonagenarians from Northern Ireland reported a significant difference in PON1 192 genotypes in Italian centenarians compared to younger controls and a similar but nonsignificant trend between octo/nonagenarian and young subjects from Northern Ireland. Using logistic regression analysis on the combined Italian and Irish data sets, there was a small survival advantage for centenarian and octo/nonagenarian subjects heterozygous for PON1 192 R allele, with a stepwise increase for RR homozygous subjects compared to QQ subjects. These data indicate a modest association between the 192R allele and longevity in two very elderly European populations [72] (Table 2).

The codon 405 isoleucine-to-valine (I405V) variation in the cholesteryl ester transfer protein gene, which is involved in regulation of lipoprotein, and its particle sizes have been assessed in a case–control study including Ashkenazi Jewish probands with exceptional longevity (mean age: 98.2 years), their offspring (mean age: 68.3 years), and an age-matched group of Ashkenazi Jews and participants from the Framingham Offspring Study as controls. Probands and offspring had a 2.9- and 3.6-fold (in men), and 2.7- and 1.5-fold (in women) increased frequency of homozygosity for the 405 valine allele of cholesteryl ester transfer protein (VV genotype), respectively, compared with controls. Probands with the VV genotype had increased lipoprotein sizes and lower serum cholesteryl ester transfer protein concentrations. Individuals with exceptional longevity and their offspring had significantly larger high-density lipoprotein and LDL particle sizes. These findings indicate that lipoprotein particle sizes are heritable and promote a healthy aging phenotype [73].

2.3.2. *Apolipoprotein Genes*

In addition to APOE, genetic polymorphisms in other apolipoprotein genes have been shown to modulate lipid metabolism and risk for CAD [74, 75] and thus have been examined for their potential role in human longevity. In Italian centenarians, the 3′APOB-VNTR polymorphism, a marker located less than 100 bp downstream of the second transcription signal of the APOB gene, has been reportedly associated with longevity [76, 77]. *APOB* is the main protein in LDL-C and plays a major role in cholesterol homeostasis [78]. In both Northern and Southern Italian populations the frequency of the 3′APOB-VNTR alleles with fewer than 35 repeats (small alleles) was significantly smaller in centenarians than in 20–60-year-old subjects [76, 77] (Table 2). These findings were not replicated in Danish centenarians and younger controls (20–64 years), but a significant gene–sex interaction relevant to alleles having more than 37 repeats (long alleles) was found [79]. Furthermore, a significant difference was observed in the frequencies of the small alleles in younger controls between Denmark and Italy that was not present in centenarians [79]. These findings support the phenomenon of regional differences from Northern to Southern Europe in allele and genotype frequencies of various genes (such as APOE and ACE1), as reported in our studies in centenarians and younger controls [23, 33].

Genes encoding for apolipoprotein C-I (APOC-I), apolipoprotein A1 (APOA-I), apolipoprotein C3 (APOCIII), and apolipoprotein A4 (APOA-IV) have all been subject to studies on longevity. In a small British study conducted in subjects aged 84 years and older, the authors examined APOE and APOC-I genes, as they are less than 10 kb apart, and APOC-I genotype and allele frequencies can be used to confirm associations at the APOE locus. This study reported a significant difference in APOC-I allele and genotype frequencies in elderly women compared to the younger sample, while no age based difference was observed in the men [52]. Similar differences in APOE genotype distribution were also observed between the elderly and young women; however, the APOE $\epsilon2$ allele was not overrepresented in either elderly women or men [52].

The APOA-I, APOCIII, and APOA-IV genes are organized in tandem in a cluster on the long arm of the human chromosome 11 and have been extensively investigated in lipoprotein disorders and vascular disease [74]. An analysis of the distribution of APOA1-MspI (−75 nt from the transcription starting site) A/P allele (absence/presence of the restriction site) frequency in three age classes (18–45 years, 46–80 years, and 81–109 years) revealed that the frequency of the P allele tends to increase in the oldest males (81–109 years), but not in the female group [80]. Thus, APOA-1 P allele appears to be linked to longevity in a sex-specific way. Because the P allele is associated with higher serum LDL-C at middle age and is overrepresented in patients affected by cardiovascular disease with high LDL-C serum levels, these results give additional evidence of an intriguing genetic paradox in centenarians. These apparently contradictory findings confirm that genetic risk factors known to act in a detrimental way at middle age can be present or even overrepresented in centenarians [24, 66]. Similarly, alleles deemed to be protective at young age, as reflected by their prevalence in the young age group, may have properties that change and become disadvantageous in older people. Analysis of APOCIII-SstI (3′UTR 3238 nt) and APOA-IV-HincII (Asp_{127}/Ser_{127}) polymorphisms in the same three age classes showed no age-related variation either in males or females [80]. Conversely, determination of APOCIII-SstI polymorphism in almost all Finnish centenarians alive in 1991 showed that the APOC-III S2 allele (SstI restriction site present) occurred more often in the centenarians than in the youngest controls [22] (Table 2). Another polymorphism of the APOC-III gene located in the 5′UTR (T-455C) within a functional insulin element was examined in Russian population and found to be associated with longevity [81]. The greater frequency of the APOC-III-455C allele was observed in subjects older than 80 years [81] (Table 2). Analysis of common APOA-IV genetic variant at codon 360 (Gln→His) revealed an overrepresentation of the 360/His allele in the aged healthy subjects compared with the younger group, indicating a role for this polymorphism in successful aging and longevity [82]. However, these studies are unique, and a role of the APOAI-CIII-AIV gene cluster in human longevity has to be further substantiated and verified through attempts to replicate these data.

3. Inflammation and Immune Response in Human Longevity

Aging is not synonymous with illness. However, aging does increase the risk for certain illnesses. Overall, elderly people have an increased rate of chronic disorders, infections, autoimmune disorders, and cancer. A significant part of this increased risk seems to be related to aging changes in the immune system. With age, the number of immune cells may decrease slightly. More important, the functioning of these cells declines. The cells are often less able to control illness than in earlier years. As a person ages, the immune system produces fewer antibodies. It also responds more slowly to injury, infection, or disease. Cells of the immune system can also lose their ability to tell the difference between normal and abnormal tissue. Accordingly, several studies have shown an association between *in vitro* T cell function and longevity, suggesting that a well-preserved immune system may be associated with extended longevity [83]. Therefore, centenarians are the best examples of successful aging and longevity, as not only do they escape death but, by and large, they escape illness for most of their lives.

Some years ago came the Genetic Theory of Aging, which suggests that all stages of life from conception until death are genetically programmed [84]. The senescence is caused by perturbations in switching on and off the capacity of genes to translate the messages whose metabolic products are necessary for tissue functions and integrity. For example, complex immune traits related to lifelong immune responsiveness and immunosenescence, such as level of circulating immunoglobulins, and the peripheral CD4/CD8 T lymphocyte ratio, seem to be under strict genetic control [85, 86].

Another theory, the "network theory," suggests that a variety of defense functions or antistress responses, globally acting as antiaging mechanisms, indirectly control the aging process [87]. The stressors include different physical (heat, ultraviolet), chemical (oxygen-free radicals and reducing sugars), and biological (viruses, bacteria) agents. This theory argues that a global reduction in the capability to cope with a variety of stressors and a concomitant progressive increase in the proinflammatory status are major characteristics of the aging process and are provoked by a continuous antigenic load and stress [88]. In fact, assuming that the major characteristic of human immunosenescence is the filling of the immunological space by memory and effector T cell clones as a consequence of the chronic antigenic stress, another major consequence of chronic exposure to antigens is the progressive activation of macrophages and related cells in most organs and tissue [88]. This phenomenon, called "inflamm-aging," was a theoretical extension of the network theory, suggesting that the immune response, the stress response, and inflammation constitute an integrated, evolutionary conserve defense network [88]. Thus, an inflammatory status is compatible with extreme longevity in good health, and "paradoxically," proinflammatory characteristics have been documented in healthy centenarians [89]. The rate of reaching the threshold of proinflammatory status over which diseases/disabilities ensues, and the individual capacity to cope with and adapt to stressors, are assumed to be complex traits having a genetic component [88].

On this basis, identification of genes and processes that set the aging rate has been attempted by several approaches in different species. Among the possible immune system and inflammation genes likely involved in longevity, class I and class II HLA genes, and genes encoding for cytokines, the hormones of the immune system that regulate the interaction among the different elements of the immune system [90] have been considered, with conflicting results.

3.1. Role of IL-6 and Other Interleukin Genes in Human Longevity

Systemic inflammation, represented in large part by the production of proinflammatory cytokines, is the response of humans to the assault of the nonself on the organism.

However, apart from disease states, there is a natural evolution of cytokine production with aging, including decreased T cell proinflammatory IL-2 and interferon-γ (type II pro-inflammatory IFN-γ), decreased non–T cell antiinflammatory type I IFN (IFN-α and IFN-β), and increased non–T cell proinflammatory IL-1, IL-6, and tumor necrosis factor α (TNF-α) [90]. The IL-6 is one of the pathogenic elements of inflammatory and age-related diseases (AD and atherosclerosis) and has been defined as the "cytokine for gerontologists" [91].

The IL-6 gene in humans is located in the short arm of chromosome 7 and has a-174 G/C polymorphism in its promoter region. Fishman *et al.* [92] reported that the IL-6-174 G/C promoter polymorphism is associated with reduced IL-6 gene expression and plasma levels. An age-related increase of IL-6 concentration has been found in serum, plasma, and mononuclear blood cell culture among elderly people and centenarians [89, 93]. Recent population-based studies identified among the elderly the magnitude of IL-6 serum levels as a predictor of mortality and functional disability, in regard to mobility or selected activities of daily living [94–96]. Bonafè *et al.* [97] found that subjects who are genetically predisposed to produce high levels of IL-6 (i.e., IL-6*G/*G homozygous men) have a reduced capacity to reach extreme longevity. In fact, the researchers noted a marked reduction of the IL-6*G/*G genotype in male, though not female, Italian centenarians compared with elderly (60–80 years) and long-lived (81–99 years) subjects. However, they found that the age-related increase of IL-6 serum levels in women was independent from the-174 G/C promoter polymorphism [97]. Furthermore, Olivieri *et al.* found that cultured peripheral blood mononuclear cells from C allele carrier individuals produced smaller amount of IL-6 than noncarriers, and that IL-6 production by peripheral blood mononuclear cells *in vitro* increased with age in C allele carriers but not in noncarriers. The authors also found that C allele carriers had lower plasma levels of IL-6 than noncarriers, but this phenomenon was significant only in men [98].

In our recent study, we did not find statistically significant differences in IL-6-174 G/C promoter genotype or allele frequencies between centenarians and middle-aged subjects from Southern Italy [99], providing no evidence for the suggested role of the IL-6-174 G/C promoter polymorphism in human longevity [97]. These findings were supported by the previously reported associations of the IL-6*G/*G genotype with elevated circulating IL-6 plasma levels at all ages: young adults, elderly adults, and centenarians [92, 97, 98, 100]. In fact, high IL-6 concentrations were shown to be a credible predictor of increased disability and mortality in the elderly [94–96], and the IL-6*G/*G genotype was associated, probably through increased IL-6 plasma levels, with higher risk of AD [101], MID [102], and vascular diseases [99, 103, 104].

Another two studies have been conducted on this polymorphism in octogenarian and nonagenarian subjects [105, 106]. In 250 Finnish community living and hospitalized nonagenarians and 400 healthy blood donors, aged 18–60 years, there was a lack of association between human longevity and gene variants of IL-1 cluster, IL-6-174 G/C promoter, IL10, and TNF-α, alone or in combination; however, a reduction of 2.5% of the Il-6*G/*G frequency between nonagenarians and widely age-selected younger control group was observed [106]. On the contrary, Rea *et al.* [105], in the Belfast Elderly Longitudinal Free-Living Ageing Study, found a trend for a reduced frequency of the IL-6*G/*G genotype in octogenarian and nonagenarian subjects compared with younger subjects of the local population, aged 20–45 years. This trend appeared more marked in elderly males compared with the females [107], but both groups showed an almost 10% decrease frequency in the oldest subjects.

Explanation for the conflicting findings could be linked to variability in the strength of association between the IL-6-174 G/C promoter polymorphism and longevity, which could be modified by European regional differences in allele frequencies, as seen above [20, 21, 23, 33]. In fact, interestingly, the IL-6*C allele frequency appears to vary across Europe, with an increasing trend from south to north in the very old populations.

Beyond the supposed role of the IL-6, some other cytokine genes, such as IL-10, IL-2, IL-1, and IFN-γ show polymorphisms that may be relevant in human longevity, as some studies revealed a different expression of cytokines in elderly compared to young people [108, 109]. In particular, a study performed on Italian centenarians reported that the-1082G IL-10 single-nucleotide polymorphism, associated to elevated production of the cytokine, was increased in male centenarians [110, 111]. Differently, the +847T single-nucleotide polymorphism at the IFN-γ gene, part of the 12CA/+847T haplotype responsible for increased production of the cytokine, was less frequent in female Italian centenarians [112]. These findings show that different alleles at cytokine genes coding for pro-(IFN-γ) or antiinflammatory (IL-10) cytokines may affect individual life span expectancy, influencing the type and the level of immune-inflammatory response to environmental stressors. The gender-related effects are difficult to explain. However, men and women have claimed to follow different strategies to reach longevity, and female centenarians always outnumber male centenarians, with ratios ranging from 4/1 to 7/1, with the exception of Apulia, where it is 2.7/1 [23], and Sardinia, where it is 2.1/1 [113]. In another study, the polymorphic variants of IL-1α (C-T transition at position-889), IL-1β (C-T transition at position-511), and IL-1 receptor antagonist (Ra) (86-bp repeated sequence in intron 2) were analyzed in 1131 subjects from Northern and Central Italy, including 134 centenarians. The results indicate that no one particular polymorphism in the IL-1 gene cluster yields an advantage for longevity [114]. In the AKEA study on Sardinian centenarians, no significant differences were observed in IL-6, IL-10, and IFN-γ polymorphism frequencies among centenarians and controls [113].

Other studies analyzed the role of different cytokine polymorphisms in the aging, but they examined somewhat younger populations, which are less selective than centenarians [106, 107, 115, 116]. In particular, in Finnish nonagenarians compared with healthy blood donors (18–60 years) no evidence of association of IL1A-889, IL1B+3953, IL1B-511, IL1RN VNTR, IL6-174, IL10-1082, and TNFA-308 polymorphisms with aging was found [106]. Similar results were obtained for a range of cytokine polymorphisms (IL-2, IL-6, IL-8, IL-10, IL-12, and IFN-γ) in healthy aged Irish population, also with respect to gender [107]. Moreover, no age-related allele or genotype frequencies were observed in the Irish for two polymorphic nucleotides in the TNF cluster (G-308A TNF-α and G+252A TNF-β), associated with increased TNF-α production [116]. A recent study in healthy elderly Bulgarians did not reveal any statistically significant allele and haplotype frequency differences between the elderly and control groups, but in families with at least two generations of longevity members in their pedigrees (IL-10 genotypes-1082G/A, -819 C/C and -592 C/C), related to the intermediate production, were positively associated, whereas genotypes-1082A/A, -819 C/T, -592 C/A, related to a low level of production, were negatively associated with longevity in Bulgarians. This effect was modulated by IL-6 and IFN-γ genotypes associated with the low level of these proinflammatory cytokines [115].

All of these findings indicate that cytokine/longevity associations may have a population-specific component, being affected by the population-specific gene pool as well as by gene-environment interactions and behaving as survival rather than longevity assurance genes [113].

3.2. Genes Linked to Immune Response: The HLA System

Studies performed in mice and humans indicated the effect of polymorphic class I and class II HLA genes on life span and human longevity. In particular, studies on experimental animals argued that certain MHC genes may be associated with shorter life and others with a longer life span; however, the effects depend both on genetic background and on the environmental conditions [117–119]. Results of analogous studies in man are confusing and contradictory. Most studies would indicate that the effect of HLA on human longevity might be rather small; however, it is not allowed at present to reach clear and definitive conclusions. Thus, when comparing HLA antigen frequencies between groups of young and elderly subjects, the same HLA antigens are increased in some studies and decreased or unchanged in others [120]. The association between heterozygosity at HLA loci and human longevity have also been addressed because it was shown that heterozygote individuals are more fit for survival than homozygote ones. However, conflicting findings have been obtained, as well; most studies found an increase in heterozygosity at HLA loci in elderly compared to young groups, whereas others did not observe any change [120]. It can be supposed that the different results observed in the various studies are a result of bias. Apart from class I and II HLA loci, the MHC system include many class III genes with the potential involvement in modulating immune response that can play a role in the genetic control of longevity. Particularly, TNF genes show strong linkage disequilibrium with HLA class I and II genes, and structural or functional defects in TNF genes may contribute to pathogenesis of MHC-associated disease [121]. A combination of specific immunogenetic and immunophenotype profiles could contribute to the successful aging and to maintaining healthy status in elderly.

4. Conclusions

Genetic analysis of pertinent genes likely to be related with CAD, CVD, or AD may provide insights into the human longevity. In fact, human longevity–enabling genes should influence aging at its most basic levels, thus affecting a broad spectrum of genetic and cellular pathways. Cytokine genes may play a role in longevity by modulating an individual's responses to life-threatening disorders as well. In fact, the release of cytokines is of crucial importance in the regulation of the type and magnitude of the immune response in the elderly. Therefore, genetic determinants of longevity should reside in those polymorphic genes that regulate immune responses or are related to illnesses in the life.

However, the real role that genetic traits play in reaching a far advanced age has not been properly and sufficiently addressed, and further studies are needed. In particular, longitudinal studies specifically designed for addressing the effect of these gene polymorphisms on extreme longevity could provide more robust evidence for further understanding their role in the reaching of human extreme longevity. Moreover, it should keep in mind the multiplicity of mechanisms regulating aging at the molecular, cellular, and systemic levels, and the whole array of environmental factors, such as diet, lifestyle, and economic circumstances, that it is not possible to leave out of consideration in determining the very complex trait of longevity.

Acknowledgments

This group is supported by Italian Longitudinal Study on Ageing (Italian National Research Council–CNR-Targeted Project on Ageing, grants 9400419PF40 and 95973PF40), and by AFORIGE (Associazione per la Formazione e la Ricerca in Geriatria). We thank Dr. Adriana Rafaschieri, Dr. Giovanni Castellaneta, and Dr. Amalia Terralavoro, Pfizer Inc., for their skillful assistance.

References

[1] Cournil A, Kirkwood TB. If you would live long, choose your parents well. Trends Genet 2001; 17:233–235.

[2] Gonos ES. Genetics of aging: Lessons from centenarians. Exp Gerontol 2000; 35:15–21.

[3] Kirkwood TB. Genetic basis of limited cell proliferation. Mutat Res 1991; 256:323–328.

[4] Kirkwood TB. Evolution of ageing. Nature 1977; 270:301–304.

[5] Kirkwood TB, Holliday R. The evolution of ageing and longevity. Proc R Soc Lond B Biol Sci 1979; 205:531–546.

[6] Barzilai N, Shuldiner AR. Searching for human longevity genes: The future history of gerontology in the post-genomic era. J. Gerontol. A Biol Sci Med Sci 2001; 56:M83–M87.

[7] Abbott MH, Abbey H, Bolling DR, Murphy EA. The familial component in longevity—a study of offspring of nonagenarians: III. Intrafamilial studies. Am J Med Genet 1978; 2:105–120.

[8] Sorensen TI, Nielsen GG, Andersen PK, Teasdale TW. Genetic and environmental influences on premature death in adult adoptees. N Engl J Med 1988; 318:727–732.

[9] Herskind AM, McGue M, Iachine IA, Holm N, Sorensen TI, Harvald B, Vaupel JW. Untangling genetic influences on smoking, body mass index and longevity: A multivariate study of 2464 Danish twins followed for 28 years. Hum Genet 1996; 98:467–475.

[10] Ljungquist B, Berg S, Lanke J, McClearn GE, Pedersen NL. The effect of genetic factors for longevity: A comparison of identical and fraternal twins in the Swedish Twin Registry. J Gerontol A Biol Sci Med Sci 1988; 53:M441–M446.

[11] Perls TT, Bubrick E, Wager CG, Vijg J, Kruglyak L. Siblings of centenarians live longer. Lancet 1998; 351:1560.

[12] Robine JM, Allard M. The oldest human. Science 1998; 279:1834–1835.

[13] Finch CE, Tanzi RE. Genetics of aging. Science 1997; 278:407–411.

[14] National Center for Health Statistics. Vital statistics of the United States. Washington, DC: Public Health Service, 1994.

[15] U. S. Bureau of the Census. Census of population, general population characteristics. Washington, DC: Department of Commerce, 1990; 1992.

[16] Perls T, Terry D. Understanding the determinants of exceptional longevity. Ann Intern Med 2003; 139:445–449.

[17] Miller RA. A position paper on longevity genes. Sci. Aging Knowledge Environ. 2001; 2001:vp6.

[18] Perls T, Kunkel L, Puca A. The genetics of aging. Curr Opin Genet Dev 2002; 12:362–369.

[19] Puca AA, Daly MJ, Brewster SJ, Matise TC, Barrett J, Shea-Drinkwater M, Kang S, Joyce E, Nicoli J, Benson E, Kunkel LM, Perls T. A genome-wide scan for linkage to human exceptional longevity identifies a locus on chromosome 4. Proc Natl Acad Sci USA 2001; 98:10505–10508.

[20] Panza F, D'Introno A, Colacicco AM, Capurso C, Capurso S, Kehoe PG, Capurso A, Solfrizzi V. Vascular genetic factors and human longevity. Mech Agein Dev 2004; 125:169–178.

[21] Panza F, Colacicco AM, Kehoe PG, Capurso A, Solfrizzi V. Vascular genetic factors and lipoprotein metabolism in human longevity. In: Panza F, Solfrizzi V, Capurso A, editors. Diet and Cognitive Decline. New York: Nova Science, 2004: 149–172.

[22] Louhija J, Miettinen HE, Kontula K, Tikkanen MJ, Miettinen TA, Tilvis RS. Aging and genetic variation of plasma apolipoproteins. Relative loss of the apolipoprotein E4 phenotype in centenarians. Arterioscl Thromb 1994; 14:1084–1089.

[23] Panza F, Solfrizzi V, Torres F, Mastroianni F, Del Parigi A, Colacicco AM, Basile AM, Capurso C, Noya R, Capurso A. Decreased frequency of apolipoprotein E epsilon4 allele from Northern to Southern Europe in Alzheimer's disease patients and centenarians. Neurosci Lett 1999; 277:53–56.

[24] Schachter F, Faure-Delanef L, Guenot F, Rouger H, Froguel P, Lesueur-Ginot L, Cohen D. Genetic associations with human longevity at the APOE and ACE loci. Nat Genet 1994; 6:29–32.

[25] Saunders AM, Strittmatter WJ, Schmechel D, et al. Association of apolipoprotein E allele e4 with late-onset familial and sporadic Alzheimer's disease. Neurology 1993; 43:1467–1472.

[26] Eichner JE, Kuller LH, Orchard TJ, Grandits GA, McCallum LM, Ferrell RE, Neaton JD. Relation of apolipoprotein E phenotype to myocardial infarction and mortality from coronary artery disease. Am. J. Cardiol. 1993; 71:160–165.

[27] Boerwinkle E, Utermann G. Simultaneous effects of the apolipoprotein E polymorphism on apolipoprotein E, apolipoprotein B, and cholesterol metabolism. Am J Hum Genet 1988; 42:104–112.

[28] Cambien F, Poirier O, Lecerf L, et al. Deletion polymorphism in the gene for angiotensin-converting enzyme is a potent risk factor for myocardial infarction. Nature 1992; 359:641–644.
[29] Kehoe PG, Russ C, McIlory S, et al. Variation in DCP1, encoding ACE, is associated with susceptibility to Alzheimer disease. Nat Genet 1999; 21:71–72.
[30] Bladbjerg EM, Andersen-Ranberg K, de Maat MP, Kristensen SR, Jeune B, Gram J, Jespersen J. Longevity is independent of common variations in genes associated with cardiovascular risk. Thromb Haemost 1999; 82:1100–1105.
[31] Blanchè H, Cabanne L, Sahbatou M, Thomas G. A study of French centenarians: Are ACE and APOE associated with longevity? C Acad Sci II 2001; 324:129–135.
[32] Choi YH, Kim JH, Kim DK, Kim JW, Kim DK, Lee MS, Kim CH, Park SC. Distributions of ACE and APOE polymorphisms and their relations with dementia status in Korean centenarians. J Gerontol A Biol Sci Med Sci 2003; 58:227–231.
[33] Panza F, Solfrizzi V, D'Introno A, Colacicco AM, Capurso C, Capurso A, Kehoe PG. Angiotensin I converting enzyme (ACE1) gene polymorphism in centenarians: Different allele frequencies between the North and South of Europe. Exp Gerontol 2003; 38:1015–1020.
[34] Davignon J, Gregg RE, Sing CF. Apolipoprotein E polymorphism and atherosclerosis. Arteriosclerosis 1988; 8:1–21.
[35] Asada T, Yamagata Z, Kinoshita T, Kinoshita A, Kariya T, Asaka A, Kakuma T. Prevalence of dementia and distribution of ApoE alleles in Japanese centenarians: An almost-complete survey in Yamanashi Prefecture, Japan. J Am Geriatr Soc 1996; 44:151–155.
[36] Asada T, Kariya T, Yamagata Z, Kinoshita T, Asaka A. Apolipoprotein E allele in centenarians. Neurology 1996; 46:1484.
[37] Yamagata Z, Asada T, Kinoshita A, Zhang Y, Asaka A. Distribution of apolipoprotein E gene polymorphisms in Japanese patients with Alzheimer's disease and in Japanese centenarians. Hum Hered 1997; 47:22–26.
[38] Davignon J, Bouthillier D, Nestruck AC, Sing CF. Apolipoprotein E polymorphism and atherosclerosis: Insight from a study in octogenarians. Trans Am Clin Climatol Assoc 1987; 99:100–110.
[39] Kervinen K, Savolainen MJ, Salokannel J, Hynninen A, Heikkinen J, Ehnholm C, Koistinen MJ, Kesaniemi YA. Apolipoprotein E and B polymorphisms—longevity factors assessed in nonagenarians. Atherosclerosis 1994; 105:89–95.
[40] Rea IM, Mc Dowell I, McMaster D, Smye M, Stout R, Evans A. MONICA group (Belfast). Monitoring of cardiovascular trends study group, apolipoprotein E alleles in nonagenarian subjects in the Belfast Elderly Longitudinal Free-living Ageing Study (BELFAST). Mech Agein Dev 2001; 122:1367–1372.
[41] Jian-Gang Z, Yong-Xing M, Chuan-Fu W, Pei-Fang L, Song-Bai Z, Nui-Fan G, Guo-Yin F, Lin H. Apolipoprotein E and longevity among Han Chinese population. Mech Agein Dev 1998; 104:159–167.
[42] Frisoni GB, Louhija J, Geroldi C, Trabucchi M. Longevity and the epsilon2 allele of apolipoprotein E: The Finnish Centenarians Study. J Gerontol A Biol Sci Med Sci 2001; 56:M75–M78.
[43] Louhija J, Viitanen M, Aguero-Torres H, Tilvis R. Survival in Finnish centenarians in relation to apolipoprotein E polymorphism. J Am Geriatr Soc 2001; 49:1007–1008.
[44] Gerdes LU, Jeune B, Ranberg KA, Nybo H, Vaupel JW. Estimation of apolipoprotein E genotype-specific relative mortality risks from the distribution of genotypes in centenarians and middle-aged men: Apolipoprotein E gene is a frailty gene, not a longevity gene. Genet Epidemiol 2000; 19:202–210.

[45] Panza F, Solfrizzi V, Colacicco AM, Basile AM, D'Introno A, Capurso C, Sabba M, Capurso S, Capurso A. Apolipoprotein E (APOE) polymorphism influences serum APOE levels in Alzheimer's disease patients and centenarians. Neuroreport 2003; 14:605–608.
[46] Couderc R, Mahieux F, Bailleul S, Fenelon G, Mary R, Fermanian J. Prevalence of apolipoprotein E phenotypes in ischemic cerebrovascular disease. A case-control study. Stroke 1993; 24:661–664.
[47] Bohnet K, Pillot T, Visvikis S, Sabolovic N, Siest G l. Apolipoprotein (apo) E genotype and apoE concentration determine binding of normal very low density lipoproteins to HepG2 cell surface receptors. J Lipid Res 1996; 37:1316–1324.
[48] Siest G, Pillot T, Regis-Bailly A, Leininger-Muller B, Steinmetz J, Galteau MM, Visvikis S. Apolipoprotein E: An important gene and protein to follow in laboratory medicine. Clin Chem 1995; 41:1068–1086.
[49] Rigat B, Hubert C, Alhenc-Gelas F, Cambien F, Corvol P, Soubrier F. An insertion/deletion polymorphism in the angiotensin I-converting enzyme gene accounting for half the variance of serum enzyme levels. J Clin Invest 1990; 86:1343–1346.
[50] Costerousse O, Jaspard E, Wei L, Corvol P. Alhenc-Gelas F. The angiotensin I-converting enzyme (kininase II): Molecular organization and regulation of its expression in humans. J Cardiovasc Pharmacol 1992; 20(Suppl. 9):S10–S15.
[51] Eisenlohr LC, Bacik I, Bennink JR, Bernstein K, Yewdell JW. Expression of a membrane protease enhances presentation of endogenous antigens to MHC class I-restricted T lymphocytes. Cell 1992; 71:963–972.
[52] Galinsky D, Tysoe C, Brayne CE, Easton DF, Huppert FA, Dening TR, Paykel ES, Rubinsztein DC. Analysis of the apo E/apo C-I, angiotensin converting enzyme and methylenetetrahydrofolate reductase genes as candidates affecting human longevity. Atherosclerosis 1997; 129:177–183.
[53] Rahmutula D, Nakayama T, Izumi Y, et al. Angiotensin-converting enzyme gene and longevity in the Xin Jiang Uighur autonomous region of China: An association study. J Gerontol A Biol Sci Med Sci 2001; 57:M57–M60.
[54] Hessner MJ, Dinauer DM, Kwiatkowski R, Neri B, Raife TJ. Age-dependent prevalence of vascular disease-associated polymorphisms among 2689 volunteer blood donors. Clin Chem 2001; 47:1879–1884.
[55] Frosst P, Blom HJ, Milos R, et al. A candidate genetic risk factor for vascular disease: A common mutation in methylenetetrahydrofolate reductase. Nat Genet 1995; 10:111–113.
[56] Riggs KM, Spiro A, 3rd, Tucker K, Rush D. Relations of vitamin B-12, vitamin B-6, folate, and homocysteine to cognitive performance in the Normative Aging Study. Am J Clin Nutr 1996; 63:306–314.
[57] Matsushita S, Muramatsu T, Arai H, Matsui T, Higuchi S. The frequency of the methylenetetrahydrofolate reductase-gene mutation varies with age in the normal population. Am J Hum Genet 1997; 61:1459–1460.
[58] Todesco L, Angst C, Litynski P, Loehrer F, Fowler B, Haefeli WE. Methylenetetrahydrofolate reductase polymorphism, plasma homocysteine and age. Eur J Clin Invest 1999; 29:1003–1009.
[59] Faure-Delanef L, Quere I, Chasse JF, Guerassimenko O, Lesaulnier M, Bellet H, Zittoun J, Kamoun P, Cohen D. Methylenetetrahydrofolate reductase thermolabile variant and human longevity. Am J Hum Genet 1997; 60:999–1001.
[60] Harmon DL, McMaster D, Shields DC, Whitehead AS, Rea IM. MTHFR thermolabile genotype frequencies and longevity in Northern Ireland. Atherosclerosis 1997; 131:137–138.

[61] Rea IM, McMaster D, Woodside JV, Young IS, Archbold GP, Linton T, Lennox S, McNulty H, Harmon DL, Whitehead AS. Community-living nonagenarians in Northern Ireland have lower plasma homocysteine but similar methylenetetrahydrofolate reductase thermolabile genotype prevalence compared to 70–89-year-old subjects. Atherosclerosis 2000; 149:207–214.

[62] Brattstrom L, Zhang Y, Hurtig M, Refsum H, Ostensson S, Fransson L, Jones K, Landgren F, Brudin L, Ueland PM. A common methylenetetrahydrofolate reductase gene mutation and longevity. Atherosclerosis 1998; 141:315–319.

[63] Rees DC, Liu YT, Cox MJ, Elliott P, Wainscoat JS. Factor V Leiden and thermolabile methylenetetrahydrofolate reductase in extreme old age. Thromb Haemost 1997; 78:1357–1359.

[64] Tan Q, Yashin AI, Bladbjerg EM, de Maat MP, Andersen-Ranberg K, Jeune B, Christensen K, Vaupel JW. Variations of cardiovascular disease associated genes exhibit sex-dependent influence on human longevity. Exp Gerontol 2001; 36:1303–1315.

[65] Meiklejohn DJ, Riches Z, Youngson N, Vickers MA. The contribution of factor VII gene polymorphisms to longevity in Scottish nonagenarians. Thromb Haemost 2000; 83:519.

[66] Mannucci PM, Mari D, Merati G, Peyvandi F, Tagliabue L, Sacchi E, Taioli E, Sansoni P, Bertolini S, Franceschi C. Gene polymorphisms predicting high plasma levels of coagulation and fibrinolysis proteins. A study in centenarians. Arterioscler Thromb Vasc Biol 1997; 17:755–759.

[67] Benes P, Muzik J, Benedik J, Elbl L, Vasku A, Siskova L, Znojil V, Vacha J. The C766T low-density lipoprotein receptor related protein polymorphism and coronary artery disease, plasma lipoproteins, and longevity in the Czech population. J Mol Med 2001; 79:116–120.

[68] Lottermoser K, Dusing R, Ervens P, Koch B, Bruning T, Sachinidis A, Vetter H, Ko Y. The plasminogen activator inhibitor 1 4G/5G polymorphism is not associated with longevity: A study in octogenarians. J Mol Med 2001; 79:289–293.

[69] Mackness B, Mackness MI, Arrol S, Turkie W, Durrington PN. Effect of the human serum paraoxonase 55 and 192 genetic polymorphisms on the protection by high density lipoprotein against low density lipoprotein oxidative modification. FEB Lett 1998; 423:57–60.

[70] Serrato M, Marian AJ. A variant of human paraoxonase/arylesterase (HUMPONA) gene is a risk factor for coronary artery disease. J Clin Invest 1995; 96:3005–3008.

[71] Bonafe M, Marchegiani F, Cardelli M, et al. Genetic analysis of paraoxonase (PON1) locus reveals an increased frequency of Arg192 allele in centenarians. Eur J Hum Genet 2002; 10:292–296.

[72] Rea IM, McKeown PP, McMaster D, Smye M, Stout R, Evans A. Paraoxonase polymorphisms PON1 192 and 55 and longevity in Italian centenarians and Irish nonagenarians. A pooled analysis. Exp Gerontol 2004; 39:629–635.

[73] Barzilai N, Atzmon G, Schechter C, Schaefer EJ, Cupples AL, Lipton R, Cheng S, Shuldiner AR. Unique lipoprotein phenotype and genotype associated with exceptional longevity. JAMA 2003; 290:2030–2040.

[74] Groenendijk M, Cantor RM, de Bruin TW, Dallinga-Thie GM. The apoAI-CIII-AIV gene cluster. Atherosclerosis 2001; 157:1–11.

[75] Menzel HJ, Boerwinkle E, Schrangl-Will S, Utermann G. Human apolipoprotein A-IV polymorphism: Frequency and effect on lipid and lipoprotein levels. Hum Genet 1988; 79:368–372.

[76] De Benedictis G, Falcone E, Rose G, et al. DNA multiallelic systems reveal gene/longevity associations not detected by diallelic systems. The APOB locus. Hum Genet 1997; 99:312–318.

[77] De Benedictis G, Carotenuto L, Carrieri G, De Luca M, Falcone E, Rose G, Yashin AI, Bonafe M, Franceschi C. Age-related changes of the 3′APOB-VNTR genotype pool in ageing cohorts. Ann Hum Genet 1998; 62:115–122.

[78] Havel JH, Kane JP. Introduction: Structure and metabolism of plasma lipoproteins. In: Scriver CR, Blaudet AL, Sly WS, Valle D, editors. The Metabolic Basis of Inherited Disease, eighth ed. New York: McGraw-Hill, 2001: 2705–2716.

[79] Varcasia O, Garasto S, Rizza T, et al. Replication studies in longevity: Puzzling findings in Danish centenarians at the 3′APOB-VNTR locus. Ann Hum Genet 2001; 65:371–376.

[80] Garasto S, Rose G, Derango F, et al. The study of APOA1, APOC3, and APOA4 variability in healthy ageing people reveals another paradox in the oldest old subjects. Ann Hum Genet 2003; 67:54–62.

[81] Anisimov SV, Volkova MV, Lenskaya LV, Khavinson VK, Solovieva DV, Schwartz EI. Age-associated accumulation of the apolipoprotein C-III gene T-455C polymorphism C allele in a Russian population. J Gerontol A Biol Sci Med Sci 2001; 56:B27–B32.

[82] Merched A, Xia Y, Papadopoulou A, Siest G, Visvikis S. Apolipoprotein AIV codon 360 mutation increases with human aging and is not associated with Alzheimer's disease. Neurosci Lett 1998; 242:117–119.

[83] Pawelec G, Adibzadeh M, Solana R, Beckman I. The T cell in the ageing individual. Mech Agein Dev 1997; 93:35–45.

[84] Goldstein S, Gallo JJ, Reichel W. Biologic theories of aging. Am Fam Physicia 1989; 40:195–200.

[85] Amadori R, Zamarchi G, Desilvestro G, Forza G, Cavatton G, Danieli A, Clementi M, Chiecobianchi L. Genetic control of the CD4/cd8 T-cell ratio in humans. Natur Med 1995; 1:1279–1283.

[86] Di Franco P, Brai M, Misiano G, Piazza AM, Giorgi G, Cossarizza A, Franceschi C. Genetic and environmental influences on serum levels of immunoglobulins and complement components in monozygotic and dizygotic twins. J Clin Lab Immunol 1988; 27:5–10.

[87] Kirkwood TB, Franceschi C. Is aging as complex as it would appear? New perspectives in aging research. Ann N Acad Sci 1992; 663:412–417.

[88] Franceschi C, Bonafe M, Valensin S, Olivieri F, De Luca M, Ottaviani E, De Benedictis G. Inflamm-aging. An evolutionary perspective on immunosenescence. Ann N Acad Sci 2000; 908:244–254.

[89] Baggio G, Donazzan S, Monti D, et al. Lipoprotein(a) and lipoprotein profile in healthy centenarians: A reappraisal of vascular risk factors. FASE J 1998; 12:433–437.

[90] Brod SA. Unregulated inflammation shortens human functional longevity. Inflamm Res 2000; 49:561–570.

[91] Ershler WB. Interleukin-6: A cytokine for gerontologists. J Am Geriatr Soc 1993; 41:176–181.

[92] Fishman D, Faulds G, Jeffery R, Mohamed-Ali V, Yudkin JS, Humphries S, Woo P. The effect of novel polymorphisms in the interleukin-6 (IL-6) gene on IL-6 transcription and plasma IL-6 levels, and an association with systemic-onset juvenile chronic arthritis. J Clin Invest 1998; 102:1369–1376.

[93] Fagiolo U, Cossarizza A, Scala E, Fanales-Belasio E, Ortolani C, Cozzi E, Monti D, Franceschi C, Paganelli R. Increased cytokine production in mononuclear cells of healthy elderly people. Eur J Immunol 1993; 23:2375–2378.

[94] Ershler WB, Keller ET. Age-associated increased interleukin-6 gene expression, late-life diseases, and frailty. Annu Rev Med 2000; 51:245–270.

[95] Ferrucci L, Harris TB, Guralnik JM, Tracy RP, Corti MC, Cohen HJ, Penninx B, Pahor M, Wallace R, Havlik RJ. Serum IL-6 level and the development of disability in older persons. J Am Geriatr Soc 1999; 47:639–646.

[96] Harris TB, Ferrucci L, Tracy RP, Corti MC, Wacholder S, Ettinger WH Jr, Heimovitz H, Cohen HJ, Wallace R. Associations of elevated interleukin-6 and C-reactive protein levels with mortality in the elderly. Am J Med 1999; 106:506–512.

[97] Bonafè M, Olivieri F, Cavallone L, et al. A gender-dependent genetic predisposition to produce high levels of IL-6 is detrimental for longevity. Eur J Immunol 2001; 31:2357–2361.

[98] Olivieri F, Bonafe M, Cavallone L, et al. The-174 C/G locus affects *in vitro/in vivo* IL-6 production during aging. Exp Gerontol 2002; 37:309–314.

[99] Capurso C, Solfrizzi V, D'Introno A, Colacicco AM, Capurso SA, Semeraro C, Capurso A, Panza F. Interleukin 6-174 G/C promoter gene polymorphism in centenarians: No evidence of association with human longevity or interaction with apolipoprotein E alleles. Exp Gerontol 2004; 39:1109–1114.

[100] Burzotta F, Iacoviello L, Di Castelnuovo A, et al. Relation of the-174 G/C polymorphism of interleukin-6 to interleukin-6 plasma levels and to length of hospitalization after surgical coronary revascularization. Am J Cardiol 2001; 88:1125–1128.

[101] Pola R, Flex A, Gaetani E, Lago AD, Gerardino L, Pola P, Bernabei R. The-174 G/C polymorphism of the interleukin-6 gene promoter is associated with Alzheimer's disease in an Italian population. Neuroreport 2002; 13:1645–1647.

[102] Pola R, Gaetani E, Flex A, Aloi F, Papaleo P, Gerardino L, De Martini D, Flore R, Pola P, Bernabei R. -174 G/C interleukin-6 gene polymorphism and increased risk of multi-infarct dementia: A case-control study. Exp Gerontol 2002; 37:949–955.

[103] Flex A, Gaetani E, Pola R, Santoliquido A, Aloi F, Papaleo P, Dal Lago A, Pola E, Serricchio M, Tondi P, Pola P. The-174 G/C polymorphism of the interleukin-6 gene promoter is associated with peripheral artery occlusive disease. Eur J Vasc Endovasc Surg 2002; 24:264–268.

[104] Rauramaa R, Vaisanen SB, Luong LA, Schmidt-Trucksass A, Penttila IM, Bouchard C, Toyry J, Humphries SE. Stromelysin-1 and interleukin-6 gene promoter polymorphisms are determinants of asymptomatic carotid artery atherosclerosis. Arterioscler Thromb Vasc Biol 2000; 20:2657–2662.

[105] Rea IM, Ross OA, Armstrong M, McNerlan S, Alexander DH, Curran MD, Middleton D. Interleukin-6-gene C/G 174 polymorphism in nonagenarian and octogenarian subjects in the BELFAST study. Reciprocal effects on IL-6, soluble IL-6 receptor and for IL-10 in serum and monocyte supernatants. Mech Agein Dev 2003; 124:555–561.

[106] Wang XY, Hurme M, Jylha M, Hervonen A. Lack of association between human longevity and polymorphisms of IL-1 cluster, IL-6, IL-10 and TNF-alpha genes in Finnish nonagenarians. Mech Agein Dev 2001; 123:29–38.

[107] Ross OA, Curran MD, Meenagh A, Williams F, Barnett YA, Middleton D, Rea IM. Study of age-association with cytokine gene polymorphisms in an aged Irish population. Mech Agein Dev 2003; 124:199–206.

[108] Caruso C, Aquino A, Candore G, Scola L, Colonna-Romano G, Lio D. Looking for immunological risk genotypes. Ann NY Acad Sci 2004; 1019:141–146.

[109] Krabbe KS, Pedersen M, Bruunsgaard H. Inflammatory mediators in the elderly. Exp Gerontol 2004; 39:687–699.

[110] Lio D, Scola L, Crivello A, Colonna-Romano G, Candore G, Bonafe M, Cavallone L, Franceschi C, Caruso C. Gender-specific association between-1082 IL-10 promoter polymorphism and longevity. Genes Immun 2002; 3:30–33.

[111] Lio D, Scola L, Crivello A, Colonna-Romano G, Candore G, Bonafe M, Cavallone L, Marchegiani F, Olivieri F, Franceschi C, Caruso C. Inflammation, genetics, and longevity: Further studies on the protective effects in men of IL-10-1082 promoter SNP and its interaction with TNF-alpha-308 promoter SNP. J Med Genet 2003; 40:296–299.
[112] Lio D, Scola L, Crivello A, Bonafe M, Franceschi C, Olivieri F, Colonna-Romano G, Candore G, Caruso C. Allele frequencies of +874T—A single nucleotide polymorphism at the first intron of interferon-gamma gene in a group of Italian centenarians. Gerontology 2002; 44:293–299.
[113] Pes GM, Lio D, Carru C, et al. Association between longevity and cytokine gene polymorphisms. A study in Sardinian centenarians. Aging Clin Exp Res 2004; 16:244–248.
[114] Cavallone L, Bonafe M, Olivieri F, et al. The role of IL-1 gene cluster in longevity: A study in Italian population. Mech Agein Dev 2003; 124:533–538.
[115] Naumova E, Mihaylova A, Ivanova M, Michailova S, Penkova K, Baltadjieva D. Immunological markers contributing to successful aging in Bulgarians. Exp Gerontol 2004; 39:637–644.
[116] Ross OA, Curran MD, Rea IM, Hyland P, Duggan O, Barnett CR, Annett K, Patterson C, Barnett YA, Middleton D. HLA haplotypes and TNF polymorphism do not associate with longevity in the Irish. Mech Agein Dev 2003; 124:563–567.
[117] Popp DM. An analysis of genetic factors regulating life-span in congenic mice. Mech Agein Dev 1982; 18:125–134.
[118] Smith GS, Walford RL. Influence of the main histocompatibility complex on ageing in mice. Nature 1977; 270:727–729.
[119] Yunis EJ, Salazar M. Genetics of life span in mice. Genetica 1993; 91:211–223.
[120] Caruso C, Candore G, Romano GC, Lio D, Bonafe M, Valensin S, Franceschi C. Immunogenetics of longevity. Is major histocompatibility complex polymorphism relevant to the control of human longevity? A review of literature data. Mech Agein Dev 2001; 122:445–462.
[121] Dawkins R, Leelayuwat C, Gaudieri S, Tay G, Hui J, Cattley S, Martinez P, Kulski J. Genomics of the major histocompatibility complex: Haplotypes, duplication, retroviruses and disease. Immunol Rev 1999; 167:275–304.

LABORATORY FINDINGS OF CALORIC RESTRICTION IN RODENTS AND PRIMATES

Yoshikazu Higami, Haruyoshi Yamaza, and Isao Shimokawa

Department of Pathology and Gerontology, Nagasaki University Graduate School of Biomedical Science, Nagasaki 852-8523, Japan

1. Introduction: Antiaging and Life Span-Extending Actions of Caloric Restriction

In 1935, McCay and coworkers were the first to determine that caloric (dietary or energy) restriction (CR) starting at or soon after weaning increases longevity in rats. Thereafter, it was recognized that a reduction in

0065-2423/05 $35.00
DOI: 10.1016/S0065-2423(04)39008-6

caloric intake by 30%–40% from *ad libitum* (AL) levels leads to a significant extension of median and maximal life span in a wide variety of short-lived species from invertebrates to mammals, including yeast, rotifers, water fleas, nematodes, fruit flies, spiders, fish, hamsters, mice, and rats [1]. Therefore, CR has been applied as a powerful tool in aging research by many investigators for nearly 70 years. Although new genetic interventions that extend rodent life span have emerged recently, CR remains the most robust, reproducible, and simple experimental manipulation extending median and maximum life span in a variety of short-lived mammals, including mice and rats [1, 2]. In many reports, the pathology of CR rodents has been compared with that of rodents fed AL. In general, CR delays the onset of or prevents several age-related diseases, including nephropathy, cardiomyopathy, and several kinds of neoplasia [1–6]. Moreover, CR suppresses age-related cellular and molecular alterations and maintains many physiological functions at more youthful levels [1, 2].

The relationship between dietary or caloric intervention and life span has been characterized in rodents. Restriction of the protein component without CR suppresses the development of both nephropathy and cardiomyopathy [6], and restriction of the fat component without CR also suppresses nephropathy [7]; however, neither markedly retards the aging processes or extends the life span [7, 8]. Thus, restricting protein, fat, or mineral components without CR suppresses certain specific age-related pathology but does not markedly influence the aging processes or life span.

A dietary regimen replacing dietary casein with lactalbumin without CR does not influence the pathology, aging processes, or longevity [9], but replacing dietary casein with soy protein without CR retards the progression of nephropathy [10]. Moreover, this dietary regimen extends longevity but is less effective than CR. Thus, dietary regimens that use several different replacement protein sources (casein, lactalbumin, or soy protein) do not markedly influence aging processes or longevity, indicating that the beneficial actions of CR depend on the reduction of caloric intake rather than of a particular nutrient [6–10].

CR during any part of the life span enhances survival to some extent, but CR throughout life maximally extends longevity [8, 11]. It is likely that the duration of CR within a life correlates with the extent of the beneficial actions on longevity [8] and pathology [12]. Moreover, as shown in Fig. 1A [11], the extent of longevity prolongation by CR depends on the extent of the reduced caloric intake (severity of CR) without malnutrition [11]. In other words, excess food consumption above an optimal amount of food intake progressively shortens longevity and promotes aging.

Many gerontologists have postulated hypotheses to try and explain the mechanism of life prolongation and the antiaging actions of CR. However,

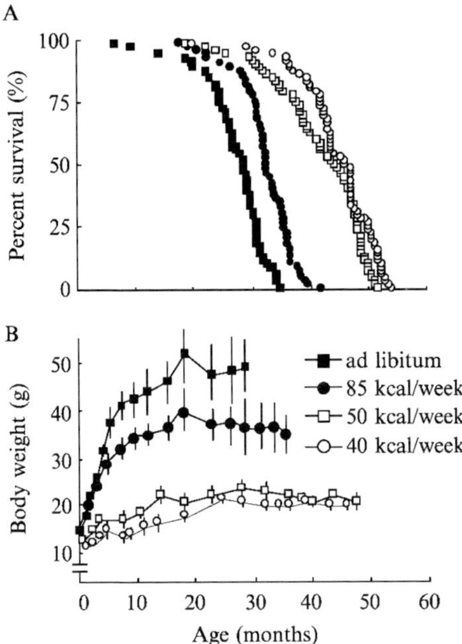

FIG. 1. Effects of caloric restriction on (A) percentage survival and (B) body weight in female C3B10 F1 mice; 85 kcal (closed circles), 50 kcal (open squares), and 40 kcal (open circles) food intake per week is equivalent to 25%, 55%, and 65% caloric restriction compared with the food intake of mice fed *ad libitum* (closed squares), respectively. Adapted from reference 11.

the exact underlying mechanism remains debatable. In general, it is believed that modulation of hyperglycemia and insulinemia, alteration of energy metabolism, and protection against internal oxidative and environmental stresses play important roles in the beneficial actions of CR [2, 13, 14]. Recently, SIRT1, an NAD^+-dependent deacetylase [15], and mitochondrial function, including production of reactive oxygen species, proton leakage, and lipid unsaturation of the mitochondrial membrane [16, 17], have been suggested to have roles in the beneficial actions of CR. Holliday explained the effect of CR from the evolutionary viewpoint that organisms have evolved neuroendocrine and metabolic response systems to maximize survival during periods of food shortage [18]. When food is plentiful, organisms grow, reproduce, and store excess energy as triglyceride in adipose tissues for later use as fuel. In contrast, on extended food shortage, animals suspend growth and reproduction to enhance survival. Thus, individuals are more likely to survive a period of food shortage and reproduce when food is once again abundant. The action of CR may derive from this adaptive response system against food shortage [18–20]. Here, we present CR-associated alterations from laboratory findings and focus on explaining these alterations on the basis of the adaptive response hypothesis.

2. Laboratory Findings in CR Rodents

2.1. Small Body

CR reduces body size and body weight [1, 2, 11, 21]. These parameters approximate those of caloric intake (Fig. 1B [11]). Growth hormone and its effecter hormone, insulin-like growth factor 1 (IGF-1), are the major regulators of body size. Normally, growth hormone is secreted from the pituitary gland in a pulsative pattern in a diurnal cycle. Growth hormone pulsative secretary dynamics are attenuated, and the mean daily level of plasma growth hormone decreases with age in AL animals [22]. Although CR suppresses the pulsative secretary dynamics of growth hormones in young animals, the pulse amplitude is increased in CR animals at 26 months of age and is indistinguishable from that in young AL animals [22]. The plasma IGF-1 level decreases significantly with age in AL rats, whereas it only slightly decreases with age in CR rats. Moreover, it is significantly higher in AL rats than in CR rats for a large part of their life, particularly from young to middle age (Fig. 2 [23]).

It is well known that dwarf animal models, including Ames mice [24] and minirats bearing an antisense GH transgene [25], live longer than their normal counterparts. Therefore, it has been suggested that the GH/IGF-1

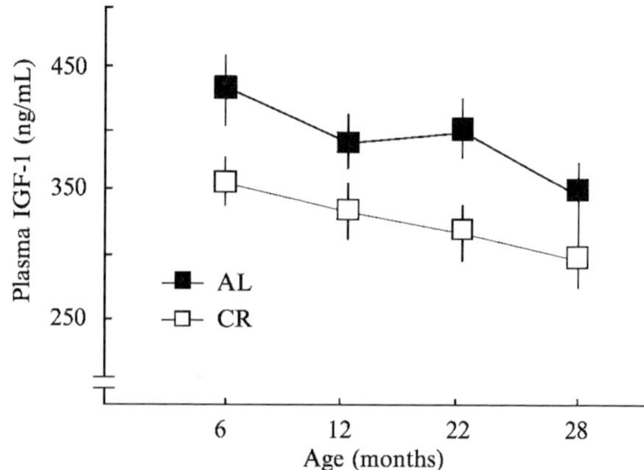

FIG. 2. Effects of aging and caloric restriction on the plasma insulin-like growth factor 1 concentration in male Brown-Norway × F344 rats. Adapted from reference 23.

axis plays a major role in the beneficial actions of CR. However, according to recent CR studies conducted on Ames mice and mini rats, CR affects aging and longevity by mechanisms other than suppression of the GH/IGF-1 axis, as it extends the life span of both Ames mice and minirats to a similar extent as it does that of their normal counterparts [26, 27].

2.2. Modulation of Hyperglycemia and Insulinemia

In F344 rats, CR significantly reduces the blood glucose concentration throughout their life (Fig. 3 [28]), and the plasma insulin levels at 6 and 12 months of age (Fig. 4 [28]). In Brown-Norway rats, the blood glucose concentration is significantly lower in CR rats than in AL rats at 11 and 17 months of age, but not at 29 months of age [29]. Furthermore, the plasma insulin levels are lower in CR rats than in AL rats at 11, 17, and 29 months of age [29]. Therefore, despite the use of different strains, CR appears to reduce both the blood glucose and plasma insulin concentrations during a large part of their life. The concentration of plasma glucagon increases with age until 24 months of age in both AL and CR F344 rats and does not differ significantly between the two dietary groups [28].

During glucose tolerance testing, the blood glucose concentration in rats increases to a peak value at 15 minutes and then returns to the basal (0 minutes) level in both AL and CR rats (Fig. 5A). However, the blood glucose concentration in AL rats gradually decreases from 15 to 90 minutes, whereas that in CR rats quickly returns to the basal level at 30 minutes [30].

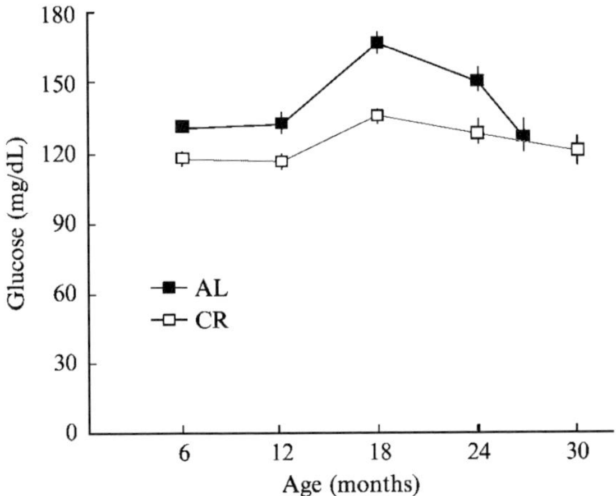

Fig. 3. Effects of aging and caloric restriction on the plasma glucose concentration in male F344 rats. Adapted from reference 28.

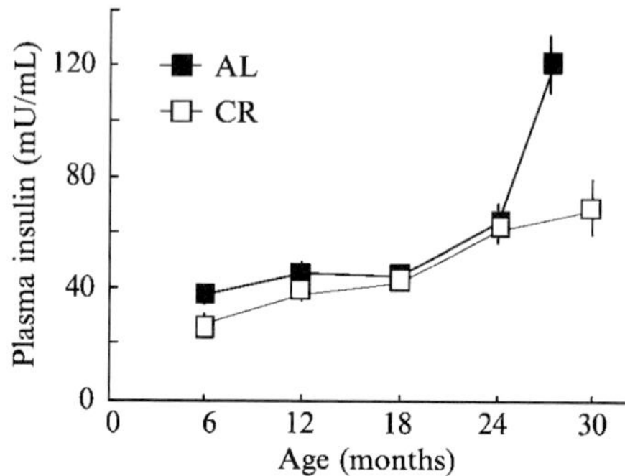

FIG. 4. Effects of aging and caloric restriction on the plasma insulin concentration in male F344 rats. Adapted from reference 28.

CR affects the serum insulin concentration during glucose tolerance testing (Fig. 5B); the concentration of insulin is transiently increased at 15 minutes in AL rats and then reduces rapidly to the basal level at 30 minutes. There is no similar surge of insulin in CR rats, indicating that CR activates non-insulin-dependent glucose disposal [30]. During insulin tolerance testing, the blood glucose concentration decreases gradually from 15 to 120 minutes after insulin injection, whereas CR reduces blood glucose concentration more rapidly and markedly (Fig. 5C), indicating that CR enhances insulin sensitivity. Taken together, CR appears to activate both insulin-dependent and non-insulin-dependent mechanisms for glucose disposal [30].

2.3. Low Body Temperature

The core body temperature of CR mice has been reported to decrease to a nearly ambient temperature (torpor) and then recover every day [31]. Koizumi *et al.* reported that the beneficial actions of CR are partially attenuated by the prevention of hypothermic episodes in CR mice by housing the animals at 30 °C [32]. The extreme hypothermia observed in CR mice is not observed in larger mammals, including rats, although CR does slightly, but significantly, reduce the body temperature in a variety of animals, including rodents [31–34]. Therefore, hypothermia may constitute a primary physiological response to CR.

2.4. Transient Reduction of Metabolic Rate

McCarter and McGee [35] examined changes in the metabolic rate immediately after starting 40% restriction of food at 6 weeks of age and continued to examine it until 24 weeks of age. As shown in Fig. 6A, the metabolic rate

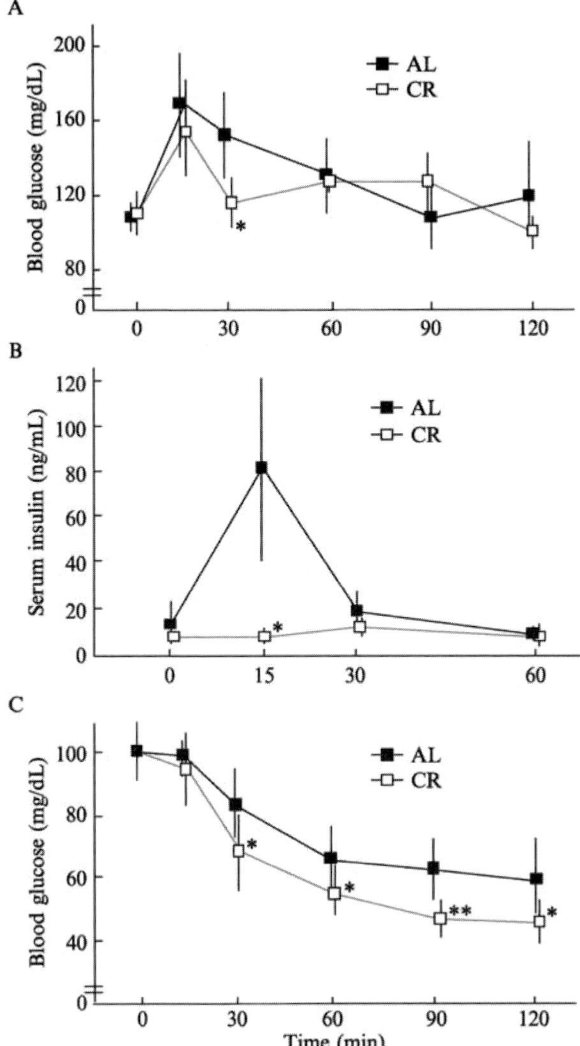

FIG. 5. (A) Blood glucose concentration during glucose tolerance testing. Dietary effect: $p = .06$, time effect: $p < .0001$, diet × time effect: $p = .0327$, *: $p < .01$. (B) Serum insulin concentration during glucose tolerance testing. Dietary effect: $p < .0001$, time effect: $p < .0001$, diet × time effect: $p < .0001$, *: $p < .0001$. (C) Blood glucose concentration during insulin tolerance testing. Dietary effect: $p < .0001$, time effect: $p < .0001$, diet × time effect: not significant, *: $p < .05$, **: $p < .01$. *Ad libitum*–fed and calorically restricted male Wistar rats were subjected to both glucose and insulin tolerance tests at 6–7 months of age after overnight fasting. Adapted from reference 30.

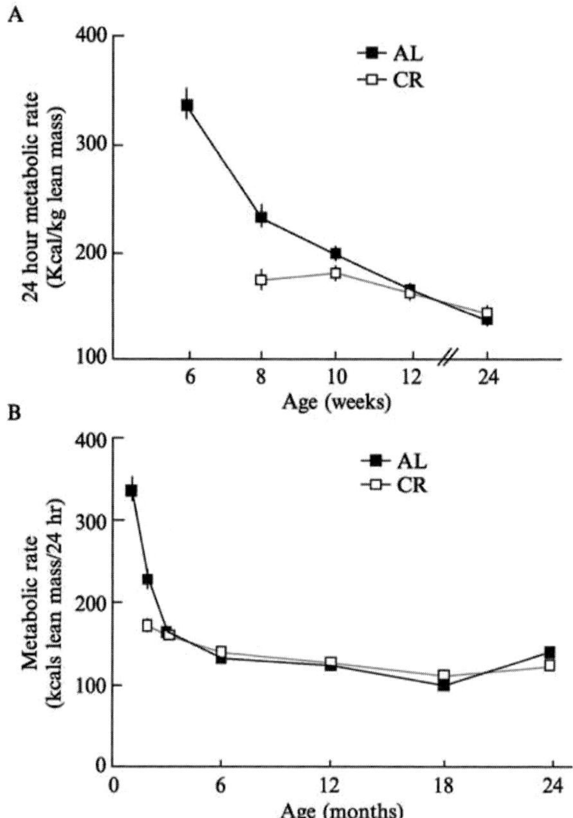

FIG. 6. (A) Variation of total daily metabolic rate for 18 weeks after initiating caloric restriction in male F344 rats at 6 weeks of age. The metabolic rate is normalized to the unit lean mass. Adapted from reference 21. (B) Metabolic rate $\times$ unit lean body mass^{-1} $\times$ 24hr^{-1} for both *ad libitum*–fed and calorically restricted male F344 rats from 6 weeks to 24 months of age. Adapted from reference 21.

per unit lean mass decreases transiently after starting CR, but after a few weeks, the metabolic rate of CR rats no longer differs from that of AL rats [35]. Moreover, both AL and CR rats exhibit variation of the metabolic rate per unit lean mass over their life span, with the rate decreasing until 18 months of age and then slightly increasing from 18 to 24 months of age (Fig. 6B [21]). These results indicate that the rate per unit lean mass of CR rats is not lower than that of AL rats [21, 35]. However, some reports have suggested that CR reduces the metabolic rate [36]. Thus, the metabolic rate results for AL and CR animals are still controversial. Recently, Blanc *et al.* suggested that the long-standing debate regarding the effects of CR on the metabolic rate may derive from the lack of consensus on how to adjust for body size and composition [37]. Therefore, further studies and a statistically suitable adjustment of the metabolic rate for body size are required.

It has been suggested that thyroid status is a possible major regulator of metabolic rate. Herlihy *et al.* [38] found that CR reduces the 24-hour mean serum triiodothyronine (T3) level, but not the thyroxine (T4) level. CR appears to have this action because it attenuates the circadian amplitude in the concentrations of both hormones. However, Snyder *et al.* [39] found no effect of CR on either T4 or T3 in young rats and no consistent effect of CR on age-related changes in the concentrations of these two hormones.

An adrenergic state is another regulator of metabolic rate. Plasma norepinephrine and epinephrine concentrations were measured in AL and CR rats at 12 months of age [40]. CR does not appear to alter either basal (resting) or stimulated (hypotension) plasma norepinephrine and epinephrine levels; however, the recovery of plasma norepinephrine and epinephrine to their basal levels after hypotensive stress is enhanced in CR rats, indicating the possible enhancement of transmitter uptake by nerve endings [40]. Plasma catecholamine concentration in aging has been measured in rodents, but the results are variable. It has been reported that aging both increases and has no effect on rat plasma catecholamine concentrations. Kingsley *et al.* [41] examined the adrenal state, including the adrenal gland weight, and contents and concentrations of dopamine, norepinephrine, and epinephrine in adrenal glands in 6-, 18- and 30-month-old AL or CR rats. Most of the parameters increase with age, and CR slightly suppresses these age-related changes; however, the adrenal weights increase in some rats at 30 months of age because of hyperplasia. After excluding the data of rats with hyperplastic adrenal glands, the data of aged rats are similar to those of younger ones. Thus, it is likely that neither aging nor CR markedly influences the adrenergic state in rats [41].

2.5. Altered Energy Metabolism

Studies of the respiratory quotient or respiratory exchange ratio (RER) indicate a metabolic shift in CR rats depending on food availability, while there is little change in AL rats [21, 42]. Figure 7 illustrates daily variations of RER with dietary regimen. Analysis of the food composition indicates an expected average value of RER of 0.89. RER in AL rats is 0.89 ± 0.02 over a 24-hour period and is relatively more constant than that in CR rats, indicating that AL rats metabolize a constant ratio of carbohydrate and lipid over 24 hours [21, 42]. In contrast, CR animals metabolize more carbohydrate

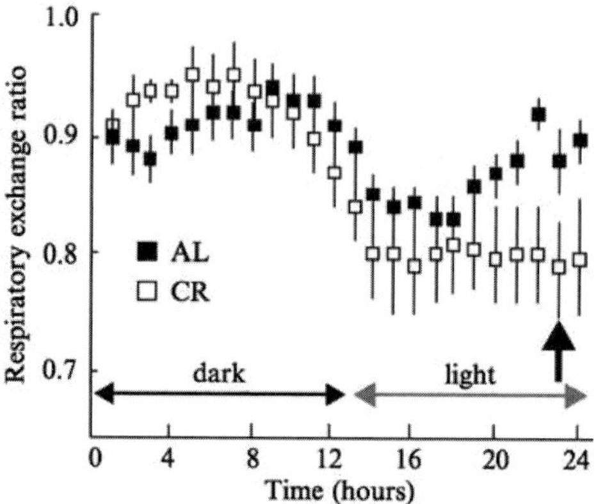

FIG. 7. Respiratory exchange ratio of male F344 rats at 18 months of age. Calorically restricted rats were fed just before turning off the light (arrow). The light was off from hours 0 to 12 and on from hours 12 to 24. Adapted from reference 21.

(RER > 0.9) immediately after feeding. When glycogen reserves are depleted, whole-body fuel utilization shifts almost exclusively to fat (RER = 0.8) before feeding [21, 42].

2.6. Alteration of Lipid Metabolism

The serum triglyceride concentration is higher in AL rats than in CR rats throughout their life. Moreover, that in AL rats increases markedly from 6 to 18 months of age, with the level remaining elevated until 24 months of age, followed by a tendency to decrease with further increasing age, whereas that in CR rats is almost constant throughout their life (Fig. 8A [28]). The serum total cholesterol concentration is higher in AL rats than in CR rats throughout their lives; in AL it increases markedly from 18 to 24 months of age, with a tendency to decrease with further increasing age, whereas that in CR increases slightly up to 30 months of age (Fig. 8B [28]).

In Wistar rats at 6 months of age, the serum levels of total lipid, triglyceride, total and free cholesterol and phospholipid are significantly higher in AL rats than in CR rats before or after feeding. The serum levels of free fatty acids are significantly higher in AL and CR rats before feeding than in CR rats after feeding. The serum levels of total ketone bodies are higher in CR rats before feeding than in AL and CR rats after feeding. The levels of the fractions, serum acetoacetate and D-3-hydroxybutyrate, show the same trend as the total ketone bodies (Table 1, unpublished data).

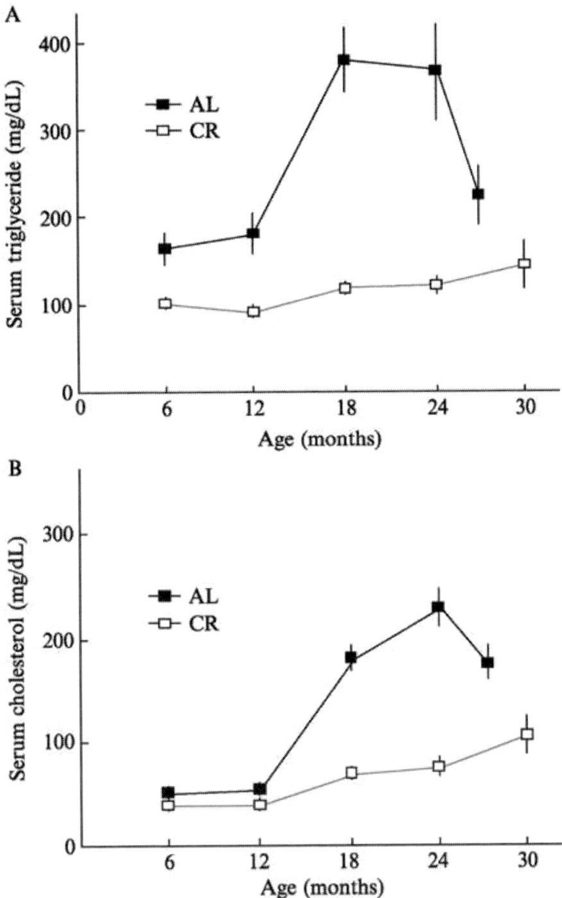

FIG. 8. (A) Effects of aging and caloric restriction on the serum triglyceride concentration in male F344 rats. Adapted from reference 28. (B) Effect of aging on the serum total cholesterol concentration in male F344 rats. Adapted from reference 28.

2.7. Lower Fat Mass (Low Adiposity)

CR-associated reductions of body size and body weight are associated with a marked decline of fat mass [21]. In general, aging is associated with an increase of adiposity, but CR suppresses this age-associated change [21, 43]. Barzilai *et al.* measured the effect of aging and CR on body weight, weight of fat mass, and weight of visceral fat (Fig. 9), and their data indicate that CR reduces fat mass more markedly than body weight and is more effective on visceral fat [43].

TABLE 1
BODY WEIGHT AND SERUM LIPIDS IN MALE WISTAR RATS AT 6 MONTHS OF AGE

	Ad libitum	Caloric restriction	
Body weight (g)	479 ± 34	342 ± 17*	
Serum lipid	*Ad libitum*	After feeding	Before feeding
Total lipid (mg/dL)	724 ± 52	438 ± 31*	468 ± 53*
Triacylglycerols (mg/dL)	380 ± 46	145 ± 44*	195 ± 46*
Total cholesterol (mg/dL)	94 ± 1	86 ± 7*	80 ± 3*
Free cholesterol (mg/dL)	20 ± 1	14 ± 2*	14 ± 1*
Phospholipid (mg/dL)	203 ± 7	164 ± 13*	152 ± 6*
Free fatty acids (mEq/L)	560 ± 81	293 ± 99**	607 ± 35
Total ketone bodies	93 ± 5	97 ± 7	161 ± 50***
Acetoacetate	36 ± 4	43 ± 2	59 ± 12***
D-3-hydroxybutyrate	57 ± 6	54 ± 6	101 ± 39***

*$p < .05$ vs. *ad libitum*.
**$p < .05$ vs. *ad libitum* and caloric restriction (before feeding).
***$p < .05$ vs. *ad libitum* and caloric restriction (after feeding).

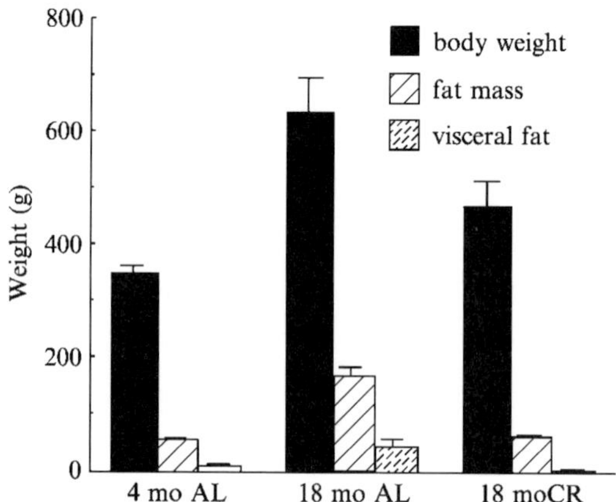

FIG. 9. Effects of aging and caloric restriction on the body weight, and weights of fat mass and visceral fat in male Sprague-Dawley rats. Adapted from reference 43.

2.8. PHENOTYPIC ALTERATION OF WHITE ADIPOSE TISSUE

Histologically, adipocytes appear smaller in CR mice [44]. A DNA microarray study containing over 11,000 genes found that CR up-regulates the expression of most genes involved in the metabolisms of glucose and amino acids and in lipogenesis and mitochondrial energy. These findings indicate that CR alters the characteristics of white adipose tissue (WAT) metabolism, resulting in a phenotypic shift reflecting energy metabolism activation and adipocyte differentiation [44].

Leptin, a peptide hormone secreted principally from adipocytes (adipocytokine), regulates appetite and energy expenditure, and a dysfunction of its signaling results in hyperphagia and obesity [45, 46]. In addition, it acts as a stress-related hormone under life-threatening conditions [47, 48]. It is well known that the fasting plasma leptin concentration is positively correlated with the percentage of fat mass [49]. Moreover, the plasma leptin level is related to feeding; the level declines rapidly within 24 hours after food deprivation, whereas feeding stimulates leptin secretion within a few hours [45]. The expression of leptin mRNA in WAT is markedly lower in CR mice than in AL mice [44]. In both AL and CR rats, the plasma leptin level shows a similar diurnal variation, with the level increasing predominantly during the dark phase when rats have food (Fig. 10A [50]). The plasma leptin concentration is consistently and significantly more than 50% lower in CR rats than in AL rats at 6, 15, and 24 months of age (Fig. 10B [50]). The concentration increases with age in AL rats but is constant in CR rats. These findings are consistent with those of previous studies showing a significant correlation between the plasma leptin concentration and fat mass in rodents and a lower fat content per unit body weight in CR rats than in AL rats [50].

Adiponectin, another adipocytokine, is predominantly secreted by smaller adipocytes and regulates energy homeostasis and glucose and lipid metabolisms [51]. The plasma adiponectin level is decreased in patients with obesity and type 2 diabetes, and its reduction seems to be associated with elevated triglyceride contents in skeletal muscle and the liver [52, 53]. Recently, it has been suggested that adiponectin reduces the blood glucose concentration and stimulates glucose utilization and fatty acid oxidation through the activation of 5′-AMP-activated protein kinase. Therefore, adiponectin signaling is one of the insulin-independent pathways involved in glucose disposal [54]. In fasted rats, the plasma adiponectin level is nearly two times higher in CR rats than in AL rats at 2 months of age, and almost three times higher in CR rats than in AL rats at 10, 15, and 20 months of age [55]. Thus, CR also modulates adipocytokine secretions from WAT.

2.9. Protection against Internal Oxidative and Environmental Stresses

Corticosterone, the principal adrenosteroid, is a so-called stress response hormone and is bound to corticosterone-binding globulin in plasma. As shown in Fig. 11 [56], the plasma concentration of corticosterone in both AL and CR rats shows a similar diurnal variation, with the level increasing in

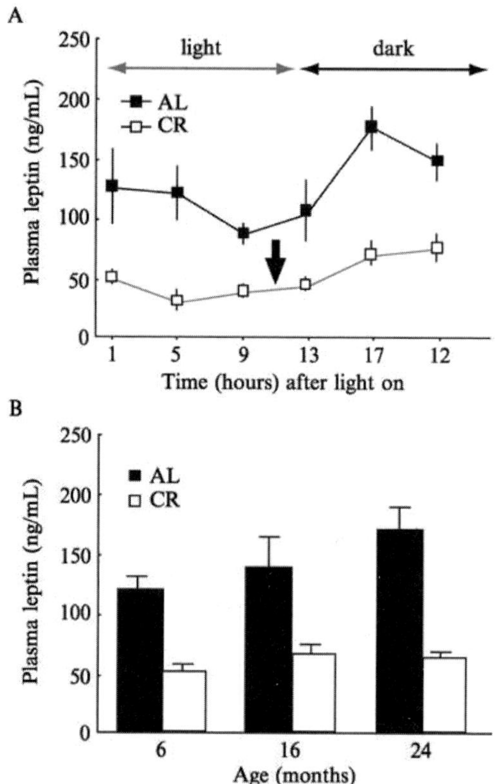

FIG. 10. (A) Diurnal variation of the plasma leptin concentration in *ad libitum*–fed and calorically restricted F344 rats at 6 months of age. Calorically restricted rats were fed just before turning off the light (arrow). Adapted from reference 50. (B) Effects of aging and caloric restriction on the plasma leptin concentration in male F344 rats. Adapted from reference 50.

the late light phase [56]. At this peak, it is slightly higher in CR rats than in AL rats [56]. The plasma concentration of free corticosterone (nonbinding form to corticosterone-binding globulin) is the biologically active fraction of the circulating hormone. The estimated plasma levels of free corticosterone in CR rats are two- to eightfold higher than those in AL rats during the daily diurnal elevation of the hormone in both young and old rats [56], indicating that CR induces mild stress conditions.

2.10. Slow Skeletal Maturation

In AL rats, bone length, weight, density, and calcium content increase rapidly with age and plateau at about 12 months of age. There is no evidence of bone loss in these animals until about 24 months of age, but by age 27

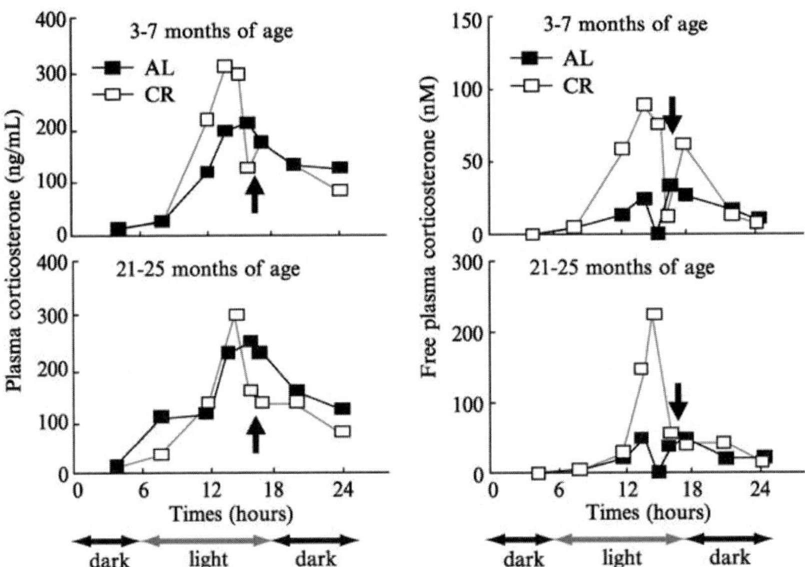

FIG. 11. Effects of aging and caloric restriction on the 24-hour plasma profiles of total and free corticosterone in young (3–7 months of age) and old (21–25 months of age) male F344 rats. Calorically restricted animals were fed just before turning off the light (arrow). The light was off from 17:00 to 05:00 and on from 05:00 to 17:00. Adapted from reference 56.

months, the animals have lost appreciable amounts of bone [57]. The plasma parathyroid hormone level of the animals increases with advancing age and rises markedly at 27 months. In CR rats, bone growth and maturation are slowed, and the animals do not experience senile bone loss or a marked terminal increase in circulating plasma parathyroid hormone up to 30 months of age [57]. Because the terminal loss of bone in aged rats is associated with severe nephropathy, the age-related loss of bone in AL rats is in part a result of the age-related decline in renal function [57].

2.11. Slow Sexual Maturation

Postweaning CR (weight was 50% of AL rats) from 21 days of age in female rats markedly delays pubertal onset. All AL rats reach puberty by 45 days of age, but only 10% of CR rats do so by 80 days [58]. Similar conclusions are found for puberty onset in male rats [59], but the susceptibility of pubertal development to CR appears to be greater in female than male rodents [60].

Male rats restricted to 50% normal food intake for 20 days exhibit lower serum LH and testosterone levels than AL rats, although the serum FSH level is unaffected [61]. In contrast, both the serum LH and FSH levels are lower in rats starved for 1 week [62]. Despite the smaller gonads of CR rats, spermatogenesis is spared, indicating that the degree of sperm development and maturation are correlated with the serum FSH level, as only starvation consistently suppresses this, while CR does not [63]. In female rats, the serum level of LH, but not FSH, is lower in CR rats than in AL rats [64].

2.12. More Fertility Later in Life

CR initiated at weaning extends the age at which animals cease to exhibit estrous cycles, with the result that CR rats are more fertile later in life [58]. Using a different experimental paradigm, middle-aged mice that were returned to AL feeding after a 7- to 8-month period of CR demonstrated a higher incidence of normal estrous cycles and fertility rates than age-matched controls that had never been under CR conditions [65]. In contrast, chronic CR of male rats, beginning at weaning, fails to suppress reproductive aging [66].

3. Laboratory Findings in CR Nonhuman Primates

The beneficial actions of CR were not tested in any animal model living longer than about 4 years until the late 1980s. Since then, studies in longer-lived species, including rhesus and squirrel monkeys, have been underway at the National Institute on Aging and the University of Wisconsin [67–69].

The longevity data in monkeys remain preliminary. However, previous reports from the National Institute on Aging indicate that CR reduces morbidity, neoplastic diseases in particular [70], and perhaps even mortality [71] in monkeys. Early in 2002 at the National Institute on Aging, the proportions of both rhesus and squirrel monkeys that had died in the CR cohorts were about one-half that in the controls [72].

The characteristics of CR monkeys are summarized in Table 2 [72] and Fig. 12A–C [73]. CR monkeys weigh less [74] and have less total and abdominal obesity than controls [75]. In addition, CR reduces body temperature and induces a transient reduction in metabolic rate [76]. Young male CR monkeys also exhibit delayed sexual and skeletal maturation [72, 74]. In addition, several lines of evidence suggest that CR improves the disease risk in rhesus monkeys [72]. CR monkeys have reduced blood glucose and insulin levels and improved insulin sensitivity [75]. CR also reduces blood pressure and lowers the serum triglyceride and cholesterol levels [77]. In

TABLE 2
SUMMARY OF PATHOPHYSIOLOGICAL EFFECTS OF CALORIC RESTRICTION (CR) IN RHESUS MONKEYS (ADAPTED FROM REFERENCE 72)

Findings	Effect of CR in monkeys	Agreement with rodent findings
Body weight	↓	Yes
Body fat	↓	Yes
Lean body mass	↓	Yes
Time to skeletal maturation (male)	↓	Yes
Time to sexual maturation (male)	↓	Yes
Transient reduction in metabolic rate	↓	Yes
Body temperature	↓	Yes
Locomotor activity (male)	→	Yes
Growth hormone/insulin-like growth factor 1	↓	Yes
IL-6	↓	Yes
Estradiol, FSH, LH	→	No
Menstrual cycling	→	Yes (rats)
Rate of DHEAS decline with aging	↓	Yes
Lymphocyte number	↓	Yes
Lymphocyte calcium response	→	No
Rate of wound closure	→	Yes
Fasting glucose and insulin	↓	Yes
Insulin sensitivity	↑	Yes
Triglycerides	↓	Yes
Cholesterol	↓	Yes
HDL2B	↑	?
Blood pressure	↓	?
Bone density (male)	→	No
Bone density (female)	↓	Yes

↓: decrease, ↑: increase, →: no change, ?: not reported, DHEAS: dehydroepiandrosterone sulfate.

general, as noted above, the ongoing monkey studies support the conclusion that most pathophysiological responses to CR in these monkeys are similar to those in rodents, possibly suggesting that CR may also extend longevity in monkeys [72].

4. Relevance to Humans

The effects of CR on longevity in humans have not been evaluated directly. Kagawa *et al.* examined a proportion of centenarians in Okinawa Prefecture compared with those in other regions of Japan [78]. Okinawa residents have a

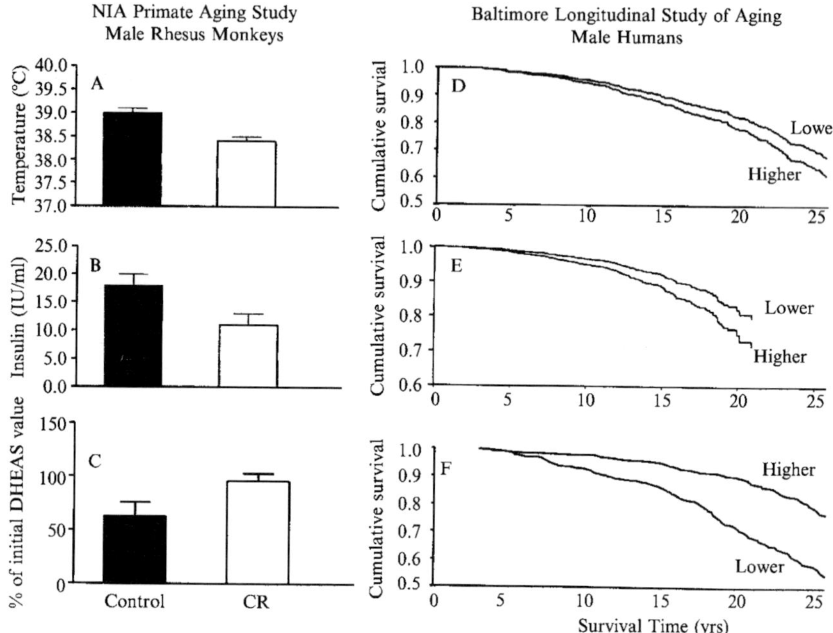

FIG. 12. (A–C) Values for male rhesus monkeys are the means ± standard errors for 20–30 animals in each group after 3–5 years of 30% CR. (D–F) The numbers of Baltimore Longitudinal Study of Aging men who died in the temperature, insulin, and dehydroepiandrosterone sulfate studies were 324, 199, and 192, respectively. Only those individuals surviving at least 3 years were followed in the dehydroepiandrosterone sulfate study, and all subjects were in good health, as evaluated by extensive medical assessment. The subject age in the three data sets ranged from 19 to 95 years. All analyses are significant to at least $p < .05$ by t-test or proportional hazard model, corrected for the initial subject age. The Baltimore Longitudinal Study of Aging data are divided into upper and lower halves for better comparisons with the two monkey groups, although all three age-adjusted variables exhibit significant and continuous effects for temperature (D), insulin (E), and dehydroepiandrosterone sulfate (F) on mortality in humans. From reference 73.

greater proportion of centenarians than the rest of Japan and boast a lower overall mortality rate as well as fewer deaths caused by vascular disease and cancer. Interestingly, Okinawa residents only take in 60% and 80% of the Japanese national average calories for children and adults, respectively [78].

Albanes *et al.* [79] examined the relationships among energy balance, body size and cancer risk using data from the first U.S. National Health and Nutrition Examination Survey and its follow-up study. The age-adjusted relative risk of cancer for men above 169 cm in height is significantly increased by 1.4- to 1.5-fold when compared to that for men under 169 cm in height. After adjustment for race, cigarette smoking, income, and body mass index, the relative risk of cancer in several organs is increased 1.5- to 1.6-fold with height. In particular, the colorectal cancer risk for men above 178.6 cm in height is significantly increased by 2.1-fold compared with that for men below 169 cm in height. Moreover, it has been suggested that a large body size and fatness, as measured by adult stature, body weight, and body mass indices, are positively related to a variety of cancer risks, including breast, colorectal, prostate, endometrial, kidney, and ovary [80]. These studies support associations among calorie intake, body size and disease, and perhaps, ultimately, longevity.

More controlled studies of a CR-like regimen have been reported for human subjects in Biosphere [81, 82]. The caloric intake was reduced between 20% and 30% from 10 weeks to over 2 years. This regimen reduced body weight, adiposity, blood pressure, blood glucose, plasma insulin, and serum lipids.

As shown in Fig. 12A–C [73], three of the most robust biomarkers of CR in both rodents and monkeys are lower body temperature, lower plasma insulin, and slow rate of decline in serum dehydroepiandrosterone sulfate (DHEAS). The survival rates of healthy men in the Baltimore Longitudinal Study of Aging in the upper and lower halves of the distributions for the corresponding markers were compared (Fig. 12D–F, [73]). Consistent with the beneficial effects of CR on aging and life span in other animals, men with lower temperature and insulin, and those maintaining a higher DHEAS level, have greater survival than their respective counterparts. Thus, CR-like effects on the physiological markers of reduced insulin, reduced body temperature, and slow decline in DHEAS would appear to be related to aging and longevity [73]. At present, conclusive evidence of the effects of CR has not been obtained in humans, although the results noted above strongly indicate that CR may also be effective in human aging and longevity.

5. Interpretation of the Beneficial Actions of CR Based On the Adaptive Response Hypothesis

The characteristics of animals subjected to long-term CR are summarized in Fig. 13. These CR effects could differ in several aspects of physiology from those of fasting (starvation), but the hormonal profile of CR (lower plasma leptin, insulin, growth hormone and gonadotropins, and higher plasma free corticosterone) is similar to that of fasting [83]. Based on studies examining the respiratory quotient or respiratory exchange ratio, carbohydrate is the major fuel after feeding in CR animals, and fuel utilization shifts from carbohydrate to fat before feeding [21, 42]. Fasting also shifts the fuel utilization

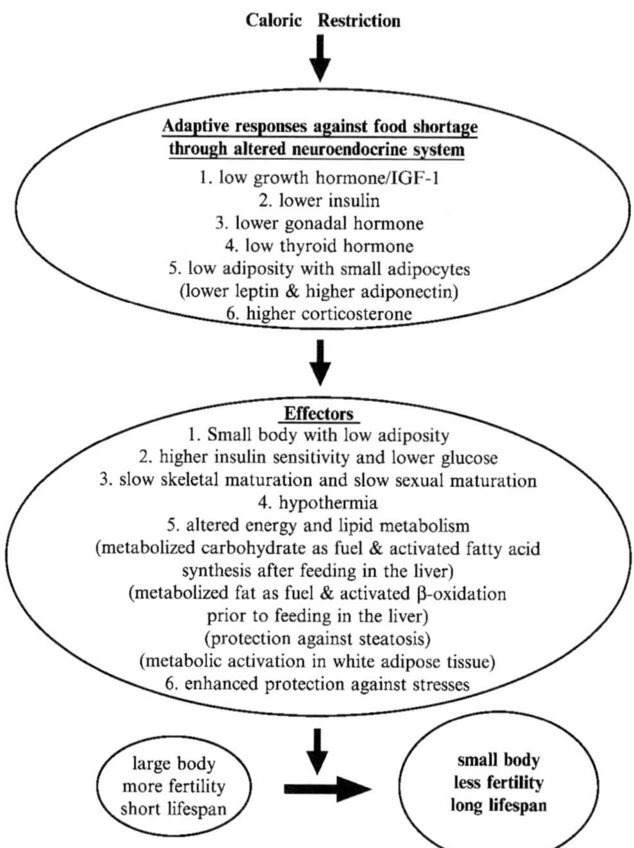

FIG. 13. Summary of the beneficial actions of caloric restriction based on the evolutional adaptive response hypothesis.

from carbohydrate to fat. The role of leptin in the altered hormonal profile and metabolic shift in CR animals has yet to be determined, but several lines of evidence indicate that it has an important role in metabolic adaptation during CR. When leptin is administered intraventricularly in fasting rats, the reduced respiratory quotient and fasting-induced hormonal profile are almost completely reversed to the control level in AL rats [20, 47, 84, 85].

Under fasting, fatty acids and glycerols derived from triglyceride in WAT are used as fuel during mitochondrial β-oxidation, and the byproduct of this oxidation generates ketone bodies, which are a major fuel source for the brain under starvation [86]. Moreover, fasting activates the mitochondrial β-oxidation, peroxisomal β-oxidation, and microsomal ω-oxidation system

[87–90]. Peroxisome proliferator-activated receptor α (PPARα). is known to be activated by both overfeeding and fasting, thereby inducing the up-regulation of the majority of genes transcriptionally involved in the mitochondrial β-oxidation, peroxisomal β-oxidation, and microsomal ω-oxidation system, and resulting in protection against hepatocyte steatosis [87–90].

In CR rats sacrificed before feeding, the genes involved in mitochondrial β-oxidation, PPARα, and several mitochondrial β-oxidation enzymes are up-regulated in the liver (unpublished data), and fat is predominantly metabolized (RER $= 0.8$). In this condition, exogenous fuel for mitochondrial β-oxidation is deficient, but the serum levels of free fatty acids and ketone bodies are increased (Table 1). In CR rats after feeding, despite lower serum levels of total lipid, triglyceride, total and free cholesterols, phospholipid and free fatty acids (Table 1), the genes associated with fatty acid synthesis, including ADD1/SREBP1, fatty acid synthase, and acyl-CoA carboxylase, are significantly up-regulated in the liver (unpublished data), and carbohydrate is predominantly metabolized (RER > 0.9). Thus, it is likely that effective lipid utilization is involved in the accelerated fatty acid synthesis after feeding and enhanced mitochondrial β-oxidation during food shortage via PPARα. Moreover, CR appears to activate carbohydrate, amino acid, lipid, and energy metabolisms in WAT [44]. Therefore, it is possible that endogenous fatty acids mostly derived from WAT must be provided as fuel for the reactions catalyzed by the mitochondrial β-oxidation enzymes, and this oxidation generates ketone bodies in CR rats before feeding (RER $= 0.8$). The fuel supplied by feeding may be transferred to and stored in WAT or used in peripheral organs more effectively in CR rats after feeding (RER > 0.9) than in AL rats via hepatic fatty acid synthesis.

PPARα-null mice have severe hypoglycemia, hypoketonemia, hypothermia, and elevated plasma free fatty acids following starvation, indicating a lack of metabolic adaptation in the liver [88]. Interestingly, ob/ob mice, which are deficient for leptin, have a similar response to fasting [91]. PPARα expression is regulated by glucocorticoids, and the diurnal cycle of the hepatic PPARα level is parallel to the circulating corticosterone level [92]. These findings indicate that metabolic adaptation to CR requires both the leptin signaling pathway and an adrenocortical system in the liver. Moreover, adiponectin stimulates fatty acid oxidation as well as glucose utilization through the activation of 5′-AMP-activated protein kinase [54]. The above evidence indicates that CR-associated reduction of leptin signaling and enhancement of corticosterone and adiponectin signaling contribute to metabolic adaptation against food shortage, including the shift from carbohydrate- to fat-based metabolism. These adaptive responses against food shortage through altered neuroendocrine systems could result in several CR-associated changes, as shown in Fig. 13.

6. Conclusions

Recently developed molecular and genetic interventions have been successful in controlling aging and longevity. Even though these lines of investigation hold great potential, studies using CR in rodents and nonhuman primates continue to provide valuable information. Because the effects of CR on aging processes, age-associated diseases, and longevity could be universal across species, this nutritional paradigm would have similar effects in humans. We believe that the genes or signal cascades altered by CR could be the principle target for generating novel mutant animal models that show suppressed aging processes and live longer. Moreover, CR mimetics, which are applicable in humans, may represent one strategy to modulate aging processes and longevity. Therefore, we emphasize the importance of dietary and calorie interventions and the interpretation of their effects in aging research, particularly in rodents, to humans.

References

[1] Weindruch R, Walford RL. The retardation of aging and disease by dietary restriction. Springfield, IL: Charles C. Thomas, 1988.

[2] Yu BP. Modulation of aging processes by dietary restriction. Boca Raton, FL: CRC Press, 1994.

[3] Cheney KE, Liu RK, Smith GS, Meredith PJ, Mickey MR, Walford RL. The effect of dietary restriction of varying duration on survival, tumor patterns, immune function, and body temperature in B10C3F1 female mice. J Gerontol 1983; 38:420–430.

[4] Duffy PH, Leakey JE, Pipkin JL, Turturro A, Hart RW. The physiologic, neurologic, and behavioral effects of caloric restriction related to aging, disease, and environmental factors. Environ Res 1997; 73:242–248.

[5] Keenan KP, Smith PF, Hertzog P, Soper K, Ballam GC, Clark RL. The effects of overfeeding and dietary restriction on Sprague-Dawley rat survival and early pathology biomarkers of aging. Toxicol Pathol 1994; 22:300–315.

[6] Maeda H, Gleiser CA, Masoro EJ, Murata I, McMahan CA, Yu BP. Nutritional influences on aging of Fischer 344 rats: II. Pathology. J Gerontol 1985; 40:671–688.

[7] Iwasaki K, Gleiser CA, Masoro EJ, McMahan CA, Seo EJ, Yu BP. Influence of the restriction of individual dietary components on longevity and age-related disease of Fischer rats: The fat component and the mineral component. J Gerontol 1988; 43: B13–B21.

[8] Yu BP, Masoro EJ, McMahan CA. Nutritional influences on aging of Fischer 344 rats: I. Physical, metabolic, and longevity characteristics. J Gerontol 1985; 40:657–670.

[9] Iwasaki K, Gleiser CA, Masoro EJ, McMahan CA, Seo EJ, Yu BP. The influence of dietary protein source on longevity and age-related disease processes of Fischer rats. J Gerontol 1988; 43:B5–B12.

[10] Shimokawa I, Higami Y, Hubbard GB, McMahan CA, Masoro EJ, Yu BP. Diet and the suitability of the male Fischer 344 rat as a model for aging research. J Gerontol 1993; 48: B27–B32.
[11] Weindruch R, Walford RL, Fligiel S, Guthrie D. The retardation of aging in mice by dietary restriction: Longevity, cancer, immunity and lifetime energy intake. J Nutr 1986; 116:641–654.
[12] Higami Y, Yu BP, Shimokawa I, Masoro EJ, Ikeda T. Duration of dietary restriction: An important determinant for the incidence and age of onset of leukemia in male F344 rats. J Gerontol 1994; 49:B239–B244.
[13] Masoro EJ. Caloric restriction and aging: An update. Exp Gerontol 2000; 35:299–305.
[14] Sohal RS, Weindruch R. Oxidative stress, caloric restriction, and aging. Science 1996; 273:59–63.
[15] Koubova J, Guarente L. How does calorie restriction work? Genes Dev 2003; 17:313–321.
[16] Ramsey JJ, Harper ME, Weindruch R. Restriction of energy intake, energy expenditure, and aging. Fre Rad Biol Med 2000; 29:946–968.
[17] Merry BJ. Molecular mechanisms linking calorie restriction and longevity. Int J Biochem Cell Biol 2002; 34:1340–1354.
[18] Holliday R. Food, reproduction and longevity: Is the extended life span of calorie-restricted animals an evolutionary adaptation? Bioessays 1989; 10:125–127.
[19] Masoro EJ, Austad SN. The evolution of the antiaging action of dietary restriction: A hypothesis. J Gerontol A Biol Sci Med Sci 1996; 51:B387–B391.
[20] Shimokawa I, Higami Y. Leptin signaling and aging: Insight from caloric restriction. Mech Ageing Dev 2001; 122:1511–1519.
[21] McCarter RJ, Palmer J. Energy metabolism and aging: A lifelong study of Fischer 344 rats. Am J Physiol 1992; 263:E448–E452.
[22] Sonntag WE, Xu X, Ingram RL, D'Costa A. Moderate caloric restriction alters the subcellular distribution of somatostatin mRNA and increases growth hormone pulse amplitude in aged animals. Neuroendocrinol. 1995; 61:601–608.
[23] Breese CR, Ingram RL, Sonntag WE. Influence of age and long-term dietary restriction on plasma insulin-like growth factor-1 (IGF-1), IGF-1 gene expression, and IGF-1 binding proteins. J Gerontol 1991; 46:B180–B187.
[24] Brown-Borg HM, Borg KE, Meliska CJ, Bartke A. Dwarf mice and the ageing process. Nature 1996; 384:33.
[25] Shimokawa I, Higami Y, Utsuyama M, et al. Life span extension by reduction in growth hormone-insulin-like growth factor-1 axis in a transgenic rat model. Am J Pathol 2002; 160:2259–2265.
[26] Bartke A, Wright JC, Mattison JA, Ingram DK, Miller RA, Roth GS. Extending the life span of long-lived mice. Nature 2001; 414:412.
[27] Shimokawa I, Higami Y, Tsuchiya T, et al. Life span extension by reduction of the growth hormone-insulin-like growth factor-1 axis: Relation to caloric restriction. FASE J 2003; 17:1108–1109.
[28] Masoro EJ, Compton C, Yu BP, Bertrand H. Temporal and compositional dietary restrictions modulate age-related changes in serum lipids. J Nutr 1983; 113:880–892.
[29] Wang ZQ, Bell-Farrow AD, Sonntag W, Cefalu WT. Effect of age and caloric restriction on insulin receptor binding and glucose transporter levels in aging rats. Exp Gerontol 1997; 32:671–684.
[30] Yamaza H, Komatsu T, Chiba T, et al. A transgenic dwarf rat model as a tool for the study of calorie restriction and aging. Exp Gerontol 2004; 39:269–272.

[31] Koizumi A, Tsukada M, Wada Y, Masuda H, Weindruch R. Mitotic activity in mice is suppressed by energy restriction-induced torpor. J Nutr 1992; 122:1446–1453.
[32] Koizumi A, Wada Y, Tuskada M, et al. A tumor preventive effect of dietary restriction is antagonized by a high housing temperature through deprivation of torpor. Mech Ageing Dev 1996; 92:67–82.
[33] Jin YH, Koizumi A. Decreased cellular proliferation by energy restriction is recovered by increasing housing temperature in rats. Mech Ageing Dev 1994; 75:59–67.
[34] Koizumi A, Tsukada M, Hirano S, Kamiyama S, Masuda H, Suzuki KT. Energy restriction that inhibits cellular proliferation by torpor can decrease susceptibility to spontaneous and asbestos-induced lung tumors in A/J mice. Lab Invest 1993; 68:728–739.
[35] McCarter RJ, McGee JR. Transient reduction of metabolic rate by food restriction. Am J Physiol 1989; 257:E175–E179.
[36] Dulloo AG, Girardier L. 24 hour energy expenditure several months after weight loss in the underfed rat: Evidence for a chronic increase in whole-body metabolic efficiency. Int J Obes Relat Metab Disord 1993; 17:115–123.
[37] Blanc S, Schoeller D, Kemnitz J, Weindruch R, et al. Energy expenditure of rhesus monkeys subjected to 11 years of dietary restriction. J Clin Endocrinol Metab 2003; 88:16–23.
[38] Herlihy JT, Stacy C, Bertrand HA. Long-term food restriction depresses serum thyroid hormone concentrations in the rat. Mech Ageing Dev 1990; 53:9–16.
[39] Snyder DL, Wostmann BS, Pollard M. Serum hormones in diet-restricted gnotobiotic and conventional Lobund-Wistar rats. J Gerontol 1988; 43:B168–B173.
[40] Herlihy JT, Thomas J, Bertrand H, Stacy C. The aging of the cardiovascular system: Modulation by dietary restriction. Ag Nutr 1992; 3:185–191.
[41] Kingsley TR, Nekvasil NP, Snyder DL. The influence of dietary restriction, germ-free status, and aging on adrenal catecholamines in Lobund-Wistar rats. J Gerontol 1991; 46: B135–B141.
[42] Duffy PH, Feuers RJ, Leakey JA, Nakamura K, Turturro A, Hart RW. Effect of chronic caloric restriction on physiological variables related to energy metabolism in the male Fischer 344 rat. Mech Ageing Dev 1989; 48:117–133.
[43] Barzilai N, Banerjee S, Hawkins M, Chen W, Rossetti L. Caloric restriction reverses hepatic insulin resistance in aging rats by decreasing visceral fat. J Clin Invest 1998; 101:1353–1361.
[44] Higami Y, Pugh TD, Page GP, Allison DB, Prolla TA, Weindruch R. Adipose tissue energy metabolism: Altered gene expression profile of mice subjected to long-term caloric restriction. FASE J 2004; 18:415–417.
[45] Schwartz MW, Woods SC, Porte D, Jr., Seeley RJ, Baskin DG. Central nervous system control of food intake. Nature 2000; 404:661–671.
[46] Zhang Y, Proenca R, Maffei M, Barone M, Leopold L, Friedman JM. Positional cloning of the mouse obese gene and its human homologue. Nature 1994; 372:425–432.
[47] Ahima RS, Prabakaran D, Mantzoros C, et al. Role of leptin in the neuroendocrine response to fasting. Nature 1996; 382:250–252.
[48] Bornstein SR. Is leptin a stress related peptide? Nat Med 1997; 3:937.
[49] Maffei M, Halaas J, Ravussin E, et al. Leptin levels in human and rodent: Measurement of plasma leptin and of RNA in obese and weight-reduced subjects. Nat Med 1995; 1:1155–1161.
[50] Shimokawa I, Higami Y. A role for leptin in the antiaging action of dietary restriction: A hypothesis. Aging Clin Exp Med 1999; 11:380–382.

[51] Kadowaki T, Hara K, Yamauchi T, Terauchi Y, Tobe K, Nagai R. Molecular mechanism of insulin resistance and obesity. Exp Biol Med 2003; 228:1111–1117.
[52] Fruebis J, Tsao TS, Javorschi S, et al. Proteolytic cleavage product of 30-kDa adipocyte complement-related protein increases fatty acid oxidation in muscle and causes weight loss in mice. Proc Natl Acad Sci USA 2001; 98:2005–2010.
[53] Yamauchi T, Kamon J, Waki H, et al. The fat-derived hormone adiponectin reverses insulin resistance associated with both lipoatrophy and obesity. Nat Med 2001; 7:941–946.
[54] Yamauchi T, Kamon J, Minokoshi Y, et al. Adiponectin stimulates glucose utilization and fatty-acid oxidation by activating AMP-activated protein kinase. Nat Med 2002; 8:1288–1295.
[55] Zhu M, Miura J, Lu LX, et al. Circulating adiponectin levels increase in rats on caloric restriction: The potential for insulin sensitization. Exp Gerontol 2004; 39:1049–1059.
[56] Sabatino F, Masoro EJ, McMahan CA, Kuhn RW. Assessment of the role of the glucocorticoid system in aging processes and in the action of food restriction. J Gerontol 1991; 46:B171–B179.
[57] Kalu DN, Hardin RH, Cockerham R, Yu BP. Aging and dietary modulation of rat skeleton and parathyroid hormone. Endocrinology 1984; 115:1239–1247.
[58] Merry BJ, Holehan AM. Onset of puberty and duration of fertility in rats fed a restricted diet. J Reprod Fertil 1979; 57:253–259.
[59] Hamilton GD, Bronson FH. Food restriction and reproductive development: Male and female mice and male rats. Am J Physiol 1986; 250:R370–R376.
[60] Bronson FH. Food-restricted, prepubertal, female rats: Rapid recovery of luteinizing hormone pulsing with excess food, and full recovery of pubertal development with gonadotropin-releasing hormone. Endocrinology 1986; 118:2483–2487.
[61] Howland BE. The influence of feed restriction and subsequent re-feeding on gonadotrophin secretion and serum testosterone levels in male rats. J Reprod Fertil 1975; 44:429–436.
[62] Campbell GA, Kurcz M, Marshall S, Meites J. Effects of starvation in rats on serum levels of follicle stimulating hormone, luteinizing hormone, thyrotropin, growth hormone and prolactin; Response to LH-releasing hormone and thyrotropin-releasing hormone. Endocrinology 1977; 100:580–587.
[63] Glass AR, Swerdloff RS. Nutritional influences on sexual maturation in the rat. Fed Proc 1980; 39:2360–2364.
[64] Howland BE. Gonadotrophin levels in female rats subjected to restricted feed intake. J Reprod Fertil 1971; 27:467–470.
[65] Nelson JF, Gosden RG, Felicio LS. Effect of dietary restriction on estrous cyclicity and follicular reserves in aging C57BL/6J mice. Biol Reprod 1985; 32:515–522.
[66] Merry BJ, Holehan AM. Serum profiles of LH, FSH, testosterone and 5 alpha-DHT from 21 to 1000 days of age in ad libitum fed and dietary restricted rats. Exp Gerontol 1981; 16:431–444.
[67] Ingram DK, Cutler RG, Weindruch R, et al. Dietary restriction and aging: The initiation of a primate study. J Gerontol 1990; 45:B148–B163.
[68] Kemnitz JW, Weindruch R, Roecker EB, Crawford K, Kaufman PL, Ershler WB. Dietary restriction of adult male rhesus monkeys: Design, methodology, and preliminary findings from the first year of study. J Gerontol 1993; 48:B17–B26.
[69] Lane MA, Ingram DK, Cutler RG, Knapka JJ, Barnard DE, Roth GS. Dietary restriction in nonhuman primates: Progress report on the NIA study. Ann N Acad Sci 1992; 673:36–45.

[70] Black A, Tilmont EM, Handy AM, Ingram DK, Roth GS, Lane MA. Calorie restriction reduces the incidence of proliferative disease: Preliminary data from the NIA CR in nonhuman primate study. Gerontologist 2000; 40:5.

[71] Roth GS, Ingram DK, Lane MA. Calorie restriction in primates: Will it work and how will we know? J Am Geriatr Soc 1999; 47:896–903.

[72] Lane MA, Mattison J, Ingram DK, Roth GS. Caloric restriction and aging in primates: Relevance to humans and possible CR mimetics. Microsc Res Tech 2002; 59:335–338.

[73] Roth GS, Lane MA, Ingram DK, et al. Biomarkers of caloric restriction may predict longevity in humans. Science 2002; 297:811.

[74] Lane MA, Reznik A, Tilmont EM, et al. Aging and food restriction alter some indices of bone metabolism in male rhesus monkeys (*Macaca mulatta*). J Nutr 1995; 125:1600–1610.

[75] Lane MA, Ball SS, Ingram DK, et al. Diet restriction in rhesus monkeys lowers fasting and glucose-stimulated glucoregulatory endpoints. Am J Physiol 1995; 268:E941–E948.

[76] Lane MA, Baer DJ, Rumpler WV, et al. Dietary restriction lowers body temperature in rhesus monkeys consistent with a postulated anti-aging mechanism in rodents. Proc Nat Acad Sci USA 1996; 93:4159–4164.

[77] Lane MA, Ingram DK, Roth GS. Calorie restriction in nonhuman primates: Effects on diabetes and cardiovascular disease risk. Toxicol Sci 1999; 52:41–48.

[78] Kagawa Y. Impact of Westernization on the nutrition of Japanese: Changes in physique, cancer, longevity, and centenarians. Prev Med 1978; 7:205–217.

[79] Albanes D, Jones DY, Schatzkin A, Micozzi MS, Taylor PR. Adult stature and risk of cancer. Cancer Res 1988; 48:1658–1662.

[80] Albanes D. Energy balance, body size, and cancer. Crit Rev Oncol Hematol 1990; 10:283–303.

[81] Walford RL, Harris SB, Gunion MW. The calorically restricted low-fat nutrient-dense diet in Biosphere 2 significantly lowers blood glucose, total leukocyte count, cholesterol, and blood pressure in humans. Proc Natl Acad Sci USA 1992; 89:11533–11537.

[82] Velthuis-te Wierik EJ, van den Berg H, Schaafsma G, Hendriks HF, Brouwer A. Energy restriction, a useful intervention to retard human ageing? Results of a feasibility study. Eur J Clin Nutr 1994; 48:138–148.

[83] Nelson JF. Neuroendocrine involvement in the retardation of aging by dietary restriction: A hypothesis. In: Modulation of Aging Processes by Dietary Restriction. Boca Raton, FL: CRC Press, 1994: 37–55.

[84] Overton JM, Williams TD. Behavioral and physiologic responses to caloric restriction in mice. Physiol Behav 2004; 81:749–754.

[85] Overton JM, William TD, Chambers JB, Rashotte ME. Central leptin infusion attenuates the cardiovascular and metabolic effects of fasting in rats. Hypertension 2001; 37:663–669.

[86] Salway JG. Metabolism at a glance, second ed. OxfordK: Blackwell Science, 1999.

[87] Hashimoto T, Cook WS, Qi C, Yeldandi AV, Reddy JK, Rao MS. Defect in peroxisome proliferator-activated receptor alpha-inducible fatty acid oxidation determines the severity of hepatic steatosis in response to fasting. J Biol Chem 2000; 275:28918–28928.

[88] Kersten S, Seydoux J, Peters JM, Gonzalez FJ, Desvergne B, Wahli W. Peroxisome proliferator-activated receptor alpha mediates the adaptive response to fasting. J Clin Invest 1999; 103:1489–1498.

[89] Kroetz DL, Yook P, Costet P, Bianchi P, Pineau T. Peroxisome proliferator-activated receptor alpha controls the hepatic CYP4A induction adaptive response to starvation and diabetes. J Biol Chem 1998; 273:31581–31589.

[90] Rao MS, Reddy JK. Peroxisomal beta-oxidation and steatohepatitis. Semin Live Dis 2001; 21:43–55.

[91] Smith SA, Cawthorne MA, Simson DL. Glucose metabolism in the obese hyperglycaemic (C57B1/6 ob/ob) mouse: The effects of fasting on glucose turnover rates. Diabete Res 1986; 3:83–86.

[92] Lemberger T, Saladin R, Vazquez M, et al. Expression of the peroxisome proliferator-activated receptor alpha gene is stimulated by stress and follows a diurnal rhythm. J Biol Chem 1996; 271:1764–1769.

IMMUNO-PCR AS A CLINICAL LABORATORY TOOL

Michael Adler

Chimera Biotec GmbH, Biomedicinecenter Dortmund, 44227 Dortmund, Germany

1. Introduction: Immuno-Polymerase Chain Reaction—Principle of the Method

The advent of the polymerase chain reaction (PCR) was a major turning point in the history of analytical techniques for biomolecules. Since the first publication of the method in 1985, the exponential signal amplification power of PCR was mirrored by an almost similar exponential increase of applications [1–4]. Using a simple and effective signal amplification by repetitive cyclic duplication of a template nucleic acid strand, the signal-enhancing

0065-2423/05 $35.00
DOI: 10.1016/S0065-2423(04)39009-8

power of a PCR is superior to all conventional linear enzymatic signal amplifications. If PCR is performed under optimized conditions, the specific detection of a single nucleic acid molecule in a given sample is possible [5–7]. An inherent disadvantage of this technique is, of course, the limitation to the detection of nucleic acids resulting from the principle of the amplification.

However, for clinical applications, nucleic acids are only a fraction of a broad and very inhomogeneous range of possible analytes, starting with large proteins like antibodies and also including small molecule hormones and markers for various diseases.

For all of these targets the versatility of enzyme-linked immunosorbent assays (ELISAs) qualifies this method as an almost universal detection platform, using the highly specific recognition potential of antibodies in conjugation with a detection enzyme and various signal-generating substrates [8, 9]. In contrast to the enormous amplification power of the PCR, these enzymatic methods normally have a detection limit of several millions of molecules.

Therefore, an approach to combining the signal amplification power of the PCR with the flexibility of the ELISA is highly interesting in regard to the improvement of the sensitivity of conventional antigen detection. In theory, this could be easily achieved by a simple exchange of the antibody–enzyme conjugate with an antibody–DNA conjugate (Fig. 1) and subsequent amplification of the DNA marker. This approach, called immuno-PCR (IPCR), was introduced in 1992 by Sano *et al.* and was from then on continuously modified as a valuable research tool [11]. Recently, a number of application papers and methodic variations have underlined the potential and suitability of the IPCR method for routine applications (e.g., in clinical analysis). With IPCR, a typical 100–10,000-fold increase in ELISA sensitivity was observed in almost all applications.

However, to adapt the remarkably elegant and evident idea behind the IPCR method to the challenges of daily usage, three main methodological aspects are the keys to successful method development and optimization: the conjugation of DNA and antibody (see Section 2.1.), the detection of the amplified DNA-marker (see Section 2.2), and the suppression of nonspecific background signals in ultrasensitive analysis, combined with the adaptation of the assay to different antigens/haptens (see Sections 2.3. and 3).

In the following sections, different methods of establishing functional DNA–protein conjugates are discussed, following a short historic overview of IPCR method development. As IPCR offers additional advantages in addition to the obvious enhancement of sensitivity, the unique potential for multiplex detection and an increased linear quantification range of the method is introduced. We show how various ELISA types suitable for

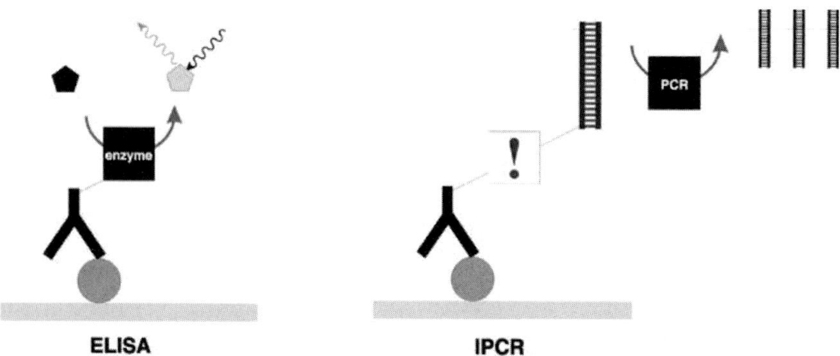

FIG. 1. Comparison of enzyme-linked immuno sorbent assay (ELISA, left) and immuno–polymerase chain reaction (IPCR, right). During ELISA, an antibody–enzyme conjugate is bound to the target antigen. The enzyme converts a substrate in solution to a detectable product. In IPCR, the antibody–enzyme conjugate is replaced by an antibody–DNA conjugate. The subsequent addition of a DNA polymerase enzyme (e.g., Taq), nucleotides and a specific primer pair uses the antibody-linked DNA marker sequence as a template for amplification of the DNA. The PCR product is finally detected as an indicator of the initial amount of antigen.

different target antigens were adapted to IPCR. The practical problems encountered while using this ultrasensitive assay—and methods for circumventing them—are covered in depth, including, for example, the validation and application of an IPCR assay for pharmacokinetic experiments in a clinical phase 1 study. Finally, a short evaluation of the continuous quest for improving the IPCR protocol and adapting it to new questions of clinical relevance is given.

2. Assay Set-Up: Key Steps and Developments of the IPCR Method

During the last few years, many applications of the IPCR method were described. A short overview, including detection limits compared to conventional ELISA, is given in Table 1. The flexibility, robustness, and sensitivity of the assay are obvious from the multitude of applications studied so far. It is also obvious, however, that the number of IPCR studies is still very small compared to, for example, approximately 200,000 PubMed-listed publications for PCR in the same timeframe. The majority of IPCR applications are research applications with protocols considered to be very demanding to the inexperienced experimenter. With an increasing number of research

TABLE 1

Overview of Immuno–Polymerase Chain Reaction (IPCR) Applications

Antigen	Year	ELISA set-up (see Fig. 3)	Comparison IPCR/ELISA	Source/short commentary
Tumor markers and disease associated proteins (see also Chapter 3.1)				
Carbohydrate tumor marker: tumor-related antigen GM3	1998	A	Detection level of <10 cells, melanoma cells diluted in 2 million healthy blood lymphocytes	Zhang *et al.* [96]
Carcinoembryonic antigen (CEA)	2001	A	Circulating CEA in human sera: 92.3% of patients with high CEA levels and 50% of patients with "normal" CEA levels were found positive	Ren *et al.* [18], use of a bispecific CEAScFv-streptavidin fusion protein-based immuno-PCR technique (see Section 2.1.1)
	2003	C	Factor 1000 (IPCR: 10 pg/mL [0.05 amol/μL], ELISA: 10 ng/mL)	Niemeyer *et al.* [60], DNA-directed immobilization–IPCR (see Fig. 4 and Section 2.1.4)
Gastritic cancer associated antigen McAb MG7-Ag	1994	A	Factor 10,000 (IPCR: 3.8×10^{-14} mol, ELISA 3.0×10^{-11} mol)	Ren *et al.* [79]
	2000	A	Factor 10,000 (IPCR: 3.8×10^{-14} mol, ELISA 3.0×10^{-11} mol)	Ren *et al.* [32]
Human proto-oncogen ETS1	1993	B	Factor 100,000 (IPCR: 9.6×10^{-21} mol, ELISA 960 amol)	Zhou *et al.* [24], introduction of the Universal-IPCR method: Sequential coupling of biotinylated antibody, STV, and biotinylated DNA (see Fig. 2B and Section 2.1.2)
PIVKA-II, tumor marker of hepatocellular carcinoma (HCC)	1995	B	Factor 1,000,000,000 (IPCR: 10–10 dilution sample of PIVKA antigen)	Tsujii *et al.* [97]
Prostate-cancer serum-tumor marker (PSA)	2000	C	Factor 1000 Immuno-RCA: 0.3 amol (0.1 pg/ml) ELISA: 300 amol	Schweitzer *et al.* [44], Immuno-RCA (see Section 2.1.3)

	2002	C	Factor 100 IPCR: $2.4 \times 10e6$ molecules PSA ELISA: 300 amol	Lind *et al.* [69], real-time IPCR (Section 2.2.3), iCyclerIQ system (Biorad)
	1995, 1999	A	Factor 100 Expression-immunoassay: 1, 1 amol (30 pg/ml)	Christopoulos *et al.* [77], Chiu *et al.* [78], expression immunoassay (see Section 3.1)
Tumor necrosis factor alpha (TNF-alpha)	1995	D	Factor 16 (IPCR: 6,25 pg/ml ELISA 100 pg/ml)	Sanna *et al.* [33]
	1999	D	Factor 50,000 (IPCR: 1 pg/L, ELISA: 5 ng/L)	Saito *et al.* [34]
	2001	D	Factor 50,000 (IPCR: 1 pg/L, ELISA: 5 ng/L)	Komatsu *et al.* [39]
Vascular endothelial growth factor	2000	C	Factor 5 (IPCR: 0.2 pg/ml fluorescence-ELISA: 1 pg/ml):	Sims *et al.* [49], real-time IPCR (see Section 2.2.3): detection of IPCR-amplificate was carried out online during PCR according to the TaqMan-principle, AbiPrism7700 sequence detector (Applied Biosystems)
			Virus antigens (3.2)	
Bovine-herpesvirus I (BHV-1)	1996	B	Antigen: Factor 1,000,000 Antibody: Factor 10,000 (IPCR: detection at 19 days following infection)	Mweene *et al.* [28, 98]
Hepatitis B surface-antigen (HBsAg)	1995	C	Factor 100 (IPCR: 0.5 pg RIA 50 pg)	Maia *et al.* [30], sandwich IPCR with subsequent PCR–enzyme linked oligonucleotide sorbent assay (see also Fig. 6 and Section 2.2.2)
	2000	A	*In situ*-IPCR: 14 of 17 liver-samples positive ELISA: 6 of 17 liver-samples positive	Cao *et al.* [82], *in situ* IPCR
HIV core protein p24	2004	C	Factor 25* (IPCR: ca. 2 copies, RT-PCR 50 copies)	Barletta *et al.* [84]

(*continues*)

TABLE 1 (*Continued*)

Antigen	Year	ELISA set-up (see Fig. 3)	Comparison IPCR/ELISA	Source/short commentary
Influenza viral antigen nonstructural protein 1 (NS 1)	2001		Factor 10* (*IPCR 10-fold more sensitive than RT-PCR (!) or virus isolation)	Ozaki *et al.* [83]
Mumps-specific IgG in serum	2002	B	Factor 1 [Compared to ELISA, IPCR was 98.6% (71/72) sensitive and 92.9% (13/14) specific]	McKie *et al.* [50, 68], real-time IPCR (Section 2.2.3), covalent antibody–DNA conjugates (Section 2.1.3), Light-Cycler (Roche)
Recombinant hepatitis B surface-antigen (HBsAg), mouse/rabbit IgG	1997	A, C, D	Factor 700 (HBsAg: IPCR: 70 fg ELISA: 49 pg) Factor 1000 (IgG: IPCR: 15 fg ELISA: 15 pg)	Niemeyer *et al.* [37], development of a fluorescence-PCR-ELISA for sensitive detection of the IPCR-amplificate (see Fig. 6 and Section 2.2.2) and identical polyclonal capture and detection antibodies
Pharmacokinetics (3.3)				
Recombinant mistletoe-lectin (rML, *rViscumin, Aviscumine*), rabbit-IgG	1999	C	Factor 1000 (rML: IPCR: 200 fg/ml ELISA: 200 pg/ml) Factor 10,000 (IgG: IPCR: 1.5 fg ELISA: 15 pg)	Niemeyer *et al.* [26], purified aggregates from biotinylated antibody, STV, and double biotinylated DNA as reagents for IPCR (see Fig. 2D and Section 2.1.4)
	2003	C	Factor 1000–10,000 [IPCR: 50 100 fg/ml (ca. 50 zeptomol) rViscumin]	Adler *et al.* [66], [73], commercial antibody–DNA conjugates, method validation
Pathogens and toxins (3.4)				
Clostridium botulinum neurotoxin type A	2001	C	Factor 1000 (IPCR: 3.33×10^{-17} mol, ELISA: 3.33×10^{-14} mol)	Wu *et al.* [48], covalent antibody–DNA conjugates (see Fig. 2C and Section 2.1.3)

	2004		Factor 100,000 (IPCR: 50 fg; ELISA: 5 ng)	Chao *et al.* [88]
Fish-pathogen (*Pasteurella piscicida*) in infected yellowtails	1996	B	Factor 10,000 (IPCR: 3.4 cfu/ml ELISA: 3.4×10^4 cfu/ml)	Kakizaki *et al.* [25], usage of double biotinylated DNA in a sequential Universal IPCR
Gliadin (food allergen)	2003		Factor 30 IPCR: 0.16 ng/ml (0.16 ppm)	Henterich *et al.* [51], real-time IPCR (Section 2.2.3) comparison of direct conjugates (Section 2.1.3) with sequential incubation, AbiPrism 7700 sequence detector (Applied Biosystems)
Pathogenic Group A Streptococcus (Strep A)	2003	A	IPCR: approximately 10 cells	Liang *et al.* [74], covalent antibody–DNA conjugates (see Section 2.1.3) vs. Universal-IPCR (see Section 2.1.2)
Embryonal growth and surface proteins, regulating hormones (3.5)				
Cell surface Qa-2 protein expression (major histocompatibility complex class Ib protein of the preimplantation embryo development)	2000	A	No comparison carried out, used in addition to RT-PCR for the study of expression in mouse embryos	Ke and Warner [14], use of the STV protein A chimera (see Fig. 2A and Section 2.1.1)
Human thyroid-stimulating hormone, chorionic Gonadotropin and β- Galactosidase (hTSH, hCG, β-Gal)	1994	C	Factor 100 (IPCR: hTSH 1×10^{-19} mol, hCG 1×10^{-17} mol, β-Gal: 1×10^{-17} mol; ELISA: hTSH 1×10^{-17} mol, hCG 1×10^{-15} mol, β-Gal: 1×10^{-15} mol)	Hendrickson *et al.* [41], covalent antibody–DNA conjugates (see Fig. 2C and Section 2.1.3.) and simultaneous application in a multiplex-IPCR, electrophoretic separation of the amplificates according to their length
	1995	C	Factor 100–1000 (IPCR: hTSH 1×10^{-19} mol, hCG 5×10^{-18} mol)	Joerger *et al.* [38], covalent antibody–DNA conjugates (see Fig. 2C and Section 2.1.3)

(*continues*)

TABLE 1 (*Continued*)

Antigen	Year	ELISA set-up (see Fig. 3)	Comparison IPCR/ELISA	Source/short commentary
Murine major histocompatibility complex (MHC)	1997	A	Factor 10 (IPCR: one embryo, ELISA: 10–15 embryos)	Quijada *et al.* [13], use of the STV protein A chimera (see Fig. 2A and Section 2.1.1)
Qa-2 antigen (product of the Preimplantation embryo development gene, mouse)	1998	A	No comparison carried out, used in addition to RT-PCR for the study of blastocystes	McElhinny *et al.* [15, 16], use of the STV protein A (see Fig. 2A and Section 2.1.1)
Other applications: Microorganisms, parasites, heart failure associated proteins, cytokines, small molecule targers, etc. (3.6)				
Antigen to Angiostrongylus cantonensis circulating fifth-stage worms	2004	C	Factor 100–1000 IPCR: 0.1 ng/L	Chye *et al.* [64]
Antihuman atrial natriuretic peptide (ANP) in serum	1997	C	Factor (time) 10 (IPCR: 5 hours, IRMA: 2–3 days)	Numata *et al.* [99]
Beta-glucuronidase (GUD) from *E. coli*	1997	C	Factor 100,000,000 (IPCR: 1×10^{-17} g/mL = 10 attog/mL; ELISA: 1 ng/mL)	Chang *et al.* [31], identical first and second polyclonal antibodies
Bovine serum albumin (BSA)	1992	A	Factor 100,000 (IPCR: 9.6×10^{-22} mol ELISA 96 amol)	Sano *et al.* [10], initial publication of the IPCR: STV-protein A chimera for coupling biotinylated DNA with an antibody (see Fig. 2A and Section 2.1.1)
Fluorescein	2000	A	Factor 1000 cIPCR: 0.3 amol/μL	Niemeyer *et al.* [92], competitive IPCR (see Fig. 10A and Section 3.6.4)
Homodimeric osteoprotegerin (rhOPG)	2001	D	Factor 25,000 IPCR: 5 pg/L homodimer	Furuya *et al.* [35]
Human angiotensinogen	2000	C	Factor 250,000 (IPCR: 0.1 ng/L, ELISA: 25 μg/L)	Sugawara *et al.* [36], identical polyclonal capture and detection antibodies

Human interleukin 18 (hIL-18)	2000	D	Factor 16,000 (IPCR: 2.5 pg/L ELISA: 40 ng/L)	Furuya *et al.* [27]
IgG from several species (mouse, rabbit, goat, and human), Aviscumin, research antibody in cell culture media	2003	A, B, C	Factor 100–1000 (IPCR: 0.1–0.01 amol [50–500 fg/mL] IgG, 40 pg/mL Aviscumin in serum, 100 pg/mL research antibody)	Adler *et al.* [62], real-time IPCR (Section 2.2.3), internal competitor (Section 2.3.6), commercial antibody–DNA conjugates, AbiPrism 7000 sequence detector (Applied Biosystems)
Intercellular adhesion-molecule 1 (sICAM-1)	1995	C	Factor 1000	Suzuki *et al.* [100]
	1996		Factor 1000	Itoh *et al.* [101]
Mouse antibody against Apoliprotein E	1993	A	IPCR: 0.5 μg, no comparable ELISA data given	Ruizicka *et al.* [55]
Oligomeric pyruvat-dehydro-genase-complex (PDC), promutagenes DNA base adduct O(6)-Methylguanosin, O(6)-Methylguanin-DNA Methyltransferase (MGMT), N-terminal peptide of MGMT	1999	B	Factor 100,000,000 (IPCR: 1 molecule, ELISA: 10–100 amol)	Case *et al.* [29, 40]
p185 (her2/neu) receptor from the crude lysate of T6-17 cells	2001	C	Factor 1,000,000,000 (10^{-13}) dilution of cell solution	Zhang *et al.* [102], IDAT (immuno detection amplified by T7 RNA polymerase)
Serotonin	2004	B	Factor 10,000 (IPCR: 0.4 pg/mL antigen, conventional ELISA: 4 puncsp;ng/mL)	Chimera Biotec GmbH/LDN [94, 93], competitive IPCR (see Fig. 10B and Section 3.6.4.)
Soluble murine T-cell receptor (sTCR)	1995	D	Factor 125 (IPCR: 0.8 pg/mL ELISA: 100 pg/mL)	Sperl *et al.* [103]
Transmembrane FcgRIIIa (CD16), (a low-affinity IgG Fc receptor)	2003	C	IPCR used in a study for comparison of FcgRIIIa level in healthy and RA patients	Masuda *et al.* [104]

Unless otherwise stated in the "commentary" column, all assays were carried out according the Universal-IPCR protocol (see Section 2.1.2), combined with gel electrophoresis for IPCR amplificate detection (see Section 2.2.1).

groups applying IPCR, there is nevertheless a remarkable improvement of assay handling and method performance. Although selected results obtained by IPCR are discussed in depth in Chapter 3, the development of the method is described in the following sections. A combination of the different methodological improvements will be essential for successful routine IPCR application.

2.1. Antibody–DNA Coupling

Providing the needed protein–DNA conjugates for IPCR is the main challenge in establishing a workable IPCR assay. Either nontrivial conjugation techniques or elaborate multistep incubation protocols of several reagents are necessary for linking antibody and DNA. Figure 2 schematically summarizes the four main strategies employed in the synthesis of the IPCR reagents.

2.1.1. *Protein Chimeras: The Origin of IPCR*

Immuno-PCR was first time introduced in 1992 by Sano, Smith, and Cantor [10] during their work on protein–protein chimeras. The researchers described an application for a highly innovative combination of the universal antibody binding "protein A" and the tetravalent biotin-binding protein streptavidin (STV; see also Section 2.1.2). This protein chimera, artificially designed by gene fusion [12] was successfully used for linking an anti-bovine serum albumine (BSA) antibody to a biotinylated DNA marker and was thereby able to detect solid-phase immobilized BSA (Figs. 2A, 3A) in an amount of 9.6×10^{-22} mol. The usefulness of this fast and elegant approach is, however, limited by several drawbacks: The STV/protein A chimera is not easily available, requiring a laboratory with access to the necessary fusion technology. In addition, this conjugate binds nonspecifically to any antibody immobilized to the microplate surface and is therefore not suited for sandwich applications using capture-antibody-coated microplates (see Fig. 3C, D).

Nevertheless, Quijada *et al.* [13] described a very sensitive application of the IPCR method, using protein-chimeras for the preimplantation detection of histocompatiblity. In this assay, single murine blastocystes as analytical targets were handled without immobilization to microplate surfaces, and therefore with no need for capture antibodies. The authors reported that the broad binding ability of protein A was turned to an advantage for the flexible detection of nonfunctionalized antibodies. Further application of the STV/protein A chimera in IPCR was described for a similar expression

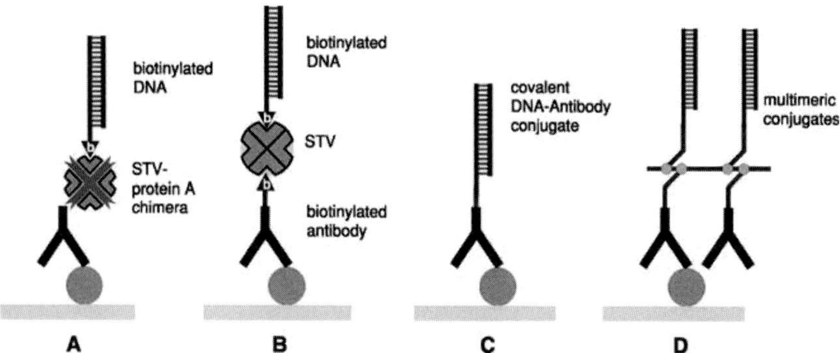

FIG. 2. Comparison of different techniques for antibody–DNA coupling. (A) In the first immuno–polymerase chain reaction (IPCR) protocol developed by Sano *et al.* [10], a protein chimera of antibody-binding protein A and biotin-binding streptavidin (STV) was preincubated with biotinylated DNA. Subsequently, this protein–DNA conjugate was immobilized against antigen-bound antibody to establish a link between antigen and DNA (see Section 2.1.1). (B) In the Universal-IPCR approach [24], the protein chimera was substituted by a sequence of three incubation steps: biotinylated antibody, tetravalent biotin-binding protein streptavidin (STV), or avidin as a linker and biotinylated DNA were coupled subsequently to the antigen (see Section 2.1.2). (C) In contrast to the manifold of incubation steps in Universal-IPCR, covalent antibody–DNA conjugates were used in a single-step assay setup. This multiplex compatible protocol requires cross-linking chemistry, synthesis, and purification for each conjugate [38, 41] (see Section 2.1.3). (D) An additional gain in sensitivity was achieved by simultaneously increasing the avidity of antibody–DNA conjugates and the amount of DNA linked to each antigen with multimeric polyvalent conjugates, each consisting of several antibodies and DNA markers (see Section 2.1.4). These conjugates are accessible, for example, by supramolecular self-organization [26, 57]. Nowadays, they are also commercially available in ready-to use kits (e.g., IMPERACER, Chimera Biotec).

analysis in blastocystes and embryos [14–17] (see Section 3.5). A variation of this technique, using a bispecific fusion protein of a carcinoembryonic antigen (CEA) binding protein and streptavidin, was successfully applied in the study of circulating CEA by Ren *et al.* [18].

Especially useful is the possibility of preparing conjugates of protein/STV chimeras and biotinylated DNA previous to their application in the assay, thus minimizing the amount of incubation steps required for IPCR (see also Section 2.1.4). Therefore, it could be concluded that if a custom-made protein chimera is accessible as a laboratory tool, or if the facilities and experience for preparing such a reagent are available, and additionally, no capture antibodies were needed, these conjugates allow for a very smart and robust approach to ultrasensitive protein detection by IPCR.

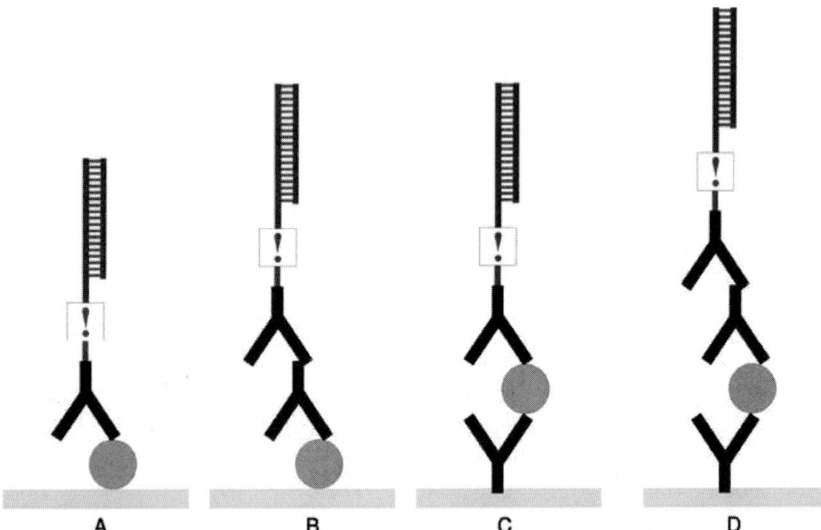

FIG. 3. Comparison of different enzyme-linked immuno sorbent assay (ELISA) methods adapted for immuno–polymerase chain reaction (IPCR). Dependent on the purification grade of the sample to be analyzed and the availability of specific and functionalized antibodies, several typical ELISA protocols were adapted to IPCR. In the direct approach (A), the pure antigen is immobilized to the microplate surface and subsequently detected by a labeled specific antibody. If no labeled antibody is available (e.g., because of unpurified ascites fluid containing the antibody or loss in activity following labeling), a standardized labeled secondary species-specific antibody is used for detection of the primary antigen-specific antibody (B). For the detection of the antigen from matrices such as serum, plasma, tissue homogenate, and so on, a capture antibody immobilized to the microplate surface was used either in a direct (C) or indirect (D) sandwich approach, with the latter one additionally including a secondary species-specific detection antibody. For different methods of coupling antibody and DNA, abbreviated by "!" in this figure, compare Fig. 2. Note that protein A chimeras (Fig. 2A) are not compatible with capture antibodies (Fig. 3C, D).

2.1.2. *Universal IPCR—Flexible Antibody–DNA Coupling with Biotinylated DNA*

The concept of linking PCR and ELISA for ultrasensitive protein analysis soon sparked the imagination of laboratory scientists, and they developed a method that separates the concept of antibody–DNA linkage from hard-to-obtain protein chimeras. An easy way to establish the desired coupling was accessible with the streptavidin/biotin binding system: The tetravalent protein STV, from *Streptomyces avidii* [19], derived, for example, as a recombinant core protein with optimized properties [20], or as its egg-yolk complement avidin [19], binds four molecules of biotin and is therefore able to link two biotinylated compounds in stepwise incubation [21–23]. Thereby, STV is often favored as coupling reagent for biotinylated compounds [19, 23] because of the additional glycosylation of avidin, which is reported for possible nonspecific interactions. This is especially important for ultrasensitive signal amplification methods such as IPCR, where the avoidance of any nonspecific binding is essential for background suppression.

A blueprint for the majority of the following IPCR applications was introduced in 1993 by Zhou *et al.* [24] with the appositely termed "Universal-IPCR" protocol. In this technique, the antigen to be detected is coupled subsequently with an antigen-specific primary antibody, a species-specific biotinylated detection antibody, STV, and a biotinylated DNA in four incubation steps, each separated by a washing step for removal of unbound reagents (see Figs. 2B and 3B).

Even from this short description, advantages and disadvantages of this approach are obvious. The major advantage of this method is the complete avoidance of any uncommon reagents or complicated synthesis steps. Biotinylated antibodies are either commercially available by a number of suppliers or easily prepared in-house with a number of standardized and robust kit reagents in a single-step reaction [23]. The synthesis of biotinylated marker DNA could be performed by preparative PCR, using commercially available biotinylated primer [25–27]. As an alternative, terminal biotin labeling of a DNA-marker was carried out using, for example, Klenow fragment [13, 24, 28, 29] or Sequenase polymerase [30]. In addition, photoactive biotin was used for synthesis of a biotinylated DNA marker for IPCR, too [31, 32].

Whereas almost any DNA sequence is a possible tool for IPCR, interestingly, a large number of publications described the usage of a standard plasmid sequence from pBluescript/M13 [25, 27, 33–36] or pUC19 [10, 13, 26, 28, 37] as a kind of standard IPCR marker derived from DNA ready available in the typical molecular biology laboratory. Maia *et al.* emphasize the importance of a DNA-marker not present in the routine work of the laboratory carrying out IPCR to avoid cross-contamination. In their work, they used a unique BTV–virus–DNA sequence only found in ruminants for the detection of HbsAg [30].

As the DNA marker is the template for signal amplification, it is surprising that only a few studies systematically evaluate the properties of different DNA length or sequence [26, 38]. Although IPCR signal amplification was found to be very robust regarding different DNA markers, ranging in size from whole plasmids [32] to short, 100-bp fragments [26, 38], optimization of the DNA marker revealed that the amplification potential of the marker could be additionally improved by further tailoring to the IPCR application;

for example, by artificially designed marker and competitor DNA sequences (see Section 2.3.6).

The selection of the marker DNA is one of the most significant methodical differences between PCR and IPCR. For PCR, the PCR protocol and polymerase applied for amplification has to be adapted painstakingly to the intended target. In contrast, IPCR gives much more freedom in assay development with the free choice of an optimized pair of a marker DNA and an amplification enzyme. No special long-range, proofreading, or otherwise modified polymerases were necessary for IPCR, as the marker could be kept as simple as possible.

Furthermore, the protocol of the Universal-IPCR simplifies the exchange of the DNA marker if several biotinylated DNAs were prepared. This flexibility in exchange of biotinylated DNA or biotinylated antibody for adaptation of a common protocol to separate applications marks the Universal-IPCR method as an ideal research tool, underlined by the multitude of applications carried out with this method (see Table 1).

However, the advantage of flexibility is inevitably linked to the drawbacks of the stepwise protocol: The sheer number of time-consuming and labor-intensive incubation steps is not well suited to routine application. Indeed, five separate incubation steps were necessary in an indirect sandwich assay (Fig. 3D), as carried out for the detection of tumor necrosis factor alpha (TNF-α) [33, 34, 39], HbsAg [37], or human interleukin 18 [35], starting with the immobilization of the antigen and finishing with the coupling of the DNA marker. In addition, appropriate washing steps flanking the incubation steps are necessary to remove reagents nonspecifically bound to the surface or other reagents. To achieve maximum performance, a systematic optimization of reagent concentration has to be carried out for each step in combination with each other parameter variation [24, 29, 36, 40] (see also Section 3.6.3). As each operation additionally reduces coupling efficiency and is a possible source for error or contaminations, a simpler approach to antibody DNA linking became desirable.

Furthermore, a simultaneous multiplex detection of several antibodies is impossible with the sequential incubation of biotinylated compounds and STV, as there is no selective linkage between a specific antibody and an individual DNA marker.

These methodical disadvantages of the multistep approach, however, could be compensated for by careful adaptation of the method. The broad number of successful implementations of the very adjustable Universal-IPCR protocol underlines the potential of the method (see Table 1). The quality of the antibodies and the amount of optimization invested in specifically fine-tuning experimental conditions such as reagent concentrations are mainly important for the efficiency of the Universal-IPCR in each new application.

Following this pathway, even for the above-mentioned "complicated" five-step assay (see Fig. 3D), an impressive sensitivity with an up to 50,000-fold ELISA improvement [34, 35, 39] was obtained, compared to an initial sensitivity-gain of "only" 16-fold for this approach [33].

In summary, the simple and flexible Universal-IPCR, with or without nonfunctionalized primary antibody, is well suited for a proof of principle of the general feasibility of the IPCR for a given problem in clinical laboratory applications, as successfully shown in several examples. It allows for an easy access to ultrasensitive research studies—if the necessary time and expertise for some fine-tuning of assay conditions is available. Under these conditions, a typical 1000- to 10,000-fold improvement of ELISA sensitivity is accessible in the research laboratory (see Table 1).

For routine applications, however, a meticulous proceeding according to a highly standardized and demanding protocol with several steps is necessary for reproducible results. However, the sometimes high method error, even in standardized and optimized protocols [29], and the sheer effort of multiple-step protocols still calls for alternative methods of DNA–protein linkage.

2.1.3. *Covalent Coupling—A Tool for Multiplex Protein Detection*

The above-mentioned potential of the IPCR assay for multiplex detection inspired Hendrickson and coworkers to the synthesis of covalent conjugates of DNA and antibodies [38, 41]. In contrast to fluorescence probes, where the number of simultaneously detectable wavelengths is limited by the availability of labels and instrument performance, DNA labels could be used as a "molecular barcode" label with a nearly unlimited resource of different sequences for the DNA marker. Each DNA marker will be assigned to a specific antigen and thereby allows for high-parallel detection of several antigens. With an increased demand for applications in proteomics, this IPCR advantage proposes to be of increasing importance. Limitations of this approach are coupled to general limitations and possibilities of multiplex PCR [42, 43], but approximately 10 antigens should be simultaneously detectable [41].

Covalent conjugates of antibodies and short, single-strand oligonucleotides were also synthesized for a combination of rolling circle amplification (RCA) and ELISA in the IPCR-related immuno-RCA technology [44–47]. In this approach, the primer covalently attached to the antibody is enzymatically elongated using an added circular, single-strand DNA as template. Although linear isothermal elongation followed by hybridization with fluorescent oligonucleotide probes is theoretically not as effective as semiexponential template amplification in PCR and requires the specifically designed circular DNA, a significant gain in sensitivity compared, for example, to direct fluorescent labeling of the antibodies was observed. This feature, combined with the permanent linkage of the elongated DNA to the antibody, makes this approach well suited to multiplex analysis on the biochip platform. Immuno-RCA is therefore described as the method of choice to transfer the enzymatic signal enhancement known from ELISA to biochips or related surfaces. A successful parallel detection of 11 analytes was reported with this method in a bead-based assay technique and using fluorescent probes for detection [46].

In the multiplex-IPCR assays carried out by Joerger and Hendrickson, DNA of different lengths was used for separation of the PCR amplificates (see Fig. 4). Results of their experiments are discussed below in Section 3.5. The ability to discriminate between different DNA markers by length is nevertheless limited by the separation capabilities of the gel. For a large number of different DNA probes, a sequence-specific detection carried out, for example, by PCR–enzyme-linked oligonucleotide sorbent assay (ELOSA; Section 2.2.2) should be preferable.

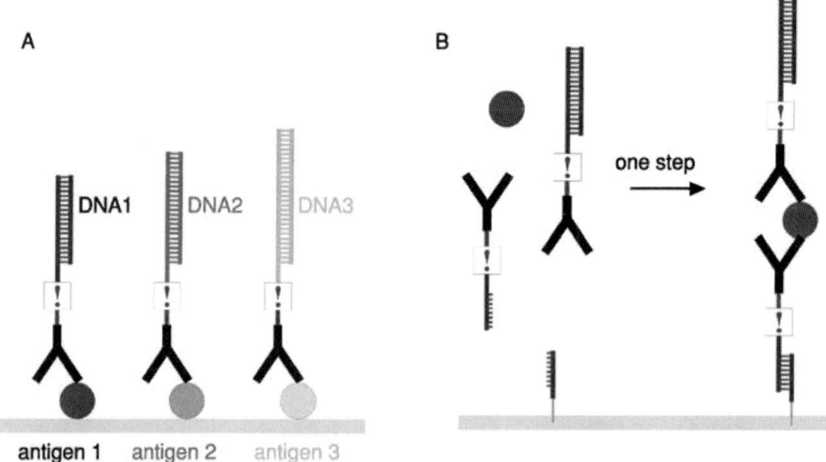

FIG. 4. Advances in immuno–polymerase chain reaction (IPCR) design. (A) Multiplex-IPCR: Different DNA labels attached to specific antibodies allow for the simultaneous parallel detection of several antigens [38, 41]. For different methods of specifically coupling antibody and DNA, abbreviated by "!" in this figure, compare Fig. 2C, D. (B) DNA-directed immobilization-IPCR: A mixture of oligonucleotide-tagged capture antibody and marker DNA–labeled detection antibody is added to the antigen-containing sample and incubated in a single step in a microwell coated with immobilized capture oligonucleotide. The antibody–antigen complex formed in solution is immobilized by hybridization to the complementary capture oligonucleotide (DNA-directed immobilization [60, 61]). The DNA–antibody coupling, abbreviated by "!", was carried out with biotin–streptavidin interaction.

Covalent coupling of antibody and DNA was carried out with a four-step protocol: 5′-amino-modified oligonucleotides and antibodies were activated separately (step 1 and 2) with heterobifunctional cross-linkers (acethylthioacetyl DNA and maleimide-modified antibodies), subsequently mixed and incubated (step 3), and finally purified using high-performance liquid chromatography / gel chromatography (step 4) [38, 41]. In contrast to the double-stranded DNA-marker discussed above, initially, single-strand DNA was used for covalent coupling. It was found, however, that double-stranded DNA has several advantages as an IPCR marker: improved stability, release of the uncoupled strand in solution from covalent conjugates and therefore easier access by the amplification enzyme, and better availability of double-strand DNA by plasmids or PCR synthesis [38].

Alternative protocols for covalent coupling were described by Wu *et al.* [48], Sims *et al.* [49], and McKie *et al.* [50]. In the latter approach, single-stranded DNA was linked in a multistep procedure with a short primer covalently coupled to the antibody. The single-stranded DNA-marker also included a Hind III restriction site. By adding the restriction enzyme previous to the PCR-step, the DNA-marker sequence was released from the immobilized immuno-complex to the supernatant liquid phase and subsequently transferred to capillary vessels (see Section 2.2.3).

In the work of Henterich *et al.* [51], a synthesis combining thiolated DNA and a sulfo-succinimidy-4-[p-maleimidi-phenyl]butyrate-activated antibody was introduced, followed by an protein-G affinity column purification of the conjugate. In this study, a direct comparison between the sequential Universal-IPCR approach and covalent conjugates was carried out for the detection of gliadin, revealing an improved IPCR sensitivity by usage of the covalent conjugate.

Although the use of covalent conjugates combines the advantages of a drastically shortened and simplified assay protocol with increased sensitivity, it is also obvious from the above-listed protocols that the conjugation procedure is not as trivial as, for example, a biotinylation. A reliable cross-linking protocol and the equipment necessary for product purification are required, whereby cross-linking, purification, and activity control has to be carried out for each new antibody intended for IPCR usage. This is a major obstacle for routine studies with changing antigens or antibodies. In addition, the typical gain in sensitivity as compared to the "standard" ratio between Universal-IPCR and ELISA was found to be rather low, ranging from 100-fold improvement [41] to 30-fold [51] or even no improvement at all [50]. If these findings are related to a decrease in antibody activity resulting from the chemical cross-linking, the general quality of the antibodies or other causes must be studied in further experiments.

Although covalent coupling is attractive for multiplex detection and significantly simplifies the general assay protocol, most single-analyte clinical applications today will probably not benefit from this type of conjugate because of the challenging coupling protocol. Once again, advantages and disadvantages as discussed for protein chimeras (Section 2.1.1) apply: If the necessary conjugates are available for the laboratory intending to perform IPCR, this approach is quite preferable compared to the multistep approach because of the efficiency of coupling and simplicity of the IPCR assay.

The development of easy accessible complete antibody DNA conjugates for single-step incubation is nevertheless the most promising approach for adapting IPCR to the needs of robust and fast clinical routine application.

2.1.4. *Polyvalent Conjugates, Specialized Reagents, and Ready-to-Use Kit Systems*

A different strategy of improving the protein–DNA conjugates was introduced by the synthesis of conjugates consisting of several antibodies and DNA markers. In these compounds a higher avidity for the antigen caused by the larger amount of antigen-binding molecules is combined with an additional signal enhancement by linking the antigen simultaneously with many DNA markers.

The latter effect was already postulated for the coupling of tetravalent STV with polybiotinylated antibodies and DNA, whereby studies of the interaction of biotinylated DNA and STV [52] indicated that the true coupling efficiency is rather low. A coupling yield of only approximately 0.1% for linking antigen to DNA by stepwise incubation of antibody, STV, and monobiotinylated DNA was calculated in a comparison of PCR and IPCR [37]. Considering the four binding pockets of STV, it appeared attractive to prepare preconjugated aggregates from biotinylated antibodies, STV, and biotinylated antibodies in analogy to the two-component avidin-biotinylated enzyme complex known in ELISA [53, 54] and to incubate these aggregates in a single-step IPCR [55]. Unfortunately, because of the low coupling efficiency and statistical distribution, a large number of nonfunctional aggregates containing only STV and DNA or STV and antibodies are to be expected in a three-component mixture [56].

A logical next step toward a larger number of markers per antigen was carried out by Kakizaki *et al.* [25]. A double-biotinylated DNA marker was implemented in the stepwise Universal-IPCR method to enhance binding efficiency between avidin and DNA. Although this approach eliminated separate incubation steps for avidin and DNA and allowed for an impressive 10,000-fold gain in sensitivity compared to ELISA for the detection of fish pathogen, it still required the *in situ* preparation of the detection complex from avidin and DNA in the microplate well and, previously, a separate incubation step for the biotinylated antibody.

A systematic study of preincubated conjugates consisting of STV and double-biotinylated DNA [26] revealed that an optimized DNA length and ratio of STV/DNA has a large effect on conjugate formation and the sensitivity of the IPCR assay. By coupling these STV–DNA conjugates with additional biotinylated antibody, a supramolecular approach to the synthesis of storable antibody–DNA conjugates was realized [26]. As reported for covalent conjugates (Section 2.1.3), however, a chromatographic purification step of the antibody–DNA conjugates for the removal of nonfunctional educts and intermediates was unavoidable for this approach, too.

An even more "molecular-engineering" approach to linking antibody and DNA was introduced by Niemeyer *et al.* [57], using covalent single-strand oligonucleotide–STV conjugates ("biomolecular adapters") [58, 59] as construction material for networks with double-biotinylated DNA. Biotinylated antibody was subsequently bound to adapter conjugates containing a complementary oligonucleotide sequence and hybridized against the DNA adapter networks. Interestingly, these well-defined smaller antibody–DNA conjugates were successfully used in IPCR model studies without additional chromatographic purification (see Section 2.1.3 and above), revealing superior performance compared to sequential incubation of biotinylated antibody, STV, and biotinylated DNA (see Section 2.1.2). Alas, as previously stated for protein chimeras and covalent conjugates, this advantage in assay handling is inevitably linked with the need for more unusual assay compounds, in this case, biomolecular STV–oligonucleotide adapters.

Sequence-specific DNA hybridization was also described by Niemeyer *et al.* [60] as a tool in a novel one-step approach to IPCR. In this work, a mixture of oligonucleotide-tagged capture antibody and a multimeric detection antibody–DNA conjugate was used to couple the antigen in solution with both antibodies. The coupling step was performed in a microwell coated with oligonucleotides complementary to the sequence linked to the capture antibody, thus merging assembly and immobilization of the conjugate in a single incubation step (see Fig. 4B). Coupling of capture antibody and oligonucleotide was performed by use of the biomolecular adaptors described above. This application of highly effective DNA-directed immobilization (DDI [61]) establishes a very fast and robust protocol. Although this approach needs two DNA-labeled antibodies, the increase in reagent complexity again simplifies assay handling and robustness.

To circumvent the need for separate functionalization for each specific antibody, the concept of species-specific secondary reagents was developed [62] according to the principle introduced for Universal-IPCR [24]: By combination of a nonfunctionalized antigen-specific primary antibody with a DNA-coupled species-specific secondary antibody (see Fig. 3B,D), only a small range of species-specific antibody–DNA conjugates is needed for IPCR. In contrast to covalent antibody–DNA conjugates, which could also be used in this fashion, the loss in coupling efficiency caused by the additional coupling step of the primary antibody is compensated by the generally higher signals of the multimeric conjugates.

The synthesis of antigen-specific primary conjugates from detection antibody and DNA is also facilitated by supramolecular conjugation. These conjugates reduces the amount of incubation steps necessary in IPCR.

Nowadays, ready-to-use multimeric conjugates are commercially available (IMPERACER conjugates, Chimera Biotec), thus liberating the scientist from the need to provide labeled DNA markers and to perform the necessary synthesis optimization steps or the actual coupling/purification of the conjugate by him- or herself. Still, nonspecific cross-reactivity of the IPCR reagents should be checked in a proof-of-principle experiment for each new application. In addition, as reported for Universal-IPCR (Section 2.1.2), the adaptation of the reagent concentrations is recommended for maximum sensitivity, whereby the reduction of incubation steps and assay parameters significantly facilitates method optimization.

The conjugates were found to be very robust and are available as secondary species-specific reagents against several common primary antibodies, such as mouse Immunoglobulin G (IgG), rabbit IgG, or goat IgG [62]. Custom-made conjugates including primary antibodies are provided by manufacturing companies such as, for example, Chimera Biotec GmbH.

2.2. DNA Detection: Readout of the IPCR Method

In addition to successful linking of target antigen and DNA marker, as discussed in the previous chapter, the subsequent amplification of the DNA is the second key factor for efficient IPCR. Similar to many protocols developed for quantitative PCR [2], the DNA amplification product has to be converted into a detectable signal. Typically, a simple yes/no decision on the presence of the DNA marker is not sufficient, and a quantitative readout dependent on the antigen concentration is needed. Therefore, in many IPCR applications the cycle number in PCR-amplification is limited to the exponential phase of the amplification; for example, 30 or fewer cycles [10, 24–26, 29, 31, 33, 37]. Alternatively, successful applications of 40 cycles were also reported [34–36, 38, 39, 41], underlining the relative flexibility of PCR conditions for the amplification step. The need for an optimized cycle number is only important for end point determinations such as gel electrophoresis (Section 2.2.1) or PCR-ELISA (Section 2.2.2). Recently, the standard approach to quantitative PCR has changed to on-line measurement of DNA amplification in real-time PCR (Section 2.2.3). This method combines superior reproducibility with the complete elimination of any additional analytic steps subsequent to PCR, thereby making the assay much easier and faster [63].

However, real-time detection requires access to a special real-time PCR cycler, which is able to detect the increase/decrease of added fluorescence labels during DNA amplification. Although these machines are more and more common for quantitative DNA analysis, their availability in clinical laboratories is still limited. Therefore, the following subsections also include a detailed overview of the "classical" approaches to quantitative (I)PCR amplificate, analysis which exchanges less demanding PCR equipment for additional hands-on time. The sensitivity of real-time or end-point IPCR detection is quite similar. A comparison of the influence of different endpoint detection methods to the overall sensitivity of IPCR is given in Fig. 5.

2.2.1. *Gel Electrophoresis and Direct Detection of DNA-Specific Labels*

Starting with the first IPCR study, gel electrophoresis retains its potential as a fast and easy method for end-point determination of DNA amplificate for IPCR assays [10, 24, 25, 29, 31, 35, 36, 38, 39, 64]. Readout is performed by intercalation fluorescence markers (e.g., ethidium bromide) and photometric/densitometric quantification of band signal intensities. The direct addition of a double-strand specific intercalation marker to the PCR amplificate and subsequent measurement of fluorescence in microwells proved to be of insufficient sensitivity for the quantification of IPCR amplificate [37]. Alternative approaches, such as radioactive labeling during PCR and subsequent imaging [33], were carried out but are not well suited for routine clinical application because of additional methodological requirements. An advantage of gel electrophoresis is the possibility of simultaneous amplificate detection for multiplex IPCR [41] and the ability to detect nonspecific amplification products.

With an increasing amount of samples, however, handling of large-scale gel electrophoresis becomes cumbersome. Although the ELISA-compatible microplate format of IPCR easily allows for simultaneous performance of 96 assays, handling, loading, and analysis of large gels for 96 or more samples, especially in multiple determinations for statistical analysis of the data, is not desirable for routine or automatic readout.

Moreover, because of sample handling and densitometric quantification, a high error was observed for double determination of gel electrophoresis

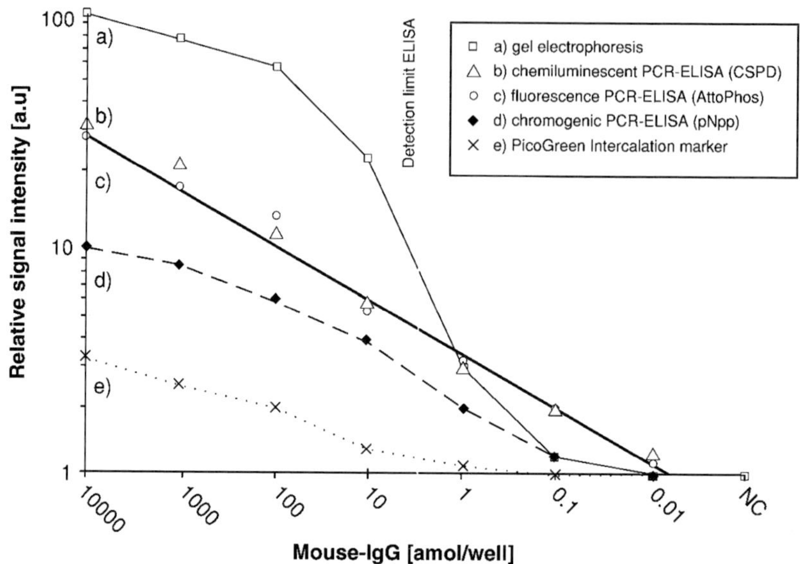

FIG. 5. Several endpoint detection methods were compared for the detection of immuno–polymerase chain reaction (IPCR) amplificate from a direct IPCR (Fig. 3A) of mouse-IgG. Although all IPCR/DNA-detection combinations were able to improve the detection limit of a comparable enzyme-linked immunosorbent assays (ELISA) of approximately 10 amol IgG in a 30-fL sample volume, several differences were observed in actual detection limit, and the linearity of the concentration/signal ratio dependent on the DNA quantification was applied. Best results were obtained for PCR-ELISA (see also Fig. 6) in combination with fluorescence- or chemiluminescence-generating substrates (b, c). With photometric substrates (d) or gel electrophoresis and subsequent spot densitometry (a), a 10-fold decrease in sensitivity was observed. In addition to the more sigmoid curve in gel electrophoresis, an enhanced overall error of 20% compared to 13% in PCR-ELISA was observed for two independent assays. The simple addition of a double-strand sensitive intercalation marker to the PCR-amplificate and measurement in a fluorescence spectrometer further decreased sensitivity (e) and appears therefore to be unsuited for IPCR amplificate quantification. (Figure modified according to references 37 and 65.)

signal readout [29, 37], particularly when compared to other quantification techniques.

2.2.2. *Microplate-Based Analysis: PCR-ELISA and PCR-ELOSA*

The addition of hapten-labeled primers or nucleotides allowed for the synthesis of modified PCR amplificate during PCR amplification. For immobilization of the functionalized amplificate, either hybridization to complementary solid-phase bound capture oligonucleotides ELOSA [30, 65–67] or coupling of twofold-labeled amplificate to solid-phase-bound STV

(PCR-ELISA [37]) were described. The immobilized hapten-labeled amplification product was subsequently detected by an antibody–enzyme conjugate similar to conventional ELISA. By choice of a chemiluminescence- or fluorescence-inducing substrate, the sensitivity of the IPCR essay is further enhanced [37]. A comparison of different detection methods is given schematically in Fig. 6; typical results are compared in Fig. 5.

In addition to the further increase in sensitivity, the advantages of microplate-based detection methods are manifold: The direct digital readout of a conventional microplate-reader allowed for easy processing of multiple samples; the compatibility to standard ELISA procedures ensures a robust and simple handling, and therefore also a significant decrease in method

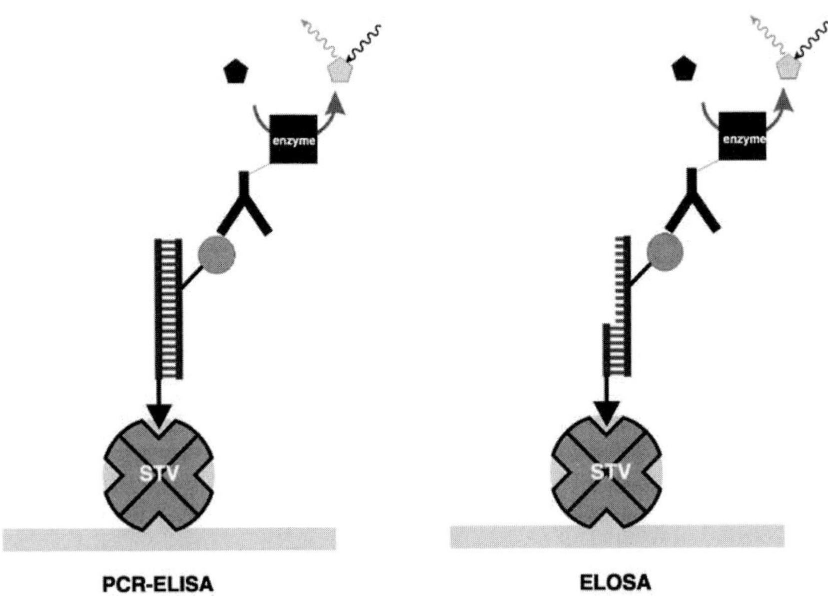

FIG. 6. Polymerase chain reaction–enzyme-linked immunosorbent assay (PCR-ELISA) and enzyme-linked oligonucleotide sorbent assay (ELOSA). During PCR, additional haptens were included in the amplification product, thus enabling the immunological detection of the amplificate following PCR. In PCR–ELISAs, biotinylated primers and a hapten-labeled nucleotide (e.g., Digoxigenin-dUTP) were used in combination. The double-labeled PCR product was immobilized using streptavidin-coated microplates and detected with a hapten-specific antibody-enzyme conjugate [37]. In an alternative approach, only hapten-labeled nucleotide or primer was used and the immobilization was carried out using hybridization against solid-phase bound capture oligonucleotides. This approach, termed ELOSA additionally discriminates between amplificates with different recognition sequences and is therefore well suited for multiplex IPCR or the parallel detection of DNA marker and competitor [30, 66, 67, 76].

error; and as automatic or semiautomatic ELISA processing equipment is widespread, this procedure is especially suited for automation.

Furthermore, the sequence-specific identification abilities of the hybridization step in IPCR-ELOSA allow for separation between different amplification products in multiplex analysis or individual detection of a marker and competitor DNA for internal standardization (see Section 2.3.6) [66].

Disadvantages of this approach are, of course, the additional time- and reagent-consuming handling steps following PCR amplification.

2.2.3. *Real-Time PCR*

A major breakthrough in the development of quantitative PCR was the invention of real-time detection methods for DNA amplification product during PCR [63]. In this technique, a fluorescence-generating probe (e.g., a Taq-man probe, see Fig. 7A) is added to the PCR mix. During amplification of the DNA template, the probe is modified/degraded, so that an initially quenched fluorescence increases parallel to the increased amount of amplified DNA. As a less specific alternative, an intercalation marker could be added to the PCR mix, which incorporates in the double-stranded PCR product and thereby increases fluorescence during PCR.

With different kinds of real-time PCR-thermocyclers, different combinations of IPCR and real-time amplificate detection were studied. In the work of Sims *et al.* [49], describing the first proof of principle of real-time IPCR, an AbiPrism7700 sequence detector (Applied Biosystems) was used in combination with a TaqMan Probe for the detection of vascular endothelial growth factor (VEGF). Because of a high background (signals for IPCR negative control after 28.3 cycles while convential PCR negative control should be negative even after 40 cycles), a large error (interassay CV: 14%–28%, intraassay CV: 24%–40%) and only a fivefold sensitivity improvement compared to conventional ELISA was found in this initial proof of principle, perhaps caused by disadvantages from covalent antibody–DNA coupling. Nevertheless, a good correlation between ELISA and IPCR was observed.

McKie *et al.* [50, 68] attempted to adapt the IPCR technology to the LightCycler real-time PCR-device (Roche) for an ultrasensitive assay of mumps IgG. As this machine uses capillary vessels in a rotating heating chamber for especially fast performance of the PCR amplification, the contents of the microplate-based antigen/antibody/DNA immobilization have to be transferred to the capillaries needed for the amplification step. Naturally, this additional transfer step previous to PCR is not beneficial for the sensitivity of the assay. In addition to the properties of the covalent DNA–protein conjugate (see Section 2.1.3), a nonspecific binding of human IgG was observed as limiting factor for assay sensitivity. Because of a combination of these effects, no IPCR sensitivity improvement was observed by McKie *et al.*

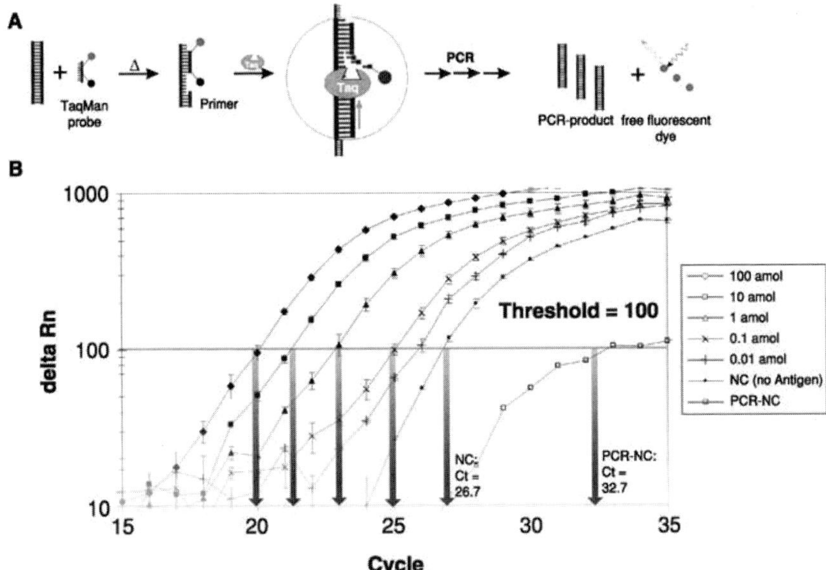

FIG. 7. A novel approach to quantification of (immuno–) polymerase chain reaction [(I)PCR] amplificate is the online detection carried out in real-time IPCR. (A) Principle of TaqMan real-time PCR (Applied Biosystems): An additional oligonucleotide probe, containing a fluorescence dye and a quencher, is added to the PCR mastermix. Although the DNA polymerase activity of the Taq enzyme elongates the PCR primer during the synthesis of the novel complementary DNA strand, the exonuclease function of the enzyme destroys the TaqMan probe bound next to the primer. The separating of fluorescence marker and quencher thereby induces a fluorescence signal for each amplified DNA strand. (B) Signal readout of a real-time IPCR experiment: Mouse-IgG immobilized on a microplate was detected with an antimouse IgG-DNA conjugate (see Fig. 3A). The fluorescence intensities (Rn) were plotted against the cycle number and normalized against a baseline using fluorescence in the first 15 cycles. A uniform threshold was set at the Rn = 100 in the linear amplification range. With a higher threshold, the PCR negative control (PCR-NC) could be excluded completely from the Ct values. The Ct value was determined as the intersection of the threshold and the signal curves. Note excellent reproducibility of the method visible in the small standard deviations (average: 0.7 ± 0.4 %) calculated from double determination. (Experimental protocol described in reference 67.)

In the work of Lind *et al.* [69], an iCycler iQ System (Biorad) was used in combination with the intercalation marker SYBR green (Molecular Probes) for the detection of prostate-specific antigen (PSA). Once again, a high background and a bad signal-to-noise ratio limited the sensitivity improvement to an 100-fold increase of detection limit in IPCR ($2.4 \times 10e6$ molecules, approximately 3 amol) compared to approximately 300 amol in conventional ELISA.

Henterich *et al.* [51] returned to the initial AbiPrism 7700/TaqMan system for the detection of gliadin in food samples, with a 30-fold gain in sensitivity compared to ELISA (see also Section 3.4). In this work, plate material and immobilization of the antigen were identified as the main source for increased nonspecific binding.

The obstacles in the development of a robust routine-suited real-time IPCR method with the typical 1000-fold IPCR sensitivity increase were finally overcome by the use of multimeric antibody–DNA reagents (see Section 2.1.4), a microplate-compatible AbiPrism 7000 real-time cycler (Applied Biosystems), and TopYield Modules (Nunc; see Section 2.3.1.). In a study carried out with several model antigens and practical examples, real-time IPCR revealed similar or better sensitivity as conventional IPCR, with significantly lowered standard derivations and a good tolerance for biological matrices such as human plasma [62, 67].

The major drawback of the real-time method IPCR is the need for a specialized real-time cycler, preferably compatible to the microplate format. With the increasing distribution of these machines, there is also an emerging range of new opportunities for real-time IPCR. The need of an additional fluorescent probe during PCR is compensated for by the fact that all materials and reagents for post-PCR processing were no longer required. However, the need for separate discriminable fluorescence probes reduces the usefulness of real-time detection in multiplex IPCR applications.

Considering the advantages resulting from reduced handling efforts and assay time, combined with the minimized contamination risk by eliminating any post-PCR handling steps of IPCR marker–DNA amplificate, these methodical improvements predestine the real-time PCR as the method of choice for IPCR routine applications (e.g., in day-to-day clinical analysis or large-scale screening tests).

2.3. Performing IPCR: A Practical Approach

Although the core steps of IPCR, DNA–antibody coupling, and DNA detection were discussed in detail earlier, the actual performance of the ultrasensitive assay requires some more considerations from the clinical laboratory scientist. A number of different protocol details and experience in IPCR assay fine-tuning were collected during the experimental work of several researchers. In the following paragraphs, the basic requirements of an optimized IPCR protocol are discussed:

2.3.1. *Plate Material*

As the standardized 96-well microplate format (and multitudes thereof) has found universal prevalence as the platform of choice in routine laboratory applications (ELISA and PCR), the full compatibility of the IPCR technique with ELISA equipment is one of the key advantages for converting well-established ELISA tests to ultrasensitive IPCR.

The requirements for microplates, micromodules, or microstrips in IPCR are nevertheless different from typical ELISA or PCR plate material. A microplate material combining antigen-binding properties with compatibility to the heating blocks in PCR thermocyclers is required to avoid the disadvantageous additional transfer step between different wells for antigen–DNA coupling and PCR amplification [33, 50, 68].

Initially, conventional PCR tubes from various suppliers consisting, for example, of polypropylene or PVC with intentionally low protein-binding capacities were used for IPCR [24, 29, 30, 41, 70]. As an alternative, polycarbonate TopYield modules (Nunc, Roskilde), specifically designed for IPCR applications with improved protein-binding properties, are increasingly popular as the material of choice for IPCR [26, 27, 34–37, 51, 62, 66, 67].

2.3.2. *Blocking of Nonspecific Interactions*

In ultrasensitive analytics it is very important to avoid nonspecific background signals. The choice of appropriate blocking reagents for nonspecific interactions is therefore an imperative for IPCR. Similar to conventional ELISA [71], a protein coating is the basic blocking reagent in IPCR. Almost all IPCR assays use either milk powder (e.g., [10, 24, 26, 29, 32, 37, 68, 70, 72]) or BSA [e.g., 28, 30, 49, 51]. Taking into account the presence of the DNA marker in each well and the extreme liability of the PCR signal amplification to a nonspecifically bound DNA marker, additional DNA (e.g., salmon sperm DNA) is included in the blocking reagent of many standardized IPCR protocols [10, 24–27, 34–37, 49, 70]. With some protocols obviously functional without the addition of (sometimes expensive) blocking DNA, the effect of the addition of DNA should be studied individually for each application. Thereby, each possibility to reduce nonspecific binding by adaptation of the blocking buffer to special assay conditions (e.g., by using animal sera as blocker for nonspecific antibody interactions [25]) should be studied during assay development.

In addition to the above-mentioned in-house preparations of common blocking reagents and variations thereof, commercial available ready-to-use blocking reagents were also successfully applied in IPCR. These reagents either includes DNA, as in Chimera Biotecs IPCR blocking reagent [62, 66, 67], or were prepared from Blockace buffer (Dainippon Pharmaceutical) and separately added DNA [34, 36].

In general it was found that best signal intensities for a sandwich IPCR technique (Fig. 3C), were obtained by using the same polyclonal antibody as a capture-and-detection antibody, probably because of minimized nonspecific interaction [31, 36, 37].

2.3.3. *Assay Duration*

In most protocols for IPCR, microplate surface coating with capture antibodies and subsequent blocking was carried out overnight, thus making it the most time-consuming step in the complete assay. Coated plates were stored at 4 °C, allowing for a reduction of actual assay time by bulk preparation in advance [67]. IPCR handling time is therefore defined by the assay duration from taking prepared plates out of storage to placing them into the PCR-cycler. Dependent on the number of steps required for the coupling of antigen and DNA, typically, assay duration is less than half a day. With a reagent incubation time of typically 30–60 minutes, IPCR processing time is therefore comparable to standard ELISA with usual 1.5–4-hour assay duration. Another 1–3 hours of post-PCR processing time has to be added for gel electrophoresis or PCR-ELISA/ELOSA quantification of the amplificates, whereas real-time IPCR is completed with the PCR step. Including the typically 90-minute to 2-hour duration of the PCR amplification, it is possible to carry out almost all standardized IPCR protocols in a single day, starting from application of the sample and ending with digitalized data readout. Hands-on time is typically 2–4 hours total and could be minimized by the usage of automatic washers (see Section 2.3.4), ready-made complete kit solutions, and real-time IPCR [62, 67].

2.3.4. *Washing Procedures*

To remove unbound reagents, washing with detergent-containing buffer (mostly Tween 20, see e.g., [24], with some exceptions; e.g., Triton X-100 [34]) is a vital part of each IPCR protocol. For enhanced stability of the reagents, Ethylenediaminetetraacetic acid (EDTA) is part of most washing buffers. In contrast to blocking reagents, which should also include DNA, IPCR washing buffer composition is similar to typical ELISA buffers [8, 9]. Dependent on the general affinity of the antibodies, DNA, or plate material for nonspecific binding, sometimes extensive numbers of washing steps (e.g., 18 [10] or even 38 [30]) were applied before PCR. In almost all standard protocols, typically five washing steps were found sufficient, followed by the important removal of EDTA previous to PCR amplification for maintaining maximum activity of the Mg-cofactor-dependent Taq enzyme. Washing could be performed automatically [29] or manually, using a multichannel pipette.

2.3.5. *Background Minimization and General Assay Sensitivity*

Independent from any strategy to minimize IPCR nonspecific binding, almost any IPCR application revealed some kind of background signal even in negative control samples without antigen. These findings are significantly different from negative PCR controls without DNA, which normally show no signal whatsoever. The key difference between PCR and IPCR is the presence of the DNA marker during incubation in any well, including negative control. Therefore, even single molecules of DNA remaining from the necessary incubation of marker DNA are able to induce a background signal in spite of all washing steps, additionally enhanced by DNA correctly bound to nonspecifically bound antibodies.

Therefore, it could be concluded that a certain background level is inevitable for IPCR because of the methodical set-up. In clinical applications, false-positive signals are therefore a potentially major problem. Great care has to be taken in method validation and choice of appropriate positive and negative controls for obtaining correct data from ultrasensitive analysis [66, 73].

This effect is also responsible for the observation that, in most IPCR applications, at least 1000 molecules of the antigen are necessary for valid and significant signals. The single-molecule sensitivity of PCR, both theoretically possible and demonstrated at several examples [5, 7], should be practically inaccessible by IPCR. Although some cases of single-molecule IPCR sensitivity are reported as exceptions from this rule [29, 31], single-cell detections or fractions thereof are entirely possible with IPCR [13, 31, 74] because of the larger number of surface or marker antigens.

As the complete removal of background seems impossible, the alternative strategy of increasing signals for an improved signal/background ratio, and therefore higher sensitivity, appeared very promising. With the introduction of multimeric conjugates [26, 62] or elaborate fine-tuning of assay parameters [25, 26, 29–31, 37], the signal-to-background ratio obtained from correct antigen binding could be noticeably enhanced, thus improving the overall performance of the IPCR.

2.3.6. *Quantification and Standardization by External and Internal Controls*

Each valid clinical assay should typically include external controls such as positive and negative controls with and without the antigen to be detected. For IPCR, an additional PCR negative control has to be included for the detection of nonspecific contamination of the amplification mix. These IPCR and PCR controls were used for the normalization of the obtained signals, thus facilitating the comparison of data between different experiments [37]. Dependent on the biological matrix to be analyzed and the quantification requirements, additional controls for assay precision and recovery are recommended according to "good laboratory practice" (GLP) procedures [75].

In an IPCR study carried out for the detection of the novel anticancer drug Aviscumin (INN name, also: rViscumin) [66, 73] (see Section 3.3), negative controls of patient samples before the application of a medicament were available. Therefore, negative controls obtained for each individual with and without antigen spiking were compared to standardized human plasma for the determination of signal recovery in each sample [66, 73]. Following normalization and quantification against an individual calibration curve for each patient, an average spike recovery of 95% ± 19% was found for 40 individual patients. Because of the normalization, it was possible to circumvent patient-to-patient variations during data analysis.

Additional standardization was achieved by including an artificially designed internal competitor DNA in the amplification mix, thus supplementing the external standards (Fig. 8A). This DNA fragment, using the same primer pair as the DNA marker of the IPCR antibody–DNA conjugate, was coamplified during PCR and induced amplification signals inverted proportional to the signals obtained from the DNA marker. By calculating the ratio of marker and competitor DNA (Fig. 8C, D), an additional signal enhancement was observed. Missing signals or an extremely high presence of both signals, marker and competitor, indicated an error in amplification. The functionality of this approach was demonstrated with gel electrophoresis [76], PCR-ELOSA [66, 76], and real-time PCR [62]. Marker and competitor DNA were individually identified either by different length (gel electrophoresis, Fig. 8B) or because of a different recognition sequence in the otherwise similar DNA sequence. This recognition sequence was detected either by hybridization with a complementary capture oligonucleotide (PCR-ELOSA, see also Fig. 6) or by a TaqMan probe (real-time PCR).

Using the example of Aviscumin, IPCR demonstrated its properties as a high-resolution quantification tool in a very narrow quantification window between 0.1 and 6 pg of the analyte, as required for the application [73]. This is a novel application for IPCR, as the method usually excels with its exceptionally broad quantification range.

For typical ELISA applications, a rather narrow detection window is observed, limited by the linear signal increase in the middle of the sigmoid enzymatic response curve. In contrast, IPCR revealed an extension of this linear detection range, often covering more than four decades of antigen concentration [26, 29, 36–38, 41, 66]. This additional quality of the IPCR

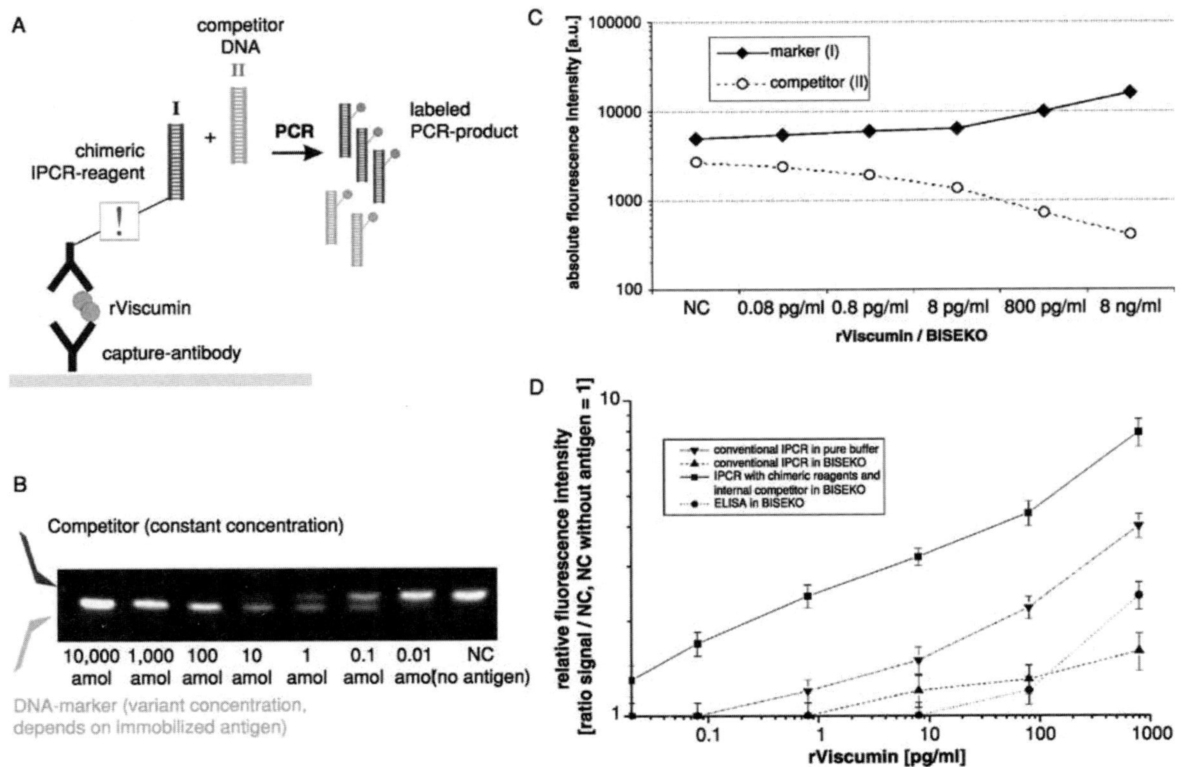
A
competitor
DNA
II
I
+
PCR
labeled
PCR-product
chimeric
IPCR-reagent
!
rViscumin
capture-antibody
B
Competitor (constant concentration)
10,000 amol
1,000 amol
100 amol
10 amol
1 amol
0.1 amol
0.01 amo
NC (no antigen)
DNA-marker (variant concentration, depends on immobilized antigen)
C
absolute flourescence intensity [a.u.]
100000
10000
1000
100
marker (I)
competitor (II)
NC
0.08 pg/ml
0.8 pg/ml
8 pg/ml
800 pg/ml
8 ng/ml
rViscumin / BISEKO
D
relative fluorescence intensity
[ratio signal / NC, NC without antigen = 1]
10
1
conventional IPCR in pure buffer
conventional IPCR in BISEKO
IPCR with chimeric reagents and internal competitor in BISEKO
ELISA in BISEKO
0.1
1
10
100
1000
rViscumin [pg/ml]

is noteworthy, as it is a very useful feature when establishing quantitative assays.

2.3.7. *IPCR Precision and Reproducibility*

As each increase in antigen sensitivity is inevitably linked to a possible enhancement of method error, the precision of the IPCR is of particular interest for clinical applications of the method. First studies indicated a rather large standard deviation for IPCR, typically being 10%–20%, and sometimes even higher [29, 30, 37, 49]. Careful optimization of the assay protocol significantly decreased the method error, especially by standardized handling, elimination of error-prone incubation and washing steps, and lately, using the highly reproducible real-time PCR detection. For real-time IPCR using standardized reagents, typical intraassay standard deviation (double determination) was found at 1.4%–2.5% in pure buffer and 3.1%–4.4% in biological matrix, compared to 8.0%–10.1% and 10%–15% for IPCR/PCR-ELISA, respectively. In three independent assays, interassay error was found to be 5.3%–9.1% for real-time IPCR and 14.1%–20.0% for IPCR/PCR-ELISA [62].

With these parameters, an optimized IPCR is comparable to conventional ELISA; the additional PCR signal amplification has no negative effect on the performance of the immunoassay. The optimized IPCR protocol is therefore well suited for all clinical applications compatible with the widespread ELISA method.

FIG. 8. Use of internal competitor in immuno–polymerase chain reaction (IPCR). (A) Supplementing the DNA–antibody conjugate containing DNA marker I, an internal DNA competitor fragment II was added in a constant concentration to the PCR mastermix and coamplified during PCR. As both DNA targets compete for primer and PCR reagents, the ratio of both PCR products depends on the amount of antigen to be detected. (B) IPCR detection of rabbit-IgG with marker and competitor DNA: Gel electrophoresis image of marker and competitor DNA PCR amplificate. (C) Comparison of the absolute signal intensities obtained by IPCR–enzyme-linked oligonucleotide sorbent assay (see Fig. 6) of the marker DNA (I) included in the PCR reagent in the presence of a constant amount of competitor DNA (II) as detected in the IPCR assay of Aviscumin [66]. Note the inverted proportional slope of the two curves. (D): Comparison of different IPCR protocols for the detection of Aviscumin [66]. With the conventional Universal-IPCR method B (see Fig. 2B), using subsequent coupling of antibody and DNA marker, approximately 1 pg/mL (0.5 amol in 30 μL sample volume) Aviscumin was detected in a buffer solution (down triangle). In serum samples, the detection limit and the signal intensity are decreased (up triangle). The use of CHIMERA reagents (IPCR method B) not only compensates the loss in detection quality but also increases the detection limit of the IPCR assay (squares). As compared to the analogous enzyme-linked immunosorbent assay, carried out in serum (circles), the use of IPCR leads to an improvement in the sensitivity of at least three orders in magnitude. (Figure 8A, C, and D reprinted from publication [66], Copyright (2003), used with permission from Elsevier.)

3. IPCR Applications: Different Target Antigens and Individual Strategies for Assay Improvement

As the first part of this review discussed how IPCR is carried out, the following part will highlight a number of interesting applications of the IPCR method with clinical relevance as well as some necessary and unique adaptations of the basic IPCR protocol.

Table 1 provides a comprehensive summary on various IPCR applications by antigen, including information about assay type, detection limits, and detection method. Different kinds of antigens revealed different properties of the IPCR method, subdivided here according to six general fields of applications.

3.1. Tumor Markers and Disease-Associated Proteins: Conventional IPCR and Methodic Alterations

The most common target antigens for IPCR were chosen from the range of tumor-/disease-associated antigens. These marker proteins were prime targets for the further development of the IPCR method: They have a potentially high impact on clinical diagnosis of widespread diseases and are inaccessible by conventional PCR because of their proteinaceous nature. Beginning with the standard Universal-IPCR protocol by Zhou *et al.*, already introduced in Section 2.1.2, several methodic alterations and improvements of the basic IPCR protocol were studied for the first time with these antigens, revealing how the basic concept of protein–DNA conjugates and subsequent DNA amplification could be varied according to different applications.

In the establishing paper of the Universal-IPCR method [24], approximately 5780 molecules of a full-length recombinant human proto-oncogene ETS-1 were detected in a 50-μL sample directly immobilized overnight against a microplate surface. As the majority of antigen molecules in this sample volume were not expected to be immobilized during overnight incubation, it was assumed that this detection limit was near the theoretically possible minimum for solid-phase bound antigen.

VEGF is a potent mitogen that also occurs at elevated levels in association with some types of cancer. Sims *et al.* [49] carried out the first real-time IPCR (see Section 2.2.3) for the detection of VEGF in human and mouse serum.

A good correlation between ELISA and IPCR was found with recovery rates between 120% and 83%, and a detection limit of 0.2 pg/mL for VEGF in 10% pooled mouse standards and 2 pg/ml in 10% individual human serum was observed. This example illustrates how matrix effects influence performance of the IPCR, an important point in adapting the IPCR protocol to novel problems. Optimization of experimental parameters, however, allowed for compensation of the differences between buffer, pooled serums, and individual serum samples [66, 73].

One of the most significant tumor markers identified so far is the prostate-cancer serum tumor marker PSA. In addition to the above-mentioned (2.2.3) real-time IPCR by Lind *et al.* [69], PSA was also target antigen in two methods related to IPCR. In the expression immunoassay [77, 78], introduced by Cristopoulos *et al.* in 1995 [77], a DNA marker encoding the luciferase enzyme was included as a label in DNA–antibody conjugates and translated following PCR amplification of the DNA. The detection of the expressed enzyme allowed for the detection of 1.1 amol (30 pg/mL) PSA by this method, using a clever if somewhat complicated nature-mimicking way to detect the marker DNA.

Another PSA detection with antibody–DNA conjugates and 0.1 pg/mL detection limit (compared to 0.1 ng/ml in ELISA) was carried out by Schweitzer *et al.* [44] using the innovative RCA for Immuno-RCA (see 2.1.3).

For the detection of CEA in human blood serum, as well as for the detection of prions (see also Section 3.4) Niemeyer *et al.* applied the one-step DDI-IPCR [60] described above (2.1.4). This method combines the advantages of coupling the antigen in solution, effective hybridization-directed immobilization, IPCR signal amplification, and real-time detection (2.2.3) into a very fast and robust protocol with a 1000-fold sensitivity increase over conventional ELISA.

Ren *et al.* reported IPCR applications for gastritic tumor–associated antigen MG7-Ag first in cell lines [79] and later on in patients [32], as well as the detection of CEA in human sera and blood stem cells [18] with a chimeric protein discussed above in Section 2.1.1. In the latter example, IPCR was able to detect a possible tumor-cell contamination of the leukophoresis blood stem cell samples from 2 of 12 patients with breast cancer. These findings are important for the evaluation of sample purity in stem cell reinfusion therapy, as highly sensitive analysis techniques for donated samples are reducing the risk of possible transfer of dangerous cell material. Furthermore, the monitoring of circulating tumor associated antigen revealed not only 92.3% recovery in samples with increased CEA levels as found by RIA but, additionally, 50% positive results in samples with an assumed normal CEA level, thus allowing for a much earlier indication of the possible risk for relapse or metastasis of resected tumors.

For MG7-Ag, similar results were obtained [32]: A direct comparison was carried out with commercially available kits for several tumor markers and an IPCR of MG7-Ag. The IPCR assay detected 81.4% seropositive samples, whereas other tests varied between 48.8% (MG7Ag with IRMA detection) and 33.7% (CEA-RIA) correct identifications of gastric carcinoma patients. IPCR additionally allowed for the discrimination between patients both with and without metastasis by signal intensity as well as the detection of other cancer subtypes. After false-positive signals were ruled out during assay development, an alarming ratio of 0.8% (2/236) supposedly healthy blood donor samples were found to be seropositive, indicating previously undetected disorders in these patients. For gastritic cancer, one of the most significant malignancies causing mortality, an early and correct diagnosis is of great importance in improving survival. The usefulness of the highly sensitive IPCR for the screening and monitoring of low-level tumor-associated antigens is therefore obvious.

TNF-α was studied by several groups as a target antigen for IPCR. As an important mediator in inflammatory and autoimmune diseases, detection of this protein was carried out in rat cerebrospinal fluid (CSF) [33] or in human serum [34, 39, 80]. IPCR allowed Sanna *et al.* for the first time to record a time course of TNF-α induction in the rat CSF following bacterial lipopolysaccharide injection. In contrast to ELISA, TNF-α was detectable 15 minutes after injection, and the kinetics of early-on antigen increase could be monitored quantitatively by IPCR. This example emphasizes how a rather small 16-fold increase of sensitivity compared to ELISA could open up a new field of kinetic studies, as ELISA only showed TNF increase after 90 minutes in a very narrow sigmoid curve [33]. Analysis of the pathophysiologic role of TNF-α in human plasma has been difficult because low concentrations were often inaccessible with classical assays, including healthy controls for reference. IPCR monitoring of TNF-α in plasma samples taken from healthy blood donors, patients with ulcerative colitis, and patients with Crohn disease revealed an increased TNF-α serum level in patients with inflammatory bowel disease and, furthermore, the ability to separate between patients with active and inactive ulcerative colitis [39]. In addition, in patients with Duchenne muscular dystrophy, an increased level of TNF-α was observed [80].

These findings illustrate how the properties and further improvements of the IPCR assay were applied for analyzing the clinical significance of marker proteins in various diseases. The ability of the IPCR assay to detect low levels or small variations in protein concentration even in biological matrices allows for the confirmation of the relationship between proposed new marker proteins and the progression of the disease.

3.2. Virus antigens: A Comparison of IPCR and Nucleic Acid Analysis

The presence of a virus is detectable by several possible marker compounds, including core RNA or DNA and an assortment of proteins. The relation between nucleic acid–based PCR or RT-PCR techniques and protein-based methods like ELISA and IPCR is of key interest for the studies of viremic infections. Although protein expression encompasses important information about the actual presence and reproduction of the complete virus, the detection is normally much less sensitive than PCR or RT-PCR. However, as there are many copies of, for example, shell or core proteins present in a single virus, but only a single copy of RNA/DNA, the increased number of targets for antigen detection is theoretically advantageous for protein detection, especially for very small amounts of virus copies in a given sample.

As observed in the following examples, improved protein detection is therefore able to improve the detection limit even below nucleic acid–based PCR methods. In addition, many factors inhibiting the PCR of nucleic acids from a given sample were easily circumvented by the specific binding and washing steps routinely included in protein-based immunological methods.

Because of an observed low-level transcription of the viral gene as one of the persistence mechanisms of the virus in its host, and because of the resulting difficulties in detection, hepatitis B surface antigen (HbsAg) presents a challenging target for IPCR method development [30, 37, 81]. Maia *et al.* [30] introduced the sandwich IPCR (termed "two-site IPCR" or "TIA") protocol for HbsAg, carried out according to the Universal-IPCR protocol. A panel of different antigen-specific monoclonal antibodies was carefully tested for their properties as capture and detection antibodies. Different biotinylation ratios were compared for the detection antibody, enabling a significant gain in sensitivity in the optimized assay. This study emphasizes the need to choose the optimal antibody for the ultrasensitive IPCR assay, as each disadvantage of the antibodies is also amplified in IPCR.

The influence of antibody selection was also described in the work of Niemeyer *et al.* [37], in which an indirect sandwich assay (Fig. 3D) with a nonfunctionalized primary detection antibody was compared to a direct sandwich assay (Fig. 3C). The removal of one incubation step resulting from the functionalized primary antibody increased coupling efficiency, thereby improving the signal-to-background ratio and, ultimately, the sensitivity of the assay.

It was found that IPCR allowed for a 100–1000-fold improvement of sensitivity while using a nonradioactive method, thus facilitating laboratory handling, waste disposal, and environmental consciousness in assay performance.

HbsAg was also the antigen of choice for the first evaluation of the properties of an *in situ* IPCR: Cao *et al.* [81, 82] introduced this method in direct similarity to the Universal-IPCR (2.1.2) and PCR-ELISA (2.2.2) protocols, respectively, for the detection of HbsAg in tissue samples. Comparing samples from 17 explanted livers of hepatitis B (HBV)-infected patients with two negative controls, an extremely good accordance of 14/17 correct positive with no false-negative identifications were found for *in situ* IPCR, and 15/17 correct positives were found for standard-IPCR. In contrast, immunohistochemistry only allowed for 5/17 (avidin-biotinylated enzyme complex) to 10/17 tyramide signal amplification (TSA) correct positive identifications. In one case of liver cirrhosis, IPCR and *in situ* IPCR allowed a correct positive identification, while even *in situ* PCR (also 14/17 positive identifications) lead to false-negative results.

These findings underline how combined information obtained from nucleic acid and ultrasensitive protein detection support each other to provide a valid diagnosis.

Bovine herpesvirus 1 was the target in an IPCR study carried out by Mweene *et al.* [28], following the Universal-IPCR protocol without methodical variations. In this work, nasal secretions of eight intranasally bovine herpesvirus 1–inoculated calves were studied over several days with IPCR. It was found that after 10 days following infection, no more virus shedding was detectable by virus isolation assay, and clinical symptoms were disappearing for some individuals after days 8–16. IPCR, however, allowed the detection of the BHV-1 even after day 20. In control samples, no false-positive signals were found. Comparing the titer calculated as log 10 reciprocal to the highest positive dilution of a cell culture supernatant of infected MDBK cells for several conventional detection methods, IPCR revealed a titer of 10.9, which was obviously superior to plaque forming assay (7.2), ELISA (3.9) or PCR (4.0).

The robust detection of antigens in a biological matrix and the prospect of monitoring an infectious illness after the disappearance of clinical or conventionally detectable symptoms are therefore powerful arguments for the application of IPCR in virus screening tests.

Other studied virus markers encompass influenza viral antigen nonstructural protein 1, detected in combination with hemagglutinin subtype–specific antigen by Ozaki *et al.* [83]. In this work, IPCR was found to be 10 times more sensitive for both antigens than either virus isolation or RT-PCR.

The same findings relating to an improved sensitivity of IPCR compared to RT-PCR were also reported by Barletta *et al.* in their study of the HIV detection by IPCR [84]. Similar to HBsAg discussed above, HIV infection is also prone to long-time low-level infections with fewer than 50 viral copies per milliliter, the actual detection limit of RT-PCR. In contrast to the amount of viral RNA, approximately 1000 times more molecules of the core protein p24 (ca. 42 zeptomol) are present in low-load samples. Although the sensitivity of conventional ELISA was found insufficient for unlocking this potential, IPCR was sufficiently sensitive to detect low concentrations, down to two viral copies per milliliter in human blood samples. In 1999,

Adler *et al.* reported an improved IPCR assay for the analogue core protein p28 of the simian immunodeficiancy virus in ape sera, thus establishing a 100-fold increase of the ELISA detection limit [76]. Although the assay at this time achieved similar sensitivity as RT-PCR and was therefore interesting for comparison of both methods, Barletta *et al.* [84] further improved the detection limit of HIV p24, finally reaching a quantification range covering three orders of magnitude between about 2 and 6500 copies of the virus. This IPCR assay was applied to HIV-infected patients whose serum virus load was considered below the detection limit of RT-PCR. In contrast, IPCR found 42% of the samples to be positive for the presence of p24.

This experiment impressively demonstrated how an improved protein assay accessible by IPCR has significantly enhanced the opportunity for the early detection of potentially dangerous false-negative samples and emphasized the value of IPCR as a screening tool for low-level infections. Although IPCR sensitivity facilitates an early therapy, actual therapy monitoring and adaptation is supported by the broad dynamic quantification range. As the actual increase/decrease of the virus load is detectable in contrast to the simple qualitative presence/absence of the virus, detailed studies of therapy and medicament influences are possible.

As there are different levels of information accessible by either proteins or nucleic acids, the direct comparison of (RT-) PCR and IPCR is leading to several conclusions: Nonspecific DNA amplification or extensive sample purification necessary for inhibitor- and interference-free PCR could be avoided because of the ELISA protocol of the IPCR method. Although the presence of nucleic acid or protein alone is probably not sufficient as a stand-alone marker for infection, the combined detection of DNA and expressed proteins—or the high sensitive monitoring of the antibody response against infection—adds valuable information for diagnosis and research of infectious diseases as well as a new level of quality control for drugs prepared from biological sources (e.g., blood banks).

3.3. Pharmacokinetics

Pharmacotoxic and pharmacokinetic studies carried out for the new antitumor drug Aviscumine (rViscumin) were supported by a robust quantitative IPCR assay developed by Adler *et al.* [66, 85]. The potency of this protein-based drug, derived from recombinant mistletoe lectine, required initial doses well below the quantification range accessible by conventional ELISA. An IPCR assay was adapted and validated for the quantification of rViscumin in standardized human serum [66, 85, 86] and subsequently modified for the detection in patient citrate plasma [87]. IPCR was carried out with an internal competitor (see Section 2.3.6) for improved quantitative analysis and control. To circumvent individual differences between the patients and enhancing the recovery, a separate calibration curve and spiking samples were included in the set of controls carried out for each patient. In summary, an overall very good recovery of 95% and a linear correlation between given dose and quantified Aviscumine was found with IPCR. The quantitative analysis of Aviscumine concentration was possible for all patients with a single exception resulting from a high background of an anomalous case [87].

A typical calibration curve for Aviscumine is given in Fig. 8D. Figure 9 shows the relation between given dose and Aviscumine found in plasma. These experiments, accompanying a 3-year clinical study [87], are reporting the first long-term and large-scale routine application of IPCR at a clinical example. Handling of the assay was carried out under standardized conditions, using stock solutions of reagents including antibody–DNA

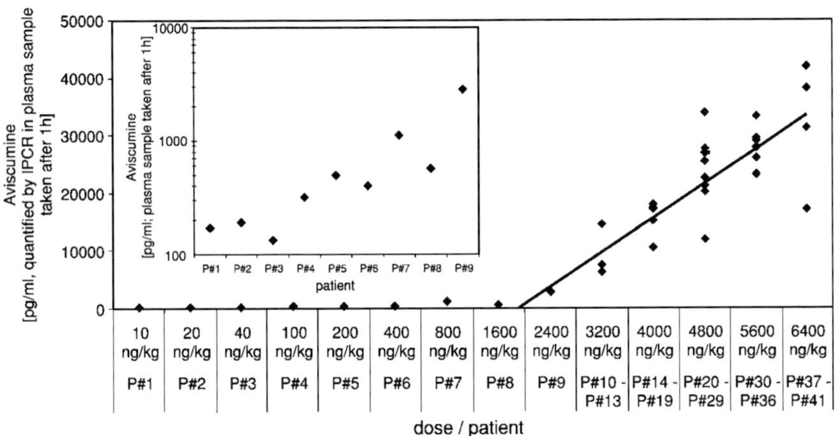

FIG. 9. Correlation between Aviscumine dose levels and Cmax concentrations in plasma as quantified by immuno–polymerase chain reaction in the sample taken 1 hour after treatment. For the application of 10 ng/kg–2400 ng/kg, only one patient was monitored for each dose (P#1–P#9, respectively). Starting with 3200 ng/kg, the calibration curve was adapted and cohorts ranging between 4 and 10 individuals were treated with the same dose (e.g., P#10–P#13 all received a treatment with 3200 ng/kg). The figure shows average concentrations for these patients. Note the linear ratio between the given dose and the concentration found in plasma. (Figure adapted from 73 and 87.)

conjugates. Under these conditions, IPCR was found to be stable and robust [73].

3.4. Pathogens and Toxins: Risk Minimizing and Prevention by IPCR of Unconventional Sample Matrices

Similar to tumor markers discussed above, a number of hazardous proteins are not immediately linked to an accompanying nucleic acid and therefore are prime targets for IPCR. These are interesting examples of how the clinical importance of IPCR is not limited to the diagnosis of diseases. IPCR was reported as a useful tool for the prevention of intoxications or infections because of highly sensitive detection of potentially dangerous compounds, especially in food analysis.

Clostridium botulinum neurotoxin, the most effective toxin known to date, with a mice lethal dose of about 50 pg/mL (330 fmol/mL) was the target antigen in IPCR assays developed by Wu *et al.* [48] and Chao *et al.* [88]. In these assays, detection limits of 5 fg (33 amol) and 50 fg (330 amol), respectively, were found.

In the work of Kakizaki *et al.*, the fish pathogen *Pasteurella piscidia* was detected in naturally infected yellowtail [25]. It was demonstrated that IPCR is not limited to blood, serum, or plasma samples as fish kidney tissue samples, suspended in formalin-PBS, were successfully used as a matrix for the assay.

The search for minuscule amounts of allergens in food to minimize the risk for susceptible patients requires great care in assay development. For individuals, for example, with coeliac disease, the control of gluten in food with reliable techniques is necessary. Henterich *et al.* [51] introduced a real-time IPCR (see Section 2.2.3) for this application as well as for the identification of the toxic peptide motif in the gluten/gliadin protein family. Covalent conjugates (see Section 2.1.3) and Universal-IPCR were used for the detection of the allergen in wheat samples. In addition, different extraction procedures and different sample sources were studied, revealing the option to use a simplified conventional ethanol extraction for some samples in IPCR compared to a complex extraction cocktail necessary for optimum ELISA performance.

As sample extraction and sample handling are of general consideration for the more "exotic" biological matrices often found in food analysis, the application of IPCR in the research project MYCOPLEX [89], founded by the European Union, promises interesting new developments. This project is dedicated to the detection by IPCR of ochra- and aflatoxins in milk and coffee, focusing on sensitivity and simplified antigen extraction by dilution of the samples.

In this context, a promising target for IPCR of recently increasing clinical importance is the potentially infectious prion protein, associated with scrapie, bovine spongiform encephalopathy, and Creutzfeldt-Jakob disease. In addition to two filed patents [90, 91], one of them using the robust one-step DDI-IPCR technology discussed above (Section 2.3.4. [60]), surprisingly, no scientific study using IPCR for prion protein detection has been published to date. One may anticipate a potentially very rewarding range of future research and development projects if the need for a highly sensitive prion detection carried out in living individuals for food control, studies on possible infection pathways, and disease monitoring is combined with the properties of the IPCR assay.

3.5. Embryonal Growth and Surface Proteins, Regulating Hormones: IPCR as a Tool for Routine Studies of Embryos

Although IPCR is normally carried out in combination with capture antibodies immobilized on microplate surfaces, Quijada and coworkers developed a variation of this technique for the highly sensitive labeling of early-stage complete embryos and blastocytes that could be manipulated for washing and labeling independent from immobilization to a surface [13]. A single embryo was found sufficient for signal generation, enabling a successful preimplantation study of the murine histocompatibility complex class I antigen.

In another application of the IPCR technology for preimplantation studies, McElhinny *et al.* determined the expression as well as the "protein painting" artificial membrane incorporation of the Qa-2 antigen, a cleavage-rate-influencing glycosylphosphatidylinositol-linked cell surface protein of the mouse major histocompatibility complex [15–17]. IPCR allowed for a single blastocyte resolution of the detection of antigen presence, thus supplementing RT-PCR and fluorescence FACScan detection.

In a study of the Qa-2-expression regulation by the "transporter associated with antigen processing" (TAP) protein, Ke and Warner [14] reported the monitoring of preimplantation embryos from TAP-1 knockout and normal mice by IPCR and RT-PCR. IPCR was carried out as described by McElhinny *et al.* [17]. Again, IPCR proved its potential as a valuable tool for the analysis of protein expression on the cell surface from small numbers (2–20) of one-cell and two-cell embryos, as well as blastocystes.

For demonstrating the properties of the multiplex–IPCR and covalent antibody–DNA conjugates (see Section 2.1.3), a number of hormones such as human thyroid stimulating hormone or chorionic gonadotropin were simultaneously detected by the workgroup of Ebersole [38, 41]. In the study of Hendrickson *et al.*, a proof of principle for this concept was carried out by the simultaneous detection of three antigens, β-Gal, human thyroid stimulating hormone, and chorionic gonadotropin, using three specific covalent antibody–DNA conjugates with DNA markers of individual length (55, 85, and 99 bp, respectively). The different amplification products, each specific for one antigen, were separated by gel electrophoresis subsequent to PCR amplification. A specific signal was obtained for 5 ng β-Gal, 100 pg human thyroid stimulating hormone, and 2.5 ng chorionic gonadotropin, incubated simultaneously, and coupled with the respective antibody–DNA conjugate mixture.

In this application, groundwork was laid out for the routine application of a set of standardized multiplex tests (e.g., in early postnatal checks or tests carried out during pregnancy). This approach is especially interesting, as the high sensitivity of IPCR allows us to perform many tests from a single small sample, thus minimizing stress for the patients in exchange for a gain in information.

3.6. Other Applications: IPCR Tailored to Specific Problems

In addition to the above-mentioned broad target groups for clinical application of IPCR, some more specialized studies are interesting examples of the versatility of IPCR.

3.6.1. *Microorganisms and Parasites: IPCR for Improvements in Sample Drawing*

The high sensitivity of the IPCR method was successfully used for the simplification of sample drawing procedures in the following examples.

For the detection of a small amount of cells in a large volume (e.g., a contamination of one cell in several liters), the inclusion of the analyte in a studied sample volume of typically 50–100 μL became a nontrivial problem of sample volume reduction without loss of the microorganism.

A highly sensitive sandwich IPCR detection of the presence of *Escherichia coli* by its soluble marker protein β-glucuronidase was developed by Chang and Huang [31]. In a comparison of the IPCR of cell extract, ELISA, and cell counting, an impressive detection limit of two molecules in a 50-μL assay volume was achieved by IPCR. The detection of a dilution of 1×10^{-4} CFU/mL in a 50-μL sample volume was the equivalent of the β-glucuronidase molecules released from a single *E. coli* cell in a 10-L sample liquid. Thereby, it was not necessary to actually find the cell in the studied volume, as the high sensitive IPCR detection of the expressed proteins enabled an indirect detection.

Although the handling of large sample volumes is typical in quality control studies, an easy sample drawing procedure is especially desirable in clinical diagnosis of human patients.

One of the most important zoonotic parasites in Taiwan, responsible for specific forms of meningitis and meningoencephalitis, is the worm *Angiostrongylus cantonensis*, living in the CSF of infected patients. Obtaining CSF samples from patients is highly stressful, requiring medically trained personnel, and is nevertheless incapable of detecting the presence of the worm in 50% of the infected patients. Therefore, Chye *et al.* [64] developed a highly sensitive sandwich IPCR technique for the detection of circulating antigens from *Angiostrongylus c.* in serum specimen. At a cutoff level of 0.1 ng/L, sensitivity and specificity for immunodiagnosis of patients with angiostrongyliasis by immuno-PCR were 98% (95% confidence interval, 91%–99%) and 100% (93%–100%), respectively. The test was positive in all parasitologically confirmed cases, thus combining a better sensitivity and specificity with a major simplification in sample drawing.

3.6.2. *Heart Failure–Associated Proteins: IPCR Allows for Significant Gain in Assay Time*

Although IPCR development is typically focused on improvement of the detection limit, the two examples discussed in 3.6.1 have already illustrated how the increase in sensitivity could be used as a means to an end instead of being the sole intended focus.

In the example of α-human atrial natriuretic peptide (ANP), found at increased plasma levels in patients with heart failure, Numata *et al.* [70] demonstrated how IPCR sensitivity accelerated conventional assay procedures. For individual treatment of the cardiac patients, a prompt detection of atrial distension by the presence of the ANP marker would be desirable. Common ANP tests, however, take 2–3 days for the quantification of plasma by radiometric or ELISA techniques. With sandwich IPCR, the assay time could be shortened to 5 hours. A good correlation between IPCR and radiometric detection was maintained, combined with an additional improvement of the detection limit to 2 ng/L ANP. The average level of ANP in plasma for 25 patients with heart failure was found to be 117 ± 100 ng/L, significantly higher than the typical level of 20 ± 14 ng/L for healthy subjects.

Reduced assay time is essential for early response in diagnosis and therapy. Because of a significant reduction in assay duration, IPCR is therefore useful even in applications in which alternative sensitive detection methods already exist.

3.6.3. *Interleukin, Angiotensinogen, and Osteoprotegerin: IPCR Assay Fine-Tuning*

Systematic optimization of the Universal-IPCR protocol revealed the full potential of the flexible technique. The research group of Watanabe and coworkers, introduced in Section 3.1 with the detection of TNF-α [34, 39, 80], has excelled in adaptation of this protocol to various antigens.

In an IPCR application developed by Furuya *et al.* [27], interleukin 18 (IL-18) was studied as an important protein in a number of immunological derangements. An indirect sandwich IPCR (Fig. 3D) was used for the detection of IL-18 in cell culture supernatants and serum samples. A quantitative study of low-level IL-18 was carried out beneath a serum concentration of typically 96–135 ng/L. In a systematic variation of assay conditions, the amount of biotinylated 227-bp DNA marker used in the Universal-IPCR protocol was identified as a source for nonspecific amplification. Following an optimization of DNA concentration and PCR cycle number, 2.5 pg/L IL-18 was detected.

In a continuation of this work with a similar Universal-IPCR protocol and the same 227-bp DNA marker, Sugawara *et al.* [36] developed an indirect sandwich IPCR for human angiotensinogen. By using identical capture and detection antibodies in combination with optimized concentrations of STV (1 mg/L; ca. 16.6 nmol/L) and biotinylated 227-bp DNA (1 ng/L; ca. 6.6 fmol/L), nonspecific binding was minimized. An optimized approximately 1:3,000,000 ratio of DNA:STV, also used for the detection of homodimeric osteoprotegerin [35], revealed the difference between the theoretical coupling efficiency of tetravalent STV and the real conditions found in successful IPCR. An increase of DNA or STV concentration induced nonspecific background, whereas lower amounts of DNA or STV decreased signal intensities. Interestingly, the actual sensitivity improvement compared to conventional ELISA was enhanced by reducing the reagent concentrations used in ELISA.

In these experiments, a uniform ratio and concentration of DNA and STV was established for the 227-bp DNA marker, indicating a successful strategy for the development of various IPCR applications by standardized assay conditions for these compounds. In contrast, the concentration of the antigen-specific biotinylated antibody was adapted for each assay.

From these findings, the need of the multistep sequential protocol for meticulous fine-tuning of assay conditions discussed above (Section 2.1.2) is obvious. The impressive and reproducible enhancement of the detection limit for several antigens, however, demonstrated with a similar IPCR

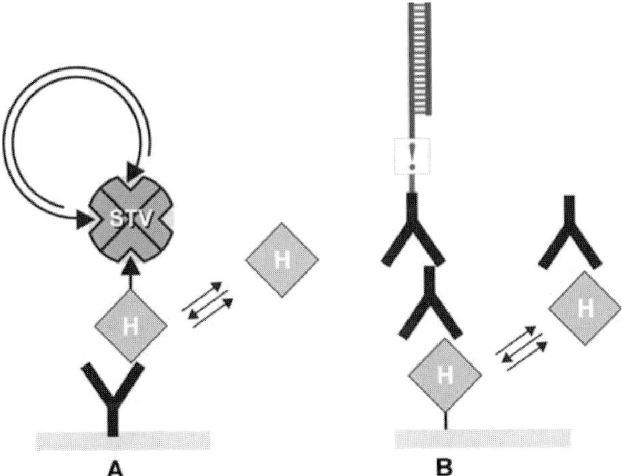

FIG. 10. Different techniques for competitive immuno–polymerase chain reaction of small-molecule compounds. (A) Using biotinylated haptens and a DNA–streptavidin nanocircle conjugate [105], hapten–DNA conjugates were synthesized and used for a competitive assay in a sample containing free hapten and capture antibody–coated surfaces [92]. (B) Hapten-coated microplates were simultaneously incubated with a sample containing free hapten and a hapten-specific antibody. Following competitive coupling, the immobilized antibody was subsequently detected by a species-specific antibody–DNA conjugate [93, 94].

protocol, strikingly exemplified the potential inherent to the method if performed under optimal conditions.

3.6.4. *Competitive IPCR for Small-Molecule Haptens: Novel Targets for IPCR*

Recently, two different strategies were developed for a competitive IPCR of small molecules, such as, for example, hormones, messengers, or transmitters without multiple antibody-binding sites necessary for sandwich assays.

In the first approach, a biotinylated analogue of the hapten to be detected was directly coupled with a DNA-STV nanocircle containing a single DNA marker (see Fig. 10A) and coincubated with the antigen-containing sample [92]. Although this approach allowed for the highly sensitive detection of the model compounds fluorescein and cholesterol, it required biotinylation of the hapten. An alternative method applied solid-phase immobilized hapten as a competitor to a supernatant sample solution containing antigen and detection antibody. Subsequent to sample incubation, the immobilized detection antibody was coupled with antispecies antibody–DNA conjugates (see Fig. 10B). A novel research ELISA of the biogenic amine serotonin was adapted to this IPCR method with only minor changes in ELISA protocol.

Thereby, an overall 10,000-fold improvement of conventional ELISA sensitivity with the detection of 0.4 pg/mL antigen (content of only 10 platelets) was achieved [93, 94].

Competitive IPCR, also tested at noradrenaline and dopamine, should therefore be a valuable tool in novel approaches to the monitoring of biogenic amines, either at levels inaccessible with conventional ELISA or in samples obtainable by noninvasive testing methods.

4. Rating of IPCR as a Tool for Clinical Applications

In summary, a typically 100–10,000-fold improvement of common ELISA sensitivity is accessible with IPCR (see Table 1).

The advantages of the IPCR technique, however, are not limited to the obvious gain in sensitivity. With the aid of increased sensitivity, it is additionally possible to develop solutions for other challenges in clinical analysis, such as, first, reduction of incubation time: a general reduction of assay duration is desirable for fast on-spot diagnosis as well as continuous monitoring of experiments [70]. For assay optimization, the elimination of overnight-incubations is especially interesting. A second solution is the enhanced robustness of the assay: most interfering compounds are removed by the immunological identification of target antigens in combination with appropriate washing steps. In addition, the increased sensitivity of IPCR compared to ELISA allowed for the detection of antigens even in impure samples, as the smallest amount of specifically bound antigen already induces a signal. Thus, elaborate sample purification steps could be reduced or eliminated completely [62].

A simple sample dilution is sufficient for removal of interfering compounds. IPCR is still able to detect the antigen in sample dilutions [66], whereby the loss in sensitivity limits the usefulness of additional dilution in conventional ELISA.

A third solution is to change the sample drawing procedures: most compounds accessible by ELISA in liquor or tissue are homogeneous, requiring stressful invasive sample drawing procedures. The ELISA detection of compounds found in liquor or tissue requires stressful invasive sample drawing procedures. These compounds are normally also present in blood, sputum, stool samples, or urine, although in much smaller concentrations. As these samples are more easily obtainable, the latter ones, even with noninvasive methods, are equally advantageous for patient and laboratory. IPCR makes it possible to transfer existing ELISA applications to these new matrices [64, 72]. This is especially interesting for sample drawing by nonmedical personal (e.g., by patients at home or monitoring during sportive activities).

A fourth solution is to develop a method that less sample material per assay: The small assay volume and high sensitivity reduces stress for patients during sample drawing and increases the number of different tests or double determinations possible with one sample. In experiments with small animals, tests of very young individuals or even developing embryos [13–16, 95], for which only small samples could be taken, IPCR is the key for specific protein analysis. In addition, for assays detecting a small number of analytes in large volumes (e.g., single pathogenic organisms [31, 74]), the assay could be simplified by reducing the amount of liquid that has to be concentrated for successful detection of the analyte. Moreover, in routine analysis of large sample numbers, the high sensitivity of IPCR enables the pooling of individual samples for quality control of whole charges.

Finally, the quantification range can be increased: The combination of ELISA and PCR techniques broadens the range of linear signal output, thus facilitating quantification [37].

The introduction of easy accessible DNA–protein conjugates and the real-time PCR protocol has eliminated most of the handling problems initially associated with IPCR, thus transforming the challenging research method to a fully-fledged and valuable tool for analysis in clinical applications. As reagents, materials, and complete kits for IPCR are nowadays commercially available, IPCR has become accessible for an even broader field of users. The combination of real-time PCR and IPCR further facilitates the assay protocol. New protocols for IPCR are combining the experience from several years of assay development. The standardized and simplified method requires only a short individual adaptation to the intended application for maximum performance. Existing publications serve as inspiring templates for several application types and assay conditions.

IPCR is not for everyone, as there are many well-established ELISA tests with sufficient sensitivity. A certain degree of handling experience and equipment needed for ultrasensitive analysis will certainly rule out applications in quick-testing, outdoor kits, or test-stripes.

For any standard ELISA application requiring enhanced sensitivity in combination with almost complete retaining of all ELISA advantages, such as microplate format, common laboratory equipment and the immunological specificity for target antigens, however, Immuno-PCR is worth consideration.

Acknowledgments

This work was supported by Chimera Biotec. The author thanks Dr. Ron Wacker and Prof. Christof M. Niemeyer for critical reading of the manuscript.

References

[1] Mullis KB, Faloona F. Specific synthesis of DNA *in vitro* via a polymerase-catalyzed chain reaction. Methods Enzymol 1987; 155:335–350.

[2] Newton CR, Graham A. Labor im Fokus. Spektrum Akademischer Verlag, 1994.

[3] Saiki RK, Gelfand DH, Stoffel S, Scharf S, Higuchi R, Horn GT, Mullis KB. Primer-directed enzymatic amplification of DNA with a thermostable DNA polymerase. Science 1988; 239:487–491.

[4] Saiki RK, Scharf S, Faloona F, Mullis KB, Horn GT, Erlich HA, Arnheim N. Enzymatic amplification of ß-globin genomic sequences and restriction site analysis for diagnosis of sickle cell anemia. Science 1985; 230:1350–1354.

[5] Nakamura S, Katamine S, Yamamoto T, Foung SK, Kurata T, Hirabayashi Y, Shimada K, Hino S, Miyamoto T. Amplification and detection of a single molecule of human immunodeficiency virus RNA. Virus Genes 1993; 7(4):325–338.

[6] Li H, Cui X, Arnheim N. Direct electrophoretic detection of the allelic state of single DNA molecules in human sperm by using the polymerase chain reaction. Proc Natl Acad Sci USA 1990; 87(12):4580–4584.

[7] Jenkins A, Kristiansen BE, Ask E, Oskarsen B, Kristiansen E, Lindqvist B, Trope C, Kjorstad K. Detection of genital papillomavirus types by polymerase chain reaction using common primers. Apmis 1991; 99(7):667–673.

[8] Crowther JR. ELISA; theory and practice. In: Walker JM, editor. Methods in Molecular Biology. Totowa, NJ: Humana Press, 1995.

[9] Price CP, Newman DJ. Principles and Practise of Immunoassay, second ed. London: Macmillan Reference Ltd, 1997.

[10] Sano T, Smith CL, Cantor CR. Immuno-PCR: Very sensitive antigen detection by means of specific antibody-DNA conjugates. Science 1992; 258(5079):120–122.

[11] Schiavo S, El-Shafey A, Jordan N, Orazine C, Wang Y, Zhou X, Chiu NH, Krull IS. Pushing the limits of detection with immuno-PCR. PharmaGenomics 2004; 36–44.

[12] Sano T, Cantor CR. A streptavidin-protein A chimera that allows one-step production of a variety of specific antibody conjugates. Biotechnology (NY) 1991; 9 (12):1378–1381.

[13] Quijada L, Moreira D, Soto M, Alonso C, Requena JM. Detection of major histocompatibility complex class I antigens on the surface of a single murine blastocyst by immuno-PCR. BioTechniques 1997; 24(4):660–662.

[14] Ke X, Warner CM. Regulation of Ped gene expression by TAP protein. J Reprod Immunol 2000; 46(1):1–15.

[15] McElhinny AS, Exley GE, Warner CM. Painting Qa-2 onto Ped slow preimplantation embryos increases the rate of cleavage. Am J Reprod Immunol 2000; 44(1):52–58.

[16] McElhinny AS, Kadow N, Warner CM. The expression pattern of the Qa-2 antigen in mouse preimplantation embryos and its correlation with the Ped gene phenotype. Mol Hum Reprod 1998; 4(10):966–971.

[17] McElhinny AS, Warner CM. Detection of major histocompatibility complex class I antigens on the surface of a single murine blastocyst by immuno-PCR. Biotechniques 1997; 23(4):660–662.

[18] Ren J, Ge L, Li Y, Bai J, Liu WC, Si XM. Detection of circulating CEA molecules in human sera and leukopheresis of peripheral blood stem cells with *E. coli* expressed bispecific CEAScFv-streptavidin fusion protein-based immuno-PCR technique. Ann NY Acad Sci 2001; 945:116–118.

[19] Green NM. Avidin and streptavidin. Methods Enzymol 1990; 184:51–67.

[20] Sano T, Pandori MW, Chen X, Smith CL, Cantor CR. Recombinant core streptavidins. A minimum-sized core streptavidin has enhanced structural stability and higher accessibility to biotinylated macromolecules. J Biol Chem 1995; 270(47):28204–28209.

[21] Bayer EA, Wilchek M. Biotin-binding proteins: Overview and prospects. Methods Enzymol 1990; 184:49–51.

[22] Diamandis EP, Christopoulos TK. The biotin-(strept)avidin system: Principles and applications in biotechnology. Clin Chem 1991; 37(5):625–636.

[23] Pierce. STV-Biotin. see www.perbio.com (1999).

[24] Zhou H, Fisher RJ, Papas TS. Universal immuno-PCR for ultra-sensitive target protein detection. Nucleic Acids Res 1993; 21(25):6038–6039.

[25] Kakizaki E, Yoshida T, Kawakami H, Oseto M, Sakai T, Sakai M. Detection of bacterial antigens using immuno-PCR. Lett Appl Microbiol 1996; 23(2):101–103.

[26] Niemeyer CM, Adler M, Pignataro B, Lenhert S, Gao S, Chi L, Fuchs H, Blohm D. Self-assembly of DNA-streptavidin nanostructures and their use as reagents in immuno-PCR. Nucleic Acids Res 1999; 27(23):4553–4561.

[27] Furuya D, Yagihashi A, Yajima T, Kobayashi D, Orita K, Kurimoto M, Watanabe N. An immuno-polymerase chain reaction assay for human interleukin-18. J Immunol Methods 2000; 238(1–2):173–180.

[28] Mweene AS, Ito T, Okazaki K, Ono E, Shimizu Y, Kida H. Development of immuno-PCR for diagnosis of bovine herpesvirus 1 infection. J Clin Microbiol 1996; 34(3):748–750.

[29] Case MC, Burt AD, Hughes J, Palmer JM, Collier JD, Bassendine MF, Yeaman SJ, Hughes MA, Major GN. Enhanced ultrasensitive detection of structurally diverse antigens using a single immuno-PCR assay protocol. J Immunol Methods 1999; 223(1):93–106.

[30] Maia M, Takahashi H, Adler K, Garlick RK, Wands JR. Development of a two-site immuno-PCR assay for hepatitis B surface antigen. J Virol Methods 1995; 52(3):273–286.

[31] Chang TC, Huang SH. A modified immuno-polymerase chain reaction for the detection of beta-glucuronidase from *Escherichia coli*. J Immunol Methods 1997; 208(1):35–42.

[32] Ren J, Chen Z, Juan SJ, Yong XY, Pan BR, Fan DM. Detection of circulating gastric carcinoma-associated antigen MG7-Ag in human sera using an established single determinant immuno-polymerase chain reaction technique. Cancer 2000; 88(2):280–285.

[33] Sanna PP, Weiss F, Samson ME, Bloom FE, Pich EM. Rapid induction of tumor necrosis factor alpha in the cerebrospinal fluid after intracerebroventricular injection of lipopolysaccharide revealed by a sensitive capture immuno-PCR assay. Proc Natl Acad Sci USA 1995; 92(1):272–275.

[34] Saito K, Kobayashi D, Sasaki M, Araake H, Kida T, Yagihashi A, Yajima T, Kameshima H, Watanabe N. Detection of human serum tumor necrosis factor-alpha in healthy donors, using a highly sensitive immuno-PCR assay. Clin Chem 1999; 45(5):665–669.

[35] Furuya D, Kaneko R, Yagihashi A, Endoh T, Yajima T, Kobayashi D, Yano K, Tsuda E, Watanabe N. Immuno-PCR assay for homodimeric osteoprotegerin. Clin Chem 2001; 47(8):1475–1477.

[36] Sugawara K, Kobayashi D, Saito K, Furuya D, Araake H, Yagihashi A, Yajima T, Hosoda K, Kamimura T, Watanabe N. A highly sensitive immuno-polymerase chain reaction assay for human angiotensinogen using the identical first and second polyclonal antibodies. Clin Chim Acta 2000; 299(1–2):45–54.

[37] Niemeyer CM, Adler M, Blohm D. Fluorometric polymerase chain reaction (PCR) enzyme-linked immunosorbent assay for quantification of immuno-PCR products in microplates. Anal Biochem 1997; 246(1):140–145.

[38] Joerger RD, Truby TM, Hendrickson ER, Young RM, Ebersole RC. Analyte detection with DNA-labeled antibodies and polymerase chain reaction. Clin Chem 1995; 41(9):1371–1377.

[39] Komatsu M, Kobayashi D, Saito K, Furuya D, Yagihashi A, Araake H, Tsuji N, Sakamaki S, Niitsu Y, Watanabe N. Tumor necrosis factor-alpha in serum of patients with inflammatory bowel disease as measured by a highly sensitive immuno-PCR. Clin Chem 2001; 47(7):1297–1301.

[40] Case M, Major GN, Bassendine MF, Burt AD. The universality of immuno-PCR for ultrasensitive antigen detection. Biochem Soc Trans 1997; 25(2):374S.

[41] Hendrickson ER, Truby TM, Joerger RD, Majarian WR, Ebersole RC. High sensitivity multianalyte immunoassay using covalent DNA-labeled antibodies and polymerase chain reaction. Nucleic Acids Res 1995; 23(3):522–529.

[42] Edwards MC, Gibbs RA. Multiplex PCR: Advantages, development, and applications. PCR Methods Appl 1994; 3(4):S65–S75.

[43] Henegariu O, Heerema NA, Dlouhy SR, Vance GH, Vogt PH. Multiplex PCR: Critical parameters and step-by-step protocol. Biotechniques 1997; 23(3):504–511.

[44] Schweitzer B, Wiltshire S, Lambert J, O'Malley S, Kukanskis K, Zhu Z, Kingsmore SF, Lizardi PM, Ward DC. Immunoassays with rolling circle DNA amplification: A versatile platform for ultrasensitive antigen detection. Proc Natl Acad Sci USA 2000; 97(18):10113–10119.

[45] Wiltshire S, O'Malley S, Lambert J, Kukanskis K, Edgar D, Kingsmore SF, Schweitzer B. Detection of multiple allergen-specific IgEs on microarrays by immunoassay with rolling circle amplification. Clin Chem 2000; 46(12):1990–1993.

[46] Mullenix MC, Dondero RS, Datta HD, Egholm M, Kingsmore SF, Perlee LT. Rolling circle amplification in multiplex immunoassays. In: Demidov VV, Broude NE, editors. DNA Amplification—Current Technologies and Applications. Norfolk, UK: Horizon Bioscience, 2004: 313–331.

[47] Liu D, Daubendiek SL, Zillmann MA, Ryan K, Kool ET. Rolling circle DNA synthesis: Small circular oligonucleotides as efficient templates for DNA polymerases. J Am Chem Soc 1996; 118(7):1587–1594.

[48] Wu HC, Huang YL, Lai SC, Huang YY, Shaio MF. Detection of *Clostridium botulinum* neurotoxin type A using immuno-PCR. Lett Appl Microbiol 2001; 32(5):321–325.

[49] Sims PW, Vasser M, Wong WL, Williams PM, Meng YG. Immunopolymerase chain reaction using real-time polymerase chain reaction for detection. Anal Biochem 2000; 281(2):230–232.

[50] McKie A, Samuel D, Cohen B, Saunders NA. Development of a quantitative immuno-PCR assay and its use to detect mumps-specific IgG in serum. J Immunol Methods 2002; 261(1–2):167–175.

[51] Henterich N, Osman AA, Mendez E, Mothes T. Assay of gliadin by real-time immunopolymerase chain reaction. Nahrung 2003; 47(5):345–348.

[52] Niemeyer CM, Adler M, Gao S, Chi L. Nanostructured DNA-protein aggregates consisting of covalent oligonucleotide-streptavidin conjugates. Bioconjug Chem 2001; 12(3):364–371.
[53] Cattoretti G, Berti E, Schiro R, D'Amato L, Valeggio C, Rilke F. Improved avidin-biotin-peroxidase complex (ABC) staining. Histochem J 1988; 20(2):75–80.
[54] Hsu SM, Raine L, Fanger H. Use of avidin-biotin-peroxidase complex (ABC) in immunoperoxidase techniques: A comparison between ABC and unlabeled antibody (PAP) procedures. J Histochem Cytochem 1981; 29(4):577–580.
[55] Ruzicka V, Marz W, Russ A, Gross W. Immuno-PCR with a commercially available avidin system [letter]. Science 1993; 260(5108):698–699.
[56] Sano T, Smith CL, Cantor CR. Technical comments. Science 1993; 260:699.
[57] Niemeyer CM, Adler M, Gao S, Chi L. Nanostructured DNA-protein networks consisting of covalent oligonucleotide-streptavidin conjugates. Bioconjug Chem 2000; 12(3):364–371.
[58] Niemeyer CM, Sano T, Smith CL, Cantor CR. Oligonucleotide-directed self-assembly of proteins: Semisynthetic DNA-streptavidin hybrid molecules as connectors for the generation of macroscopic arrays and the construction of supramolecular bioconjugates. Nucleic Acids Res 1994; 22(25):5530–5539.
[59] Kukolka F, Lovrinovic M, Wacker R, Niemeyer CM. Covalent coupling of DNA oligonucleotides and streptavidin. Methods Mol Biol 2004; 283:181–196.
[60] Niemeyer CM, Wacker R, Adler M. Combination of DNA-directed immobilization and immuno-PCR: Very sensitive antigen detection by means of self-assembled DNA-protein conjugates. Nucleic Acids Res 2003; 31(16):e90.
[61] Niemeyer CM, Boldt L, Ceyhan B, Blohm D. DNA-Directed immobilization: Efficient, reversible, and site-selective surface binding of proteins by means of covalent DNA-streptavidin conjugates. Anal Biochem 1999; 268(1):54–63.
[62] Adler M, Wacker R, Niemeyer CM. A real-time immuno-PCR assay for routine ultrasensitive quantification of proteins. Biochem Biophys Res Commun 2003; 308(2):240–250.
[63] Heid CA, Stevens J, Livak KJ, Williams PM. Real time quantitative PCR. Genome Res 1996; 6(10):986–994.
[64] Chye SM, Lin SR, Chen YL, Chung LY, Yen CM. Immuno-PCR for detection of antigen to Angiostrongylus cantonensis circulating fifth-stage worms. Clin Chem 2004; 50 (1):51–57.
[65] Adler M. Hapten labeling of nucleic acids for immuno-polymerase chain reaction applications, bioconjugation protocols: Strategies and methods. In: Niemeyer CM, editor. Methods in Molecular Biology. Humana Press, 2004: 163–180.
[66] Adler M, Langer M, Witthohn K, Eck J, Blohm D, Niemeyer CM. Detection of rViscumin in plasma samples by immuno-PCR. Biochem Biophys Res Commun 2003; 300(3):757–763.
[67] Adler M, Niemeyer CM. Enhanced protein detection using real-time immuno-PCR. In: Broude NE, editor. DNA Amplification—Current Technologies and Applications. Norfolk, UK: Horizon Bioscience, 2004: 293–312.
[68] McKie A, Samuel D, Cohen B, Saunders NA. A quantitative immuno-PCR assay for the detection of mumps-specific IgG. J Immunol Methods 2002; 270(1):135–141.
[69] Lind K, Kubista M. Real-time immuno-PCR on the iCycler iQ system. BioRadiations 2002; 109:32–33.

[70] Numata Y, Matsumoto Y. Rapid detection of alpha-human atrial natriuretic peptide in plasma by a sensitive immuno-PCR sandwich assay. Clin Chim Acta 1997; 259(1–2):169–176.

[71] Vogt RF, Jr, Phillips DL, Henderson LO, Whitfield W, Spierto FW. Quantitative differences among various proteins as blocking agents for ELISA microtiter plates. J Immunol Methods 1987; 101(1):43–50.

[72] McKie A, Vyse A, Maple C. Novel methods for the detection of microbial antibodies in oral fluid. Lancet Infect Dis 2002; 2(1):18–24.

[73] Adler M, Langer M, Witthohn K, Wilhelm-Ogunbiyi K, Schöffski P, Fumoleau P, Niemeyer CM. Adaptation and performance of an IPCR-assay for the quantification of aviscumine in patient plasma samples. Biochem Biophys Res Commun 2004; (in press).

[74] Liang H, Cordova SE, Kieft TL, Rogelj S. A highly sensitive immuno-PCR assay for detecting Group A Streptococcus. J Immunol Methods 2003; 279(1–2):101–110.

[75] ICH Topic Q2B Validation of Analytical Procedures. EMEA-European Agency for the Evaluation of Medicinal Products. Full text available at www.emea.eu.int/pdfs/human/ich/028195en.pdf, 1996.

[76] Adler M. Herstellung und Charakterisierung supramolekularer DNA-Protein-Aggregate und deen Anwendung als Reagenzien in der Immuno-PCR. Doctoral theis, Germany: University of Bremen, 2001.

[77] Christopoulos TK, Chiu NH. Expression immunoassay. Antigen quantitation using antibodies labeled with enzyme-coding DNA fragments. Anal Chem 1995; 67(23):4290–4294.

[78] Chiu NH, Christopoulos TK. Two-site expression immunoassay using a firefly luciferase-coding DNA label. Clin Chem 1999; 45(11):1954–1959.

[79] Ren J, Fan DM, Zhou SJ. Establishment of immuno-PCR technique for the detection of tumor associated antigen MG7-Ag on the gastric cancer cell line. Chung Hua Chung Liu Tsa Chih 1994; 16(4):247–250.

[80] Watanabe N. Clinical significance of measurement of circulating tumor necrosis factor alpha. Rinsho Byori 2001; 49(9):829–833.

[81] Cao Y, Kopplow K, Liu GY. *In situ* immuno-PCR to detect antigens. Lancet 2000; 356 (9234):1002–1003.

[82] Cao Y. *In situ* immuno-PCR. A newly developed method for highly sensitive antigen detection *in situ*. Methods Mol Biol 2002; 193:191–216.

[83] Ozaki H, Sugita S, Kida H. A rapid and highly sensitive method for diagnosis of equine influenza by antigen detection using immuno-PCR. Jpn J Vet Res 2001; 48(4):191–196.

[84] Barletta JM, Edelman DC, Constantine NT. Lowering the detection limits of HIV-1 viral load using real-time immuno-PCR for HIV-1 p24 antigen. Am J Clin Pathol 2004; 122 (1):20–27.

[85] Adler M, Niemeyer CM, Witthohn K, Langer M, Blohm D. Detection of subattomol amounts of rViscumin in serum by immuno-PCR, Poster auf der Biotechnology 2000 Conference and Exhibition, Berlin. 2000.

[86] Adler M, Niemeyer CM, Witthohn K, Langer M, Blohm D. High-sensitivity detection of rViscumin in blood serum samples using immuno-PCR. Biol Chem 2000; 381(Suppl.): S178.

[87] Schöffski P, Riggert S, Fumoleau P, et al. Phase 1 trial of intravenous Aviscumine (rViscumin) in patients with solid tumors. A study of the European Organization for Research and Treatment of Cancer (EORTC) New Drug Development Group (NDDG). Ann Oncol 2004; 15(12):1816–1824.
[88] Chao HY, Wang YC, Tang SS, Liu HW. A highly sensitive immuno-polymerase chain reaction assay for Clostridium botulinum neurotoxin type A. Toxicon 2004; 43(1):27–34.
[89] MYCOPLEX. Available at: http://www.biozoon.de/index_1024.htm.
[90] Tetro JA. A specific and sensitive method to detect PrPSc in peripheral blood using a capture immunosorbent assay polymerase chain reaction (PrP-Immuno-PCR). Full text available at http://v3.espacenet.com/origdoc?DB=EPODOC&IDX=CA2237739&F=0&QPN=CA2237739, 1999.
[91] Niemeyer CM, Adler M. Method for the detection of an analyte. Full text available at http://v3.espacenet.com/origdoc?DB=EPODOC&IDX=EP1249500&F=0&QPN=EP1249500, 2002.
[92] Niemeyer CM, Wacker R, Adler M. Hapten-functionalized DNA-streptavidin nanocircles as supramolecular reagents in a novel competitive immuno-PCR. *Angew Chem Int Ed* 2001; 40:3169–3172.
[93] Adler M, Wacker R, Booltink E, Manz B, Niemeyer CM. Detection of femtogram amounts of biogenic amines using self-assembled DNA–protein nanostructures. Nat Methods 2005; 2:147–149.
[94] CHIMERA, LDN, "IPCR Technology for Biogenic Amine Immunoassays," LDN Labor Diagnostic Nord GmbH & Co KG, Norhorn, Germany & CHIMERA BIOTEC GmbH, Dortmund, Germany, Poster & Tech note, MEDICA, 2003. See also: http://www.ldn-de.net.
[95] Warner CM, McElhinny AS, Wu L, Cieluch C, Ke X, Cao W, Tang C, Exley GE. Role of the Ped gene and apoptosis genes in control of preimplantation development. J Assist Reprod Genet 1998; 15(5):331–337.
[96] Zhang Z, Irie RF, Chi DD, Hoon DS. Cellular immuno-PCR. Detection of a carbohydrate tumor marker. Am J Pathol 1998; 152(6):1427–1432.
[97] Tsujii H, Okamoto Y, Nakano H, Watanabe K, Naraki T. Application of immuno-PCR assay to detect an ultra-low level plasma protein, pivka-II. International Hepatol Commun 1995; 3(1000):154.
[98] Mweene AS, Okazaki K, Kida H. Detection of viral genome in non-neural tissues of cattle experimentally infected with bovine herpesvirus 1. Jpn J Vet Res 1996; 44 (3):165–174.
[99] Numata Y, Matsumoto Y. Rapid detection of anti-human atrial natriuretic peptide in plasma by a sensitive immuno-PCR sandwich assay. Clin Chim Acta 1997; 259:169–176.
[100] Suzuki A, Itoh F, Hinoda Y, Imai K. Double determinant immuno-polymerase chain reaction: A sensitive method for detecting circulating antigens in human sera. Jpn J Cancer Res 1995; 86(9):885–889.
[101] Itoh F, Suzuki A, Hinoda Y, Imai K. Double determinant immuno-polymerase chain reaction for detecting soluble intercellular adhesion molecule-1. Artif Organs 1996; 20(8):898–901.

[102] Zhang HT, Kacharmina JE, Miyashiro K, Greene MI, Eberwine J. Protein quantification from complex protein mixtures using a proteomics methodology with single-cell resolution. Proc Natl Acad Sci USA 2001; 98(10):5497–5502.
[103] Sperl J, Paliwal V, Ramabhadran R, Nowak B, Askenase PW. Soluble T cell receptors: Detection and quantitative assay in fluid phase via ELISA or immuno-PCR. J Immunol Methods 1995; 186(2):181–194.
[104] Masuda M, Morimoto T, De Haas M, Nishimura N, Nakamoto K, Okuda K, Komiyama Y, Ogawa R, Takahashi H. Increase of soluble FcgRIIIa derived from natural killer cells and macrophages in plasma from patients with rheumatoid arthritis. J Rheumatol 2003; 30(9):1911–1917.
[105] Niemeyer CM, Adler M, Gao S, Chi L. Supramolekulare Nanoringe aus Streptavidin und DNA. Angew Chem 2000; 112:3183–3187.

INDEX

D

E

L

M

N

P

R